Vienna in the Age of Uncertainty

Vienna in the Age of Uncertainty

Science, Liberalism, and Private Life

DEBORAH R. COEN

The University of Chicago Press

CHICAGO AND LONDON

DEBORAH R. COEN is assistant professor of history at Barnard College.

The University of Chicago Press, Chicago 60637
The University of Chicago Press, Ltd., London

Printed in the United States of America

16 15 14 13 12 11 10 09 08 07   1 2 3 4 5

ISBN-13: 978-0-226-11172-8 (cloth)
ISBN-10: 0-226-11172-5 (cloth)

Library of Congress Cataloging-in-Publication Data

Coen, Deborah R.
Vienna in the age of uncertainty : science, liberalism, and private life / Deborah R. Coen.
p. cm.
Includes bibliographical references and index.
ISBN-13: 978-0-226-11172-8 (cloth : alk. paper)
ISBN-10: 0-226-11172-5 (cloth : alk. paper) 1. Exner family. 2. Vienna (Austria)—Biography. 3. Vienna (Austria)—Intellectual life. 4. Intellectuals—Austria—Vienna—Biography. I. Title.
DB844.E98C64 2007
509.2′243613—dc22
[B]
2007003723

♾ The paper used in this publication meets the minimum requirements of the American National Standard for Information Sciences—Permanence of Paper for Printed Library Materials, ANSI Z39.48-1992.

Contents

Illustrations

Acknowledgments

IT IS a pleasure to thank the many teachers, colleagues, and friends who have given me suggestions, criticism, and support in the eight years since I began this project. It started as an M.Phil. dissertation at Cambridge University under the supervision of Simon Schaffer, who has continued to provide advice and inspiration throughout. Over the years I have also benefited from conversations with former M.Phil. classmates, including Charlotte Bigg and Anna Maerker. At Harvard, where the project took shape as a doctoral dissertation, Professors Bob Brain, Everett Mendelsohn, Anne Harrington, and David Blackbourn were generous with their time and input. I owe a great deal to my fellow Ph.D. students in Harvard's History of Science Department and to our dissertation working group, as well as to Harvard's European History Dissertation Workshop. Thanks above all to Jimena Canales, Michael Gordin, Karl Hall, Jeremiah James, David Kaiser, Elizabeth Lee, Theresa Levitt, Denise Phillips, Matthew Stanley, and Katja Zelljadt, and to Orit Halpern for sharing copies of Karl Frisch's films with me. David Cohen at Swarthmore College has given me the insights of a working physicist and Robin Kirman those of a novelist. At the Max Planck Institute, I enjoyed valuable conversations with Mechthild Fend, Michael Hagner, Anke te Heesen, Ed Jurkowitz, Skuli Sigurdsson, and Friedrich Steinle. I feel fortunate to have had an inspiring cohort at the Harvard Society of Fellows, among whom Avner Ben-Zaken, Anna Henchman, and Tara Zahra gave thoughtful suggestions for this project. During my research trips to Vienna, I was grateful for advice from Dieter Bogner, Christa Hammerl, Michael Hantel, Gabrielle Koller, Elisabeth Nemeth, Walter Ogris, and Wolfgang Reiter. I was fortunate to be able to participate in the 2004 workshop "Sigmund Exner—Physiologie, Psychologie, Ästhetik und empirische Kulturforschung," organized by Veronika Hofer and Michael Hagner, and I thank the Vienna Circle Institute for providing travel funds. I am grateful to Veronika Hofer and Michael Stöltzner for numerous discussions of the "Exnerei" and for their contributions to the session on this topic at the 2001 History of Science Society Meeting. Another HSS session, "All in the Family" at the 2003 meeting, brought thoughtful feedback from many participants on the topic of science and private life; thanks in particular to Paul White. Tom Broman, Malachi Hacohen, and Norton Wise have also given helpful critiques of conference papers. More recently, I benefited from the chance to present a chapter at the University of Chicago's Fishbein Center, and I thank Robert J. Richards for inviting me. Throughout my research, it has been a unique privilege to have had encouragement from members of the Exner-Frisch family. Many thanks to Professors Christoph Exner, Herwig Frisch, and Peter Dijkgraaf for generously providing materials and recollections.

I am especially grateful to those who read this manuscript in full at various stages of its development. As dissertation readers, Lorraine Daston, Peter Galison, Joan Richards, and Charles Rosenberg provided criticism, encouragement, and wisdom. As a (formerly anonymous) reader for the University of Chicago Press, Mitchell Ash gave the manuscript the benefit of his incomparable knowledge of Austrian science. A special note of thanks to my graduate advisor, Peter Galison. He has been a patient and generous mentor to me since I was an undergraduate, and his methods shaped this book from the start.

Several institutions provided generous support for this project. I thank the National Science Foundation, the Harvard Center for European Studies, the German Academic Exchange Service, the Max Planck Institute for the History of Science (Berlin), and the Harvard Society of Fellows. I am very grateful for assistance from the staffs of these institutions and especially to Diana Morse at the Society of Fellows. The staffs of many libraries and archives have facilitated my research. Special thanks to Stefan Seinell at the Archive of the Austrian Academy of Sciences, Brigitte Kromp at the University of Vienna's Central Library for Physics, Johannes Seidl and Kurt Mühlenberger at the University of Vienna's archives, Roland Franz at the Museum for Applied Art in Vienna, and Silvia Herkt at the Oskar Kokoschka Collection of Vienna's University for Applied Art. At the University of Chicago Press, I would like to thank editor Catherine Rice and her assistant Pete Beatty.

Writing a family history has consistently reminded me of the debt I owe my own. Without their help this project would not have been possible. Many thanks to my siblings David, Michael, Jennifer, and Gwen for moral and technical support; to Naomi, Robert, David, and Jessica Tuchmann for their patience and boundless generosity; and to Amalia Tuchmann for letting her mom finish this book during her gestation and first months in the world. My husband, Paul Tuchmann, has been much more than a first-rate critic and proofreader. He has been my hero. Thank you, Paul, for letting me find you, and for staying found. To my parents, Stanley and Ruth Imber Coen, I owe the greatest debt of all. It may be that my fascination with fin-de-siècle Vienna dates from my early exposure to psychoanalysis at the dinner table. In any case, they have always been my inspiration. It is not nearly enough of a tribute, but I dedicate this book to them.

Klimt's *Philosophie,* a mural intended for the Philosophy Faculty of the University of Vienna. Only this black-and-white photograph survives. Courtesy of Galerie Welz, Vienna.

INTRODUCTION

A Scientific Dynasty

IT HAS become an icon of the modernist subversion of rationality. Gustav Klimt's mural *Philosophy* was commissioned as a tribute to the illuminating light of knowledge by the faculty of philosophy and natural sciences at the University of Vienna. But the image unveiled in the first year of the twentieth century was murky and otherworldly, its naked figures clinging to each other in despair. In a flurry of public protest, Vienna's scientists and philosophers barred the painting from the walls of the university. Instead, thanks to Carl Schorske's account in *Fin-de-Siècle Vienna*, Klimt's *Philosophy* lives on in the historical imagination as a symptom of what Schorske diagnosed as the "crisis of rationalism." To Schorske, the mural and the uproar it occasioned were tokens of the fall of Viennese liberalism, a movement that he defined by its confidence in "rational man" and in the "scientific domination of nature." Klimt and the professors seemed to Schorske to stand on opposite sides of a historical disjuncture, representatives, respectively, of the old rational, ethical culture and the new aesthetic, subjective culture.[1]

Among the scientists who figure in Schorske's account as critics of Klimt's mural and thus as proponents of "liberal rationality" were the brothers Sigmund and Franz Serafin Exner. The Exner family, Schorske pointed out, achieved renown in science as well as politics, displaying a breadth of interest characteristic of their upper middle-class culture of "rationalistic liberalism." Viewed from the perspective of the Exner family, however, the Klimt

1. Carl Schorske, *Fin-de-Siècle Vienna: Politics and Culture* (New York: Vintage, 1981); see too William McGrath, *Dionysian Art and Populist Politics in Austria* (New Haven: Yale University Press).

scandal looks quite different, as do the very categories of "rationality" and "liberalism."

If we bring the Exner family as a whole into focus, suggestive connections come into view. At the turn of the century Sigmund Exner's wife Emilie was the director of an art school for women, where she modeled the curriculum on the teachings of the Secession, Vienna's aesthetic innovators. Franz Serafin's daughter Hilde was herself a member of the Secession, as was her cousin Nora. So we should not be surprised to find, as we will in later chapters, that the ideal of rationality defended by the Exner brothers in 1900 was in fact closely aligned with the pedagogical principles of Viennese modernism, with its emphasis on the autonomous yet disciplined observation of nature. Nor were the brothers threatened by the themes of subjectivity and skepticism implicit in Klimt's painting, as became clear in the science of color they developed in the wake of this controversy.

Attending in this way to family life alongside the public sphere, this book recasts the widely held image of fin-de-siècle Vienna. Continuities will emerge between science and aesthetics, reason and subjectivity. We will find a culture in which skepticism, far from being liberalism's downfall, was in fact its core value; a culture in which the family sphere was not a retreat from rational thought and political engagement but constitutive of them. By moving between "public" and "private" life, this study identifies the anxieties that shaped Austrian liberalism, the dual threats of religious dogmatism and radical relativism. And it uncovers the source of imperial Austria's profound creativity in the sciences: a value system in which probabilistic reasoning was not a sign of tentativeness but rather a display of authority.

Perhaps no time and place have come to represent so vividly the origins of the "modern" as Vienna circa 1900. Freudian psychology and the expressionist fervor of Klimt and Schiele have become a shorthand for the self-awareness of modern individuals. If such self-knowledge was modernism's reward, its cost has seemed equally clear. While sounding the depths of the human psyche, Vienna's modernists apparently cut themselves loose from the legacy of the Enlightenment, a transformation at once intellectual and political. According to Schorske, liberalism failed in Austria because the middle class became critical of the absolute claims of science and law. They ostensibly abandoned reason for relativism and liberalism for an obsessive preoccupation with their private selves.

There is comfort to be found in Schorske's story, for it makes the failures of this archetypically "modern" society seem less arbitrary. It seems to explain why Austrian liberalism gave way to an acrid and ultimately perilous form of populist politics. In Schorske's account, in short, liberalism arose in Austria in the mid-nineteenth century as the embodiment of public reason only to be destroyed a generation later by the bourgeoisie's flight into skepticism and the cultivation of the intimate sphere. His story rests, then, on a specific conception of rationality—one to which doubt and uncertainty could be lethal and which flourished in public but withered in private. In fact, we know little as yet about what rationality really meant in the world we have taken as our paradigm for its "modern" subversion. Allan Janik writes that today the "'big' questions about Viennese culture" center on "just how 'rational' developments there have been," and to answer these questions, Janik argues, we need research on the history of natural science in Austria.[2] By the same token, Schorske took for granted the categories of public and private. Yet, as feminist political theorists alert us, to demarcate public from private is always a political act.[3] Rather than drawing this boundary a priori, we need to learn how our historical actors mapped it.[4] In order to rethink the linked dichotomies at the heart of Schorske's thesis—between reason and uncertainty, publicity and privacy—I take up the story of the Exners, Vienna's foremost scientific dynasty.

With their hands in projects spanning science, politics, and the arts, the Exner family is an ideal guide for a tour of Vienna at the turn of the twentieth century. Over the course of three generations, born between 1802 and 1886, they produced ten professors at Austrian universities. At musical evenings in Vienna and at Brunnwinkl, their lakeside retreat near Salzburg, they entertained the cream of Austria's intellectual and artistic elite. Ernst Mach and Ludwig Boltzmann were close colleagues, Sigmund Freud and Erwin Schrödinger numbered among their students, and Josef Breuer and Marie Ebner von Eschenbach were their intimate friends. Their scientific accomplishments were celebrated by contemporaries, if neglected by historians, and remain fascinating as well as relevant today. We will meet the physicist

2. Allan Janik, "Vienna 1900 Revisited," in *Rethinking Vienna 1900*, ed. Steven Beller (New York: Berghahn Books, 2001), p. 47.

3. See, e.g., Joan B. Landes, ed., *Feminism, the Public and the Private*, (Oxford: Oxford University Press, 1998); and Johanna Meehan, ed., *Feminists Read Habermas: Gendering the Subject of Discourse* (New York: Routledge, 1995).

4. Resources for such an analysis include the ten-volume series Bürgertum in der Habsburgermonarchie, edited by Hannes Stekl and published by Böhlau Verlag, Vienna; Karl-Heinz Rossbacher, *Literatur und Liberalismus. Zur Kultur der Ringstrassenzeit in Wien* (Vienna: Verlag Jugend & Volk, 1992); and Karlheinz Rossbacher, *Literatur und Bürgertum. Fünf Wiener jüdische Familien von der liberalen Ära bis zum Fin de Siècle* (Vienna: Böhlau, 2003).

Franz Serafin Exner, who made physical indeterminism plausible well before the advent of quantum mechanics, and his nephew Felix Exner, who showed how meteorological problems could be handled statistically in ways that computers have now made invaluable. Felix's father Sigmund Exner introduced the concept of neural networks that is at the heart of contemporary brain science and artificial intelligence. Those familiar with the work of Nobel Prize–winning biologist Karl Frisch might be surprised to learn that he too was an Exner on his mother's side. Frisch's studies of insect communication remain exemplary today for biologists studying the relationship between brain and behavior. Moreover, the Exners were equally active outside the laboratory. In their prominent public roles as university administrators, as advisors to the education ministry and to the upper house of parliament, the Exners shaped the content and values of the education system of a mighty European empire.

In 1840 Vienna was the capital of the largest Roman Catholic power in the world, controlling lands from the Adriatic to the Russian steppe.[5] From the birth of the first Franz Exner at the turn of the nineteenth century to the death of the grandson who bore his name in 1947, the Exners' Austrian world changed dramatically. Franz Exner I belonged to the generation of educated middle-class Austrians who first demanded liberal reforms of the conservative, Catholic regime. His children came of age in the 1860s, the golden age of Austrian liberalism. By the 1870s, economic depression and a resurgence of Catholic power had weakened liberal authority in Austria. As the third-generation Exners matured in the Vienna of the fin de siècle, the liberal party dissolved and the politics of the multiethnic empire became dominated by nationalism, anti-Semitism, and socialism.[6]

The Exners built their moral authority as liberals on their personal capacities to confront and manage uncertainty in a world where religion no longer guaranteed truth. They transmitted this authority in their teaching

5. "Austria" had multiple meanings in the nineteenth and early twentieth centuries. I use it to refer to the entirety of the Habsburg lands before 1867, to the non-Hungarian half of the empire between 1867 and 1918, and to the rump nation-state carved by the peace treaties in the interwar period. A note on names: in the Habsburg Empire individuals ennobled for state service earned the right to use the participle "von." With the abolition of the nobility in 1919, its use was banned, such that the "von Frisch" family became merely "Frisch."

6. On the history of Austrian liberalism, see Pieter Judson, *Exclusive Revolutionaries: Liberal Experience, Social Politics, and National Identity in the Austrian Empire, 1848–1914* (Ann Arbor: University of Michigan Press, 1996); Lothar Höbelt, *Kornblume und Kaiseradler, Die deutschfreiheitlichen Parteien Altösterreichs 1882–1918* (Vienna: Verlag für Geschichte und Politik, 1993); Harry Ritter, "Austrian-German Liberalism and the Modern Liberal Tradition," *German Studies Review* 7 (1984): 227–48.

and research through the mathematics of probability theory, an algorithm for quantifying and thus disciplining uncertainty. What follows is not an exhaustive biography of a family, nor of a mathematical concept. Rather, my goal is to use the story of this scientific dynasty to chart the evolution of a form of authority specific to the convergence of science and liberalism in imperial Austria.[7] Still, the lessons of this story extend to the United States of the twenty-first century, where scientists working in such controversial fields as evolutionary biology, stem cell research, and global climate change face critics eager to portray their research as merely speculative. As in the Exners' Vienna, religious fundamentalists today contrast "mere" scientific theories with the unquestionable truth of biblical texts. The evidentiary basis of politically volatile research must be ever stronger to withstand "the manufacture of uncertainty" by the political right.[8]

These charged debates have driven scientists to reflect publicly on how they deal with uncertainty. Many have emphasized that all good science is uncertain, because science is empirical knowledge that is always subject to revision based on further experience. But some have gone farther, insisting that uncertainty is a source of creativity rather than a necessary evil. "Science thrives on uncertainty," such researchers argue. "You can have gaps," one biologist says. "You don't have to fill all the gaps with supernatural explanations. If you don't have gaps, you don't have science."[9] A geophysicist declares that climate change research illustrates how scientists "work and thrive in an environment of uncertainty." Uncertainty is "akin to a glass half full, not half empty."[10] Although a century removed and a continent apart from these disputes, the Exners stand today as examples of scientists who won authority not as gatekeepers of knowledge but as stewards of uncertainty.

IN THE 1840s, education in Austria was still strictly constrained by Roman Catholic dogma. The curriculum of the empire's secondary schools remained Catholic in its principles and absolute in its presentation. In this theological framework, as one textbook declared, there could be only one true

7. Werner Michler has argued that natural science "served as Austrian liberalism's preferred medium of self-expression." *Darwinismus und Literatur: Naturwissenschaftliche und literarische Intelligenz in Österreich, 1859–1914* (Vienna: Böhlau, 1999), p. 12.

8. Chris Mooney, *The Republican War on Science* (New York: Basic Books, 2005).

9. "Debate gets down to definition of science," *St. Louis Post-Dispatch*, Nov. 13, 2005.

10. Henry N. Pollack, *Uncertain Science... Uncertain World* (Cambridge: Cambridge University Press, 2003), pp. 3, 41.

philosophy uniting science and faith, "since truth, knowledge of which is indeed the goal of all philosophy, can itself only be singular."[11]

The reorganization of Austria's education system on a primarily secular basis was one of the few lasting victories that Europe's liberals won in the revolutions of 1848. Virtually alone in Europe, liberal educators in Austria celebrated 1848 as a victory, and Franz Exner was their hero. Exner had been a professor of philosophy at Prague, where he had won his students' admiration for his daring flirtation with scientific ideas proscribed by the Catholic authorities. When ministers of the imperial government became desperate to quell student protests, they called Exner to Vienna as their advisor. Having won the ear of the administration, Exner and his liberal allies faced the monumental challenge of replacing Catholic dogma as the foundation of moral education.

What is most surprising in retrospect about Exner's proposals is his optimism. Even the man whose pedagogical philosophy he borrowed, Johann Friedrich Herbart, had denied that a uniform national education system could successfully take on the responsibility of *Erziehung*.[12] Herbart believed that *Erziehung* would remain, as it had traditionally been, the province of parents and private tutors. Exner had to tread carefully to avoid upsetting a delicate balance of power between families and schools. The saga of these reforms, then, can be read as part of a history of the renegotiation of the private and public, of family and nation-state.[13]

The reformers subverted the moral authority of the Catholic church by painting skepticism as a virtue. They wove this lesson into the new philosophy course designed for students of the *Gymnasien*, the elite secondary schools that were the sole pathway to matriculation at a university.[14] By the

11. *Lehrbuch der Philosophie* (Vienna: k.k. Schulbücher-Verschleiß-Administration, 1835), vol. 1, pp. 14–15, 5. Translations are my own unless otherwise noted.

12. *Erziehung*, literally "upbringing," has no direct equivalent in English. In the context considered here it usually designated the process of shaping a young person's character and morals. In this sense, it will be translated as "character building" or "moral education."

13. For these themes, see Michelle Perrot, "The Family Triumphant," in *A History of Private Life*, ed. Michelle Perrot, trans. by Arthur Goldhammer (Cambridge, Mass.: Harvard University Press, 1990), vol. 4, pp. 99–129; Lothar Gall, *Bürgertum in Deutschland* (Berlin: Siedler, 1989); Franz J. Bauer, *Bürgerwege und Bürgerwelten: Familienbiographische Untersuchungen zum deutschen Bürgertum* (Göttingen: Vandenhoeck and Ruprecht, 1991).

14. In 1850–51 the empire's *Gymnasien* had a total enrollment of approximately twenty thousand students, or less than 0.2 percent of the total population. A few thousand more were students at the far less prestigious *Realschulen*, secondary schools that taught modern languages instead of Latin and Greek and that placed greater emphasis on natural science. *Realschule* graduates could

1860s a sixteen-year-old student at an Austrian *Gymnasium* would have learned from his philosophy textbook that the precondition for knowledge was neither faith, nor trust, nor impartiality, but rather "doubt." He would have read that the intended effect of his last two years of secondary school was the overthrow of all his convictions, a head-first plunge into uncertainty. Somewhere along a man's path to intellectual maturity, he was told, "[t]he foundation of his knowledge sways beneath his feet and he grasps for an anchor to steady himself."[15] With this warning came the assurance that he would emerge from this experience with a strong and noble character. Through this coming-of-age story, liberal educators laid siege to the moral authority of the Catholic church.

Yet the liberals' pedagogical strategy was precarious. They risked pushing students to doubt too widely, jeopardizing their own nascent authority. Faced with the empire's centrifugal forces of nationalism and democratic radicalism, liberals had to justify their claim to represent a staggeringly heterogeneous population. Even as they discredited dogmatism, they had to speak from a position that transcended narrowly personal or class-based interests. Their challenge was thus to steel students against the twin dangers of dogmatism and radical relativism. It was the task of the new philosophy course to steer students clear of what one *Gymnasium* director in retrospect called "political and social aberrations [*Verirrungen*] of the head and heart."[16]

Their solution lay in a novel element of the new philosophy course, the theory of probability. The pre-1848 philosophy course had alluded perfunctorily to quantitative probabilities without providing a single illustration. By contrast, the philosophy texts of the 1850s and 1860s presented probability in mathematical detail and with numerous examples. An 1860 textbook,

continue their studies at technical colleges. By the late 1870s, they also obtained some of the same privileges as *Gymnasium* graduates, including eligibility for the one-year voluntary military service. As in Germany, France, and England, the students at these schools were male. Private associations in Austria began to establish separate academic secondary schools for women in the 1870s. Still, by 1910 there were fewer than three thousand female secondary students. The total enrollment in secondary schools in Austria began to grow after 1850, with more rapid growth after 1870. Compared to Prussia, secondary school enrollment in Austria was far lower at midcentury but roughly equal by 1910. Gary B. Cohen, *Education and Middle-Class Society in Imperial Austria, 1848–1918* (West Lafayette, Indiana: Purdue, 1996), chapters 2, 4.

15. Robert Zimmermann, *Philosophische Propadeutik*, 2nd ed. (Vienna: Wilhelm Braumüller, 1860), p. 356.

16. Ludwig Chevalier, "Über den Unterricht in der philosophischen Propädeutik an österreichischen Gymnasien," *Jahresbericht des k.k. Staats-Untergymnasiums in Prag-Neustadt* 4 (1885): 3–39, quotation on p. 17.

penned by one of Franz Exner's students, asked the reader to suppose that a family has three sons, of whom we know two, and these two strongly resemble each other. In this case, the text instructs, we would assign the statement "all sons in this family resemble each other" the probability 2/3. The statement is, in other words, probable. If the family had four sons the statement would be "doubtful"; if it had five sons it would be "improbable."[17]

To the philosophers of the Enlightenment, the calculus of probabilities had been an admission of defeat.[18] Pierre-Simon Laplace and his contemporaries constructed probabilistic reasoning as a crutch for mere mortal minds, whose knowledge of nature's intricate causal chains would always be imperfect. Laplace remained a convinced determinist, believing that unpredictability in nature was the result of the difficulty of specifying exact initial conditions. Although probabilities in his day were often modeled as the frequencies of events such as throws of a die, they represented states of mind, not states of the world. In a deterministic universe, states of the world could never be indeterminate. Compare, for instance, the following statements:

1. Given the testimony of the witnesses, it is probable that the defendant is guilty as charged.
2. The probability of tossing a head with this coin is 1/3.

In the first case, probability is the degree of belief accorded to a proposition (the defendant is guilty) on the basis of certain evidence (the witnesses' testimony). Today, probability in this sense is often called "subjective," "epistemic," or "belief-type." The second is a statement about the relative frequency of an event in the world: approximately one out of three times the coin will land heads up. For a fair coin, the probability of tossing a head is one half; so in this case, you have been slipped a weighted coin. Here probability is a property of an object, namely, the *frequency* with which it behaves in a certain way. Probability in the former sense was the dominant interpretation from the seventeenth until well into the nineteenth century, and this is the sense in which probability was taught in the Austrian *Gymnasien* after 1848.

17. Robert Zimmermann, *Philosophische Propadeutik,* 3rd ed. (Vienna: Wilhelm Braumüller, 1867), p. 56.

18. On early modern probability theory, see Lorraine Daston, *Classical Probability in the Enlightenment* (Princeton: Princeton University Press, 1988); and Ian Hacking, *The Emergence of Probability: A Philosophical Study of Early Ideas about Probability, Induction, and Statistical Inference* (London and New York: Cambridge University Press, 1975).

By that time, however, an alternative understanding of probability had begun to emerge, one illustrated by statement number two above.[19] As centralizing nation-states collected and published quantitative information about their populations, bureaucrats and private citizens began to notice regularities in the rates of such events as births, deaths, marriages, even suicides. A mathematical principle formulated by Simeon-Denis Poisson in 1835 suggested to many contemporaries that such regularity was characteristic of mass phenomena. Poisson's "law of large numbers" states that in a series of trials with different, independent outcomes, as the number of trials approaches infinity, the relative frequency of each possible outcome will approach a constant, namely, the true probability. Soon after, John Venn distinguished clearly for the first time between the two interpretations of probability sketched above. Venn insisted that statements of probability could only apply to empirical records of repeated events. Probability was the relative frequency of an event in a series and could not legitimately be applied to single instances. Probability in this sense was knowledge won from experience, not—as for Laplace—the product of ignorance.

By the 1860s natural scientists had begun to apply this new statistical perspective to the physical world, by direct analogy to social systems: individuals were to society as molecules were to a gas. Probabilistic descriptions of aggregate regularities allowed physicists, like social scientists, to ignore the innumerable, minute factors governing the behavior of individuals. James Clerk Maxwell in England and Ludwig Boltzmann in Austria thus borrowed freely from the tradition of social statistics, but they clung to probability's meaning in statement number one. They interpreted the less-than-perfect certainty of their results as a reflection of their own ignorance of the causal factors at work, not as signs that physical phenomena were unpredictable in some fundamental way. Maxwell maintained a religiously grounded belief in the existence of a perfectly ordered molecular world, a sort of microscopic Platonic reality. Even if humans could achieve no more than probabilistic knowledge of the world, Maxwell believed in a higher intelligence that could.[20] Like Maxwell, Boltzmann was unsatisfied with

19. The following sketch of probability's history draws on Theodore M. Porter, *The Rise of Statistical Thinking, 1820–1900* (Princeton: Princeton University Press, 1986); Ian Hacking, *The Taming of Chance* (Cambridge: Cambridge University Press, 1990); Stephen M. Stigler, *The History of Statistics: The Measurement of Uncertainty before 1900* (Cambridge, Mass.: Harvard University Press, 1986); and Alain Desrosières, *The Politics of Large Numbers* (Cambridge, Mass.: Harvard University Press, 1998).

20. James Clerk Maxwell, "Molecules," *Nature* 8 (1873): 437–41.

the alternatives of absolute necessity or pure chance. But Boltzmann's perspective was that of a secular humanist, and he accepted the limitations of empirical knowledge. He wrote of science's view of the world as being like a painting of a curtain—nothing lay behind it. For Boltzmann, unlike Maxwell, placing physics on a probabilistic basis demanded the acceptance of uncertainty.[21]

This contrast points to what was unique about probability's history in Austria. I will argue that Austrian liberals made probabilistic reasoning into a virtue rather than an admission of defeat. Probability took its place within the liberal school reforms as a guide to reasoning after the renunciation of absolute certainty. Probability would fortify young minds against the lure of religious dogmatism by providing a model for action in the face of uncertainty. And it would skirt relativism by transforming knowledge won from lived experience into a language of universal validity and transparency.

By examining the contempt of Austrian scientists for deterministic laws we can better understand the aspirations and strategies of liberals in the Habsburg Empire. Liberalism can be characterized, most generally, as an ideology, as an organized political movement, or as a "desire for any kind of reform."[22] It has been said to comprise any or all of the following commitments: to a free-market economy, to representative government based on a limited franchise, and to some set of civil rights, such as freedom of speech and of religion. For many years after the Second World War, historians sought to explain the rise of National Socialism by pinpointing ways in which liberalism in Central Europe deviated from such a general definition. This is not my aim.[23] Without writing Austrian history as an inevitable descent into fascism, however, it is still possible to isolate defining

21. Ludwig Boltzmann, "The Second Law of Thermodynamics," in *Theoretical Physics and Philosophical Problems: Selected Writings*, ed. Brian McGuinness (Dordrecht: Reidel, 1974), pp. 13–32. Blackmore shows that Boltzmann took pains to make his version of determinism compatible with free will. See John T. Blackmore, *Ludwig Boltzmann: His Later Life and Philosophy, 1900–1906*, vol. 2, *The Philosopher*, Boston Studies in the Philosophy of Science, 174 (Dordrecht and Boston: Kluwer, 1995), p. 163; Michael Stöltzner, "Vienna Indeterminism. Causality, Realism and the Two Strands of Boltzmann's Legacy" (Ph.D. diss., University of Bielefeld, 2003), pp. 95–97.

22. Dieter Langewiesche, *Liberalism in Germany*, trans. by Christiane Banerji (Princeton: Princeton University Press, 2000); James Sheehan, *German Liberalism in the Nineteenth Century* (Chicago: Humanities Press, 1978); and Georg Franz, *Liberalismus. Die deutschliberale Bewegung in der Habsburgischen Monarchie* (Munich: Georg Callwey, 1956).

23. David Blackbourn and Geoff Eley, *The Peculiarities of German History: Bourgeois Society and Politics in Nineteenth-Century Germany* (Oxford: Oxford University Press, 1984). Trying to restore a sense of "contingency" to German history, the authors stress the inadequacy of a historical model

traits of Austrian liberalism—a movement rightly known for its enormous influence in the twentieth century, due to philosophers and economists like Carl Menger, F. A. von Hayek, and Karl Popper.

Nineteenth-century Austrians' own definition of liberalism was at once far more specific and more all-encompassing than a list of political commitments. Other historians have emphasized that Central European liberals intended the qualifications for political participation to extend far beyond property holdings to include ethical qualities such as rationality and intellectual independence.[24] Liberals saw these very qualities as the products of education and upbringing, thus of experiences in both the public and private spheres. Liberal identity in Austria lay not only in an ideology but also in a character—a style of speaking, reasoning, and interacting, the product of an individual's education in the broad sense of *Erziehung*.[25] Those who displayed this character embodied the liberals' moral authority over their opponents on either side of the political spectrum. On the right stood clericals decrying the dangers of skepticism; on the far left stood democrats and socialists challenging the liberals' claim to represent interests beyond those of their own class. Poised in the center, Austrian liberals laid claim to an authority rooted in their ethical character, itself the fruit of a perpetual process of self-cultivation. Against their right-wing challengers they displayed the virtues of a flexible and independent mind. Against socialists and nationalists they vaunted their self-discipline, with which they believed themselves capable of transcending a class- or nation-based perspective. The probability calculus entered public education as a standardized tool for producing these trappings of a liberal habitus, a shorthand for the virtues upon which liberals rested their claim to authority.

Austria's culture of uncertainty was an alliance of science and liberalism related to and yet distinct from those that emerged elsewhere in Europe in the nineteenth century. Recent historiography makes possible certain tentative generalizations. In Western Europe, liberal elites tended to embrace science in the name of social intervention from above. British and French

that assumes that the emergence of a bourgeois class necessarily gives rise to a liberal ideology, then to parliamentary politics, and finally to democracy.

24. Judson, *Exclusive Revolutionaries*.

25. On *Bürgerlichkeit* as a "habitus" in Pierre Bourdieu's sense, see Ulrike Döcker, "Bürgerlichkeit und Kultur," in *Bürgertum in der Habsburgermonarchie*, ed. Ernst Bruckmüller et al., vol. 1 (Vienna: Böhlau, 1990). For "habitus" in the French academic context, see Bourdieu, *Homo Academicus*, trans. by Peter Collier (Stanford: Stanford University Press, 1984).

liberals adopted statistical reasoning, in particular, as a guide to engineering society (even as they understood the stability of statistical laws to signal the limits of government intervention). In France statistical thinking became part of the liberal tradition of Comtean positivism, which valued knowledge for the sake of prediction and control.[26] Like the Austrians, French positivists could be bitterly anticlerical and contemptuous of dogmatism, but unlike the Austrians they insisted that scientific laws were "immutable" and "universal."[27] East of the Rhine, the stress fell more heavily on science's value for character building. In Prussia and Baden, as in Austria, liberals embraced empirical science as a model of independent thinking and consensus building. In Prussia and Baden, however, with their burgeoning electrical and chemical industries, liberals concerned themselves more immediately with industrialization than their Habsburg counterparts and promoted science for practical ends. Like the French, they emphasized the predictive value of scientific laws.[28] By contrast, the stress that Austrian liberals laid on the probabilistic character of scientific knowledge was a strategy tailored to Habsburg politics. Nowhere outside of Austria in the nineteenth century did liberals face both a politically dominant church and splintering national movements. The challenge of liberal science in Austria was to define rationality in such a way as to discredit at once the absolute claims of religion while justifying their claim to knowledge that transcended a narrowly class- or nation-based perspective. Probabilistic reasoning was the solution.

In the century following Franz Exner's reforms, Austrians trained in the reformed *Gymnasien* became famous for their contributions to probability's

26. Theodore Porter, *The Rise of Statistical Thinking, 1820–1900* (Princeton: Princeton University Press, 1986); and Charles C. Gillispie, *Science and Polity in France: The Revolutionary and Napoleonic Years* (Princeton: Princeton University Press, 2004).

27. Jan Goldstein, "The Hysteria Diagnosis and the Politics of Anticlericalism in Late Nineteenth-Century France," *Journal of Modern History* 54 (1982): 209–39, quotation on p. 234.

28. Timothy Lenoir, *Instituting Science: The Cultural Production of Scientific Disciplines* (Stanford: Stanford University Press, 1997); Arleen Marcia Tuchman, *Science, Medicine and the State in Germany: The Case of Baden, 1815–1871* (Oxford: Oxford University Press, 1993); Edward Jurkowitz, "Helmholtz and the Liberal Unification of Science," *Historical Studies in the Physical and Biological Sciences* 32 (2002): 291–317; and Ian F. McNeely, *"Medicine on a Grand Scale": Rudolf Virchow, Liberalism, and the Public Health* (London: Wellcome Trust Centre for the History of Medicine at University College London, 2002). For an Austrian liberal critique of a strictly utilitarian view of science, see Michler's discussion of the Vienna philologist and Darwinist Wilhelm Scherer in *Darwinismus und Literatur*, pp. 67–68. On the chemical industry in the Habsburg Empire, see Robert W. Rosner, *Chemie in Österreich, 1740–1914: Lehre—Forschung—Industrie* (Vienna: Böhlau, 2004).

scientific applications and philosophical interpretations. Thanks to Michael Stöltzner's recent analysis, we can now recognize "Vienna indeterminism" as a coherent philosophical tradition, one in which Franz Serafin Exner played a pivotal role.[29] Ernst Mach, Franz Brentano, Richard von Mises, and Rudolf Carnap were among the Austrian philosophers who probed probability's conceptual foundations. In the physical sciences, Ludwig Boltzmann, Marian von Smoluchowski, Paul Ehrenfest, and Erwin Schrödinger extended probability's reach into new domains of empirical knowledge. Philosopher Friedrich Jodl argued that science had nothing to fear from the notion of chance, nor from the proposition that physical events were akin to throws of the die.[30] Similarly, in economics, the liberal "Austrian school" reinterpreted the concept and origin of laws to accommodate the uncertainty of a science of "highly complex phenomena, or of structures determined by a greater number of particular facts than could ever be concretely ascertained by scientific observers."[31] In the interwar period, the philosophers of the Vienna Circle sought to put science on a strictly logical foundation by reformulating empirical laws as probability statements.[32]

One could argue that in all these cases, probability was tied to a characteristically liberal and anticlerical rejection of absolute claims. Contra Schorske, uncertainty did not run wild in fin-de-siècle Vienna. Those who renounced the goal of certainty did so not in rejection of Enlightenment values but in defense of them. They "tamed" uncertainty by quantifying it.[33] Skepticism was thus not liberalism's downfall but instead a vital element of liberal culture and natural science in post-1848 Vienna.[34] In fact, the abandonment of certainty was what made Austrian science so dynamic in this period.

29. Stöltzner, "Vienna Indeterminism: Mach, Boltzmann, Exner," *Synthese* 119 (1999): 85–111.

30. Friedrich Jodl, "Der Begriff des Zufalles: Seine theoretische und praktische Bedeutung" (1904), in *Vom Lebenswege: gesammelte Vorträge und Aufsätze*, ed. Wilhelm Börner (Stuttgart and Berlin: Cotta, 1916–17), pp. 515–33.

31. F. A. Hayek, *The Fortunes of Liberalism: Essays on Austrian Economics and the Ideal of Freedom*, ed. Peter G. Klein (London: Routledge, 1992), p. 56.

32. Richard von Mises, *Wahrscheinlichkeit, Statistik, und Wahrheit* (Vienna: Springer, 1936); Rudolf Carnap, *Logical Foundations of Probability* (Chicago: University of Chicago Press, 1950). Of course, von Mises, Carnap, Schlick, and Reichenbach reached no consensus on probability's interpretation.

33. Cf. Hacking, *Taming of Chance*.

34. Hacohen makes the crucial point that "Philosophers who challenged certitude . . . often led efforts for social reform and popular science education." Malachi Hacohen, "Karl Popper, the Vienna Circle and Red Vienna," *Journal of the History of Ideas* 59 (1998): 711–50, quotation on p. 718.

This enthusiasm for less-than-certain knowledge may seem incongruous against the background of science elsewhere in nineteenth-century Europe. Since the eighteenth century, scientists had expanded the domain of mechanical explanation to include ever more complex phenomena, from hydrodynamics to meteorology to experimental psychology.[35] In Prussia, scientists were proclaiming victory in their efforts to reduce natural phenomena to causal laws. Hermann von Helmholtz, the liberal spokesman of Prussian science and a teacher to Sigmund Exner, declared in 1869 that to "comprehend natural phenomena" meant nothing else than to "ascertain their laws."[36] Half a century later, Max Planck, the leading German physicist of the next generation, insisted that determinism was an indispensable basis for scientific inquiry.[37]

Physicist Franz Serafin Exner dismissed Planck's warning out of hand. He and his students productively treated physics as the study of statistical regularities, not absolute laws. Philosophers and scientists including Philipp Frank and Erwin Schrödinger have credited Franz Serafin Exner with the first clear demonstration that physics could safely discard the causal principle.[38] Earlier physicists had used probabilities as placeholders, bracketing causes of which they were ignorant until they could obtain fuller knowledge of the microscopic world. Probabilities in Exner's sense were not second-best

35. On the "mechanical world picture" and German physical theory in general, see Christa Jungnickel and Russell McCormmach, *Intellectual Mastery of Nature: Theoretical Physics from Ohm to Einstein,* vol. 1, *The Torch of Mathematics, 1800–1870* (Chicago: University of Chicago Press, 1986); on G. T. Fechner's mathematization of psychology, see Michael Heidelberger, *Die Innere Seite der Natur: Gustav Theodor Fechners wissenschaftlich-philosophische Weltauffassung* (Frankfurt am Main: Klostermann, 1993). Kathryn Olesko's studies of the all-consuming pursuit of precision in German physics and culture provide further evidence of this German / Austrian contrast. The first of these projects is *Physics as a Calling: Discipline and Practice in the Königsberg Seminar for Physics* (Ithaca, N.Y.: Cornell University Press, 1991).

36. Helmholtz continued: "Isolated facts . . . only become valuable in a theoretical or practical point of view when they make us acquainted with the *law* of a series of uniformly recurring phenomena, or, it may be, only give a negative result showing an incompleteness in our knowledge of such a law." Helmholtz, "The Aim and Progress of Physical Science," in *Science and Culture: Popular and Philosophical Essays,* ed. David Cahan (Chicago: University of Chicago Press, 1995), p. 208. On Helmholtz's understanding of scientific laws as an expression of German liberalism, see Jurkowitz, "Liberal Unification."

37. J. L. Heilbron, *The Dilemmas of an Upright Man: Max Planck and the Fortunes of German Science* (Cambridge, Mass.: Harvard University Press, 1996), p. 66. Stöltzner argues that Serafin Exner was the implicit target of Planck's attack on indeterminism; see Stöltzner, "Vienna Indeterminism," p. 107.

38. Frank, *The Law of Causality and Its Limits,* ed. R. S. Cohen, trans. by M. Neurath and R. S. Cohen (Dordrecht: Kluewer, 1998), pp. 70, 72; Schrödinger, "What Is a Law of Nature?" in *Science, Theory, and Man* (1935; New York: Dover, 1957), pp. 133–47.

descriptions. They were the only possible form of physical knowledge about microscopic events, namely, their relative frequencies. The introduction of indeterminism into modern physics began not as a "capitulation to hostile forces," as Paul Forman famously argued, but as an empirically motivated expression of a native scientific culture.[39] It was a stance echoed by the physiologist Sigmund Exner, who described causal thinking as a "mental habit" that evolution might overcome.[40] Even the jurist Adolf Exner, evoking the ridicule heaped by French revolutionaries on the hairstyle of aristocrats, decried determinism as the "pigtail of the nineteenth century."[41] This epithet captured precisely the liberal spirit behind the Exners' affinity for probability.

Mathematical techniques alone did not set these Austrians apart from their colleagues abroad. Indeed, scientists in Britain (Maxwell), the United States (Gibbs), France (Perrin), and elsewhere in the German-speaking lands (Clausius, Einstein, Heisenberg, Born) all played key roles in the development of a probabilistic physics. Nor, as these examples show, was probabilistic reasoning in the sciences always and everywhere aligned with liberalism, the meaning of which is itself historically and geographically contingent. Take the extreme case of Werner Heisenberg: a nationalistic German working in Munich, his path to the "uncertainty principle" (a probabilistic interpretation of atomic physics) was motivated by a commitment to positivism. Such examples of probabilistic physics outside of Austria help to highlight the specificity of probability's history in imperial Austria, where liberal dominance over the politics of education was poised precariously against the forces of clericalism. In Austria, probabilistic reasoning marked a course between dogmatic absolutism on one hand and subversive skepticism on the other. It thus laid a path not only for science but for liberal judgment more broadly.

WHAT follows, then, is a study of a utopian project to train liberal minds on an imperial scale. But we will approach this vast project through a small-scale experiment. In the foothills of the Alps in the 1880s, the Exners built a liberal

39. Paul Forman, "Weimar Culture, Causality, and Quantum Theory, 1918–1927: Adaptation by German Physicists and Mathematicians to a Hostile Intellectual Environment," *Historical Studies in the Physical Sciences* 3 (1971): 1–116. For a recent critique of the Forman thesis in light of Viennese developments, see Stöltzner, *Vienna Indeterminism*, pp. 21–46.

40. Sigmund Exner, *Entwurf zu einer physiologischen Erklärung der psychischen Erscheinungen* (1894; Thun and Frankfurt: Verlag Harri Deutsch, 1999), pp. 362–70.

41. Adolf Exner, *Über politische Bildung* (Vienna: Adolf Holzhausen, 1892), p. 22.

learning environment of the kind newly threatened by a conservative administration in Vienna. Here, at their summer retreat Brunnwinkl, they hosted readings, lectures, and discussions for relatives, friends, colleagues, and students. In Schorske's framework, the move from Vienna 250 kilometers west to the lakes of the Salzkammergut would signal a retreat from the political stage. This study argues, to the contrary, that the project of education and self-cultivation that defined Austrian liberalism was as rooted in domestic life as in the public sphere, and equally at home in the countryside as in the capital.

In Vienna in the 1880s the branches of the Exner family clustered in the neighborhood of the university, their homes roughly in walking distance of one another. Family members could often be found chatting in the leafy garden of the apartment on Josefstädterstrasse where the second generation had lived as adolescents, later the home of Marie Exner von Frisch. They gathered regularly to listen to music at Franz Serafin's home, joined by friends and colleagues. Their children played together on Pelikangasse, where Adolf and several family friends resided. In later years, as the children grew and their lives diverged, it was in the summers, at Brunnwinkl, that the family reassembled. This pattern was typical of the Austrian bourgeoisie: by renting or buying neighboring cottages, vacationers maintained and expanded social networks among relatives and friends who were separated during the rest of the year.[42]

By the early twentieth century, a summer vacation in the Salzkammergut had become a staple of academic life in imperial Austria. According to a recent search of the guest books of resorts in just one corner of this region, the visitors between 1873 and 1915 included approximately one hundred fifty professors.[43] Jews were prominent among them. Indeed, Viennese Jews often found it easier to assimilate at the summer resort (*Sommerfrische*) than in the imperial capital.[44] Even after the Second World War, Hermann Broch, Friedrich Torberg, and other refugees from the Third Reich still thought longingly of the Salzkammergut as their "Heimat," their true homeland. "When I get homesick in America," wrote Gina Kaus in her recent autobiography *From Vienna to Hollywood*, "it is never for Vienna, but always

42. Hannes Haas, "Der Traum von Dazugehören—Juden auf Sommerfrische," in *Geschmack der Vergänglichkeit: Jüdische Sommerfrische in Salzburg*, ed. Robert Kriechbaumer (Vienna: Böhlau, 2002), pp. 41–58.

43. Johanna Palme, *Sommerfrische des Geistes: Wissenschaftler im Ausseerland* (Bad Aussee: Alpenpost, 1999), p. 22.

44. Haas, "Traum von Dazugehören," p. 51.

homesickness instead for Aussee, for a certain forest path, a certain view, for a scent."[45]

These thinkers themselves traced the *Sommerfrische*'s liberal atmosphere to the example set in the early nineteenth century by Archduke Johann, the brother of Emperor Franz I and a passionate alpinist, hunter, and naturalist.[46] Liberals embraced Johann for his progressive views, exemplified by his marriage to a postmaster's daughter from the Salzkammergut.[47] He was also a representative of the ascendance of science in a liberalizing Austria. He served as the first curator of the Austrian Academy of Sciences and founded an institute and museum for mining, metallurgy, and natural history in Graz. Images of Johann as a hunter and mountaineer became an early form of publicity for tourism in the Austrian Alps and an icon of the liberal culture of the *Sommerfrische*.

Emperor Franz Josef put his own stamp on the Salzkammergut when he chose Bad Ischl, near Salzburg, as his summer residence. Each year on the eighteenth of August, urban vacationers in the region celebrated the emperor's birthday with a rustic feast, a tradition that marked "simultaneously the high point and the beginning of the close of the holiday season."[48] When Franz Josef passed through St. Gilgen by train in 1893, a flock of Brunnwinkl children were waiting at the station to greet him.[49] The Exners were typical in the pride they took in their royal neighbor, lending the summer retreat an air of apolitical patriotism.[50]

Woven into the culture of the *Sommerfrische*, then, were a tolerance for diversity, an enthusiasm for natural science, and a loyalty to the supranational principle of the empire that were broadly characteristic of Austria's liberal *Bildungsbürgertum* or educated middle class. By the twentieth century, the

45. Wolfgang Kos and Elke Krasny, eds., *Schreibtisch mit Aussicht: Österrreichische Schriftsteller auf Sommerfrische* (Vienna: Verlag Carl Ueberreuter, 1995), pp. 221, 222, 224.

46. See the reminiscences of liberal economist Felix Butschek in Palme, *Sommerfrische des Geistes*, pp. 156–57; and Selma Krasa, "Erzherzog Johann und Franz Joseph I. Zur Ikonographie der Jagdporträts des Kaisers," in *Jagdzeit: Österreichs Jagdgeschichte. Eine Pirsch* (Vienna: Historisches Museum der Stadt Wien, 1997), pp. 51–53.

47. The Frankfurt parliament chose Johann as the representative of the Habsburg emperor in 1848.

48. *Schreibtisch mit Aussicht*, p. 109.

49. Frisch, *50 Jahre Brunnwinkl*, p. 25.

50. On this liberal brand of Austrian patriotism, see Hannes Stekl, "Bürgertumsforschung und Familiengeschichte," in *Bürgerliche Familien: Lebenswege im 19. und 20. Jahrhundert. Bürgertum in der Habsburgermonarchie VIII*, ed. Hannes Stekl (Vienna: Böhlau, 2000), pp. 9–34, quotation on p. 20.

Sommerfrische had become an object of nostalgia in a tumultuous age. One Salzkammergut memoirist applied Ernst Bloch's aphorism to this phenomenon: "One needs a firm piece of ground to stand on. This is true above all for those who want to tear things off their hinges."[51] The *Sommerfrische* was thus for many members of Austria's liberal *Bildungsbürgertum*—such as Sigmund Freud, Theodor Gomperz, and Arthur Schnitzler—the fixed point that levered intellectual creativity and political vision.

Karl Kraus described the *Sommerfrische* as "a constant marriage of city and country."[52] Elise Gomperz found her "quiet alpine nest" to be "far more metropolitan" than Vienna itself.[53] The Exners likewise made Brunnwinkl into a rustic version of a Vienna salon. Yet the summer retreat was never simply an urban parlor transposed to the countryside. Interactions with the natural surroundings and with the native population were vital to the *Sommerfrische* experience. Arthur Schnitzler, for instance, retained vibrant memories of the natural world in which he spent his childhood summers: "Many only fleeting impressions of smaller flora and fauna . . . adhered indelibly in me. I will never forget my first liverleaf . . . Such were the impressions—which resurge in me almost more as symbols than as memories—that meant nature to me in my childhood."[54] Nature was a school for the eyes, for artists as for scientists. Thus Peter Altenberg's aphorism: "'I know every path here,' says the tourist. 'I know every blade of grass,' says the poet."[55] Even after decades in the United States, the physicist Victor Weisskopf vividly recalled the natural scenery of the Salzkammergut, which remained for him "the ideal image of nature in all its abundance and splendor."[56]

Even in the midst of such pleasure, however, some of these holidaymakers managed to see this natural world through the eyes of those who depended on it for their livelihoods. Admiration for the "inner nobility" of the alpine peasants was common among the Viennese,[57] and relations between the summer and year-round populations were often warm, particularly before the boom in tourism at the turn of the century.[58] Hugo von Hofmannsthal

51. The musicologist Gerlinde Haid, quoted in Palme, *Sommerfrische des Geistes*, p. 161.

52. Palme, *Sommerfrische des Geistes*, p. 12.

53. Quoted in Rossbacher, *Literatur und Bürgertum*, p. 271.

54. Kos and Krasny, *Schreibtisch mit Aussicht*, pp. 34–35.

55. Kos and Krasny, *Schreibtisch mit Aussicht*, p. 82.

56. Palme, *Sommerfrische des Geistes*, p. 59.

57. Josephine von Wertheimstein, quoted in Rossbacher, *Literatur und Bürgertum*, p. 271.

58. Palme, *Sommerfrische des Geistes*, p. 98.

captured the self-consciousness of these Viennese vacationers in a tribute to the fire brigade of the village where he summered as a youth. It was a hymn to the locals, to those "who always stay, when we go":

> Who do not watch for the mysterious
> Blue beauty of this water
> Nor its fragrance and grace
> But for the meager
> Growth of their scrawny crops
> For the fruit of the small garden
> For food to live on;
> Who fear the wild, beautiful storm
> Because it could cast
> Sparks in the dry barns.[59]

Through contact with the peasants of Strobl, Hofmannsthal had come to see in the Salzkammergut scenery more than a romantic backdrop. He had come to admire his peasant neighbors for their ability to survive in such an unpredictable world.

Like Hofmannsthal and other bourgeois intellectuals, the Exners wrote and spoke frequently about their experiences of the natural world and the native inhabitants of the Salzkammergut. In popular lectures, autobiographical sketches, and even scholarly writings, they portrayed themselves against the backdrop of their lakeside hamlet. Through these exercises in self-fashioning, they derived moral and intellectual authority from Brunnwinkl's natural surroundings.[60] As we will see, they even culled models of rationality from this rural environment, holding up the figures of the hunter, the farmer, or the naturalist as rational ideals—and as prescriptions for education in the empire at large.

Brunnwinkl was thus the bridge between the Exners' public and private lives. In thinking about Brunnwinkl's hybrid character as a public-private space, it is useful to draw loosely on the psychoanalytic concept of "transitional space." In the work of D. W. Winnicott, this term denotes a figurative space between individuals, made possible by the security of a stable environment, in which play and creativity become possible.[61] "Play" was a recurring

59. Kos and Krasny, *Schreibtisch mit Aussicht,* 190.

60. Lorraine Daston and Fernando Vidal, eds., *The Moral Authority of Nature* (Chicago: University of Chicago Press, 2004).

61. For a Winnicottian interpretation of another Central European intellectual colony, see Peter Loewenberg, "The Creation of a Scientific Community: The Burghölzi, 1902–1914," in *Fantasy and*

motif in the Exners' accounts of their lives at Brunnwinkl. As a transitional space, the summer retreat belonged to the Exners' imaginations as well as to the physical world; it was simultaneously subjective and objective, internal and external, private and public.

The homes of other *bildungsbürgerlich* families have been the subjects of recent historical studies. Historians of Central Europe have argued that middle-class homes were equal in importance to civic associations as sites for the negotiation of political and economic power.[62] Through marriages and the cultivation of family networks, the educated middle class defined itself as a distinct social group.[63] Particularly among the Viennese, family ties wove intricate and tightly knotted patterns through the scientific community.[64] The Exners' daughters found husbands among their students, while their sons married colleagues' daughters. They spoke of the university as a many-branched family, of scholarly writings as their intellectual progeny, of mentors as fathers and students as sons. To his students, Franz Serafin Exner was known as "Papa." The Exners interwove domestic life and the life of science so tightly that it is impossible to understand one without the other.[65] From this perspective, we can begin to identify the contributions of

Reality in History (Oxford: Oxford University Press, 1995), pp. 46–89. My use of Winnicott's model diverges from Loewenberg's primarily in that I take Winnicott's physical metaphors seriously. I am interested in the reliable environment constituted by people *and* the inanimate world.

62. Rebekka Habermas, *Frauen und Männer des Bürgertums: Eine Familiengeschichte (1750–1850)* (Göttingen: Vandenhoek & Ruprecht, 2000), p. 183. Classic works on the history of the family in modern times include Michelle Perrot, ed., *A History of Private Life*, trans. by Arthur Goldhammer, vol. 4 (Cambridge, Mass.: Harvard University Press, 1990); Leonore Davidoff and Catherine Hall, *Family Fortunes: Men and Women of the English Middle Class, 1780–1850* (London: Hutchinson, 1987); and Michael Stürmer et al., *Wagen und Wägen: Sal. Oppenheim jr. & Cie.: Geschichte einer Bank und einer Familie* (Munich: Piper, 1989).

63. On the German *Bildungsbürgertum*, see M. Ranier Lepsius, "Richard Lepsius und seine Familie—Bildungsbürgertum und Wissenschaft," in *Karl Richard Lepsius (1810–1884)*, ed. Elke Freier and Walter Reineke (Berlin: Akademie-Verlag, 1988); Jürgen Kocka, "Bildungsbürgertum—Gesellschaftliche Formation oder Historikerkonstrukt?" in *Bildungsbürgertum im 19. Jahrhundert*, ed. Werner Conze and Jürgen Kocka (Stuttgart: Klett-Cotta, 1989), vol. 4, pp. 9–20; and Friedrich Lenger, *Werner Sombart, 1863–1941: Eine Biographie* (Munich: C. H. Beck, 1994).

64. For examples not mentioned here, see Wolfgang L. Reiter, "Zerstört und vergessen: Die Biologische Versuchsanstalt und ihre Wissenschaftler/innen," *Österreichische Zeitschrift für Geschichte* 10 (1999): 585–614.

65. On the need to historicize the concepts of "public" and "private" within the history of science, see Catherine Goldstein, "Mathematik im Frankreich des frühen 17. Jahrhunderts," in *Zwischen Vorderbühne und Hinterbühne: Beiträge zum Wandel der Geschlechterbeziehungen in der Wissenschaft vom 17. Jahrhundert bis zur Gegenwart*, ed. Theresa Wobbe (Bielefeld: transcript Verlag, 2003), pp. 41–72.

women to liberal culture in Austria and to the culture of scientific inquiry—not as helpmeets or assistants but as moral authorities in their own right.

My focus on a family is a novelty among historians of science, who tend to stage science either as a vast ensemble production or a one-man show. At the first extreme lies science in its "*public* character,"[66] a product of the rise of the public sphere. On this model, science's power rests on the potential of an open and diverse community to achieve objectivity through the multiplication of individual perspectives. At the opposite extreme is the enduring ideal of science in solitude: the scientist as lone genius, unworldly and purely cerebral.[67] Such a drama might star Darwin, gestating his *Origin* in the Kent countryside, or Einstein, dreaming of life as a lighthouse keeper. By force of will, these characters seem to have independently transcended their worldly situations and thus the confines of individual subjectivity. These two backdrops to scientific work—one crowded and boisterous, one monastically still—are more alike than they might seem. In both stagings, science appears to thrive in isolation from emotional ties, above all those of domestic spaces. For the solitary genius, the family is at best a sheltering influence. Common wisdom has it that members of the household best contribute to the scientist's work by keeping their distance. The ideal of "public reason," on the other hand, has been defined in opposition to the "tribalism" of families and friends, a state prone to taboos and mysticism. As Hannah Arendt argued, "even the richest and most satisfying family life can offer only the prolongation or multiplication of one's own position with its attending aspects and perspectives . . . [T]his family 'world' can never replace the reality rising out of the sum total of aspects presented by one object to

66. Karl Popper, *The Open Society and Its Enemies* (New York, 1950). For this theme, see Larry Stewart, *The Rise of Public Science: Rhetoric, Technology and Natural Philosophy in Newtonian Britain* (Cambridge: Cambridge University Press, 1992); Richard Yeo, *Science in the Public Sphere: Natural Knowledge in British Culture, 1800–1860* (Aldershot: Ashgate, 2001); Geoffrey Sutton, *Science for a Polite Society: Gender, Culture, and the Demonstration of Enlightenment* (Boulder, Col.: Westview, 1995); Jan Golinski, *Science as Public Culture: Chemistry and Enlightenment in Britain, 1760–1820* (Cambridge: Cambridge University Press, 1992); and Jürgen Habermas, *The Structural Transformation of the Public Sphere,* trans. Thomas Burger (Cambridge, Mass.: MIT Press, 1991). A note on terminology: in its early modern usage, "private" designated all that was not part of the state, including the market, salons, and civic associations. Habermas instead uses the term "intimate" to refer primarily to the realm of the bourgeois family. I will use "intimate" and "private" interchangeably to refer to the sphere in which interpersonal relations are dominated by marriage, birth, and friendship and will use "domestic" to refer specifically to the household.

67. For a critique of this genre, see Steven Shapin, "The Mind Is Its Own Place: Science and Solitude in Seventeenth-Century England," *Science in Context* 4 (1991): 191–218.

a multitude of spectators."[68] Scientific rationality would seem nowhere less at home than in the home.

It is only recently that historians have begun to think of the family as a resource for scientists' self-fashioning in the professional age. Janet Browne has shown how Darwin built his scientific persona in his later years around his seclusion in a country household. Mary Jo Nye has revealed that the de Broglies carved their professional identities as physicists from their family's aristocratic status.[69] Studies of domestic science have restored credit to wives and daughters for their direct contributions to scientific work. Still, they have rarely considered the indirect ways in which domestic gender roles have shaped scientific activities.[70] More fundamentally, these studies have seldom questioned the categories of public and private at the heart of their investigations.

The literature on fin-de-siècle Vienna has tended to cement this dichotomy between reason and family life. An alternative interpretation of the intimate world of the Viennese can be found in Karl-Heinz Rossbacher's recent study of the literature of Austrian modernism.[71] The hallmark of this body of writing, according to Rossbacher, was its questioning of the reality of the individual subject and its search for a viewpoint outside the self. Rossbacher reads this literary culture as an effort to navigate the modern tension between individualism and intimacy. Drawing on Norbert Elias's account of the civilizing process, Rossbacher interprets literary writing and

68. Hannah Arendt, *The Human Condition* (Chicago: University of Chicago Press, 1998), p. 57. This judgment seems to have medieval roots, for instance, in the writings of Abelard; see Gadi Algazi, "Scholars in Households: Refiguring the Learned Habitus, 1480–1550," *Science in Context* 16 (2003): 9–42, quotation on p. 13.

69. Janet Browne, *Charles Darwin: The Power of Place* (Princeton: Princeton University Press, 2002); Mary Jo Nye, "Aristocratic Culture and the Pursuit of Science: The De Broglies in Modern France," *Isis* 88 (1997): 397–421. For other examples of science in the home, see the essays in Therese Wobbe, ed., *Zwischen Vorderbühne und Hinterbühne: Beiträge zum Wandel der Geschlechterbeziehungen in der Wissenschaft vom 17. Jahrhundert bis zur Gegenwart* (Bielefeld: transcript Verlag, 2003); H. M. Pycior et al., eds., *Creative Couples in the Sciences* (New Brunswick: Rutgers University Press, 1996), Paul White, "Science at Home," in *Thomas Huxley: Making the "Man of Science"* (Cambridge: Cambridge University Press, 2002); Alix Cooper, "Homes and Households," in *The Cambridge History of Science,* volume 3*: Early Modern Science,* ed. Katharine Park and Lorraine Daston (Cambridge: Cambridge University Press, forthcoming), Joan Richards, "Parallel Universes: Natural Theology and the Power of Reason," *Science in Context* (forthcoming); Algazi, "Scholars in Households." On science and private life more broadly, see too Robert J. Richards, *The Romantic Conception of Life* (Chicago: University Chicago Press, 2002).

70. For critical attention to this question see White, "Science at Home."

71. Rossbacher, *Literatur und Bürgertum.*

discussion as channels for formalizing intimate conversation once polite society made direct language taboo. Essential to this interpretation, although Rossbacher does not emphasize it, is the context of Austrian liberalism. Opposed, on one hand, to the dogmatism of the Catholic church, and fearful, on the other, of the anarchic potential of radical skepticism, Austrian liberals performed a delicate balancing act. They were alternately enamored of and repelled by individualism, and dreaded solipsism as much as they did conformity. Rossbacher at times implies that literature was simply a means of connecting. Yet it is clear that the liberal intellectuals of his study were seeking something more multivalent, a discourse that could simultaneously establish intimacy and distance.

I argue that science performed that function. Within the close-knit circle of liberal intellectuals to whom Rossbacher introduces us, one was as likely to meet a scientist as a poet. We tend to forget that in the era between the laboratory revolution and the dawn of "big science," scientific work was often carried out in the home, with relatives as collaborators. The very rhythm of this domestic labor, with its intervals of solitude and companionship, served to moderate the extremes of individualism and conformity. Science trained the senses and the communicative faculties. To become a connoisseur of flora and fauna, to record constancy and change in the heavens, to report on the physiology of one's own senses—all of these activities involved learning to observe the world independently, yet with a mind constantly attuned to the challenges of communication. By presenting novel challenges of perception and communication, science became a means of negotiating the boundaries between interiority and intimacy. The modalities of probability in particular allowed the Exners to steer between the virtues of individualism and the dangers of solipsism. Science in the home thus became part of the conflict-ridden modern struggle to form and maintain a self that is both independent and related. And the home itself became a transitional space, bridging the public and private lives of its inhabitants.

Working at the border between the physical and sensory realms, the Exners aimed to bridge the gulfs between individual minds, to build a high road over the abyss of solipsism. In the parlance of Viennese liberalism, their goal was *Vielseitigkeit* (many-sidedness or versatility), the ability to see from perspectives beyond the personal. According to an educational philosophy that had become common wisdom among the Austrian *Bildungsbürgertum* by the mid-nineteenth century, learning proceeded from one-sidedness to many-sidedness. *Vertiefung* (absorbed contemplation) in the object of knowledge gave way to *Besinnung* (self-conscious reflection), through which the pupil related the newly acquired knowledge to other ideas and experiences.[72] To call someone many-sided was to bestow the highest compliment of this educated society. Many-sidedness expressed a liberal vision of a society whose cohesion would emerge from successful communication, not from imposed uniformity. Communication was the key to overcoming the personal and class-based conditions of perception. The ideal of many-sidedness equated the common (a normative concept) with the normal (a descriptive and statistical concept). In the context of the industrial revolution, the call for many-sidedness was a bid to overcome the specialization required by a modern division of labor and the dehumanizing effects of bureaucratization. The modern state demanded experts, however, and entrusted public institutions with producing them. The liberal bourgeoisie therefore expected the seeds of many-sidedness to be planted and nurtured within the sphere of the family.[73]

Here, then, was the relationship of the family and the state as envisioned by Austria's liberal educated middle class: neither an opposition between private and public, nor a replica of patriarchal authority on a smaller scale, the family was instead a training ground for the challenge of participating in a newly diverse society. Even in an age of public education, bourgeois families cultivated many-sided personalities in their parlors, gardens, and summer retreats. Robert Zimmermann, for decades Austria's most influential philosopher and a student of the first Franz Exner, explained how a group could arrive at a concept of human nature by distinguishing among themselves "what emerges as similar and of the same common nature from what is unique to the individual." Human nature so defined would have only a finite "degree of justification" or probability.[74] By expanding the reach of such a conversation one could, in principle, approach asymptotically to a universal perspective. In this way, the family was meant to be a first approximation to universality. It was the smallest sample size of the population that allowed one to glimpse the contours of the universally human. What was most intimate thus came to be associated with the "common," the "normal," and the average.

72. This discussion relies on Dietrich Benner's lucid analysis in his *Die Pädagogik Herbarts* (Weinheim: Deutscher Studien Verlag, 1986); and in J. F. Herbart, *Systematische Pädagogik*, Bd. 2: *Interpretationen*, ed. Dietrich Benner (Weinheim: Deutscher Studien Verlag, 1997).

73. Benner, *Systematische Pädagogik*, pp. 108–15.

74. Robert Zimmermann, *Über Hume's empirische Begründung der Moral* (Vienna: Carl Gerold's Sohn, 1884), p. 12.

Of course, the Exners were by no means a "representative" family. While their reliance on probabilistic reasoning and their devotion to the culture of the *Sommerfrische* were typical of their liberal colleagues, they were in many ways exceptional. But my goal is not to arrive at a universal epistemology of the domestic sphere. My aim is far more modest, yet potentially troubling to those who wish to draw a neat line between reason and affect. The Exners demonstrate one historically contingent set of possibilities for configuring the relationship between science and family life. For instance, I argue that their usage of such pivotal terms as "subjective," "normal," and "universal" was as conditioned by experiences of domestic intimacy as it was by "public" discourse. In such ways, their intimate lives shed light on the content of their research. Science, it has been said, is an "epistemology of *common* experience."[75] Yet where if not among family do scientists first learn the meaning of "common"?

Writing a family history accentuates certain methodological problems that are often overlooked in other forms of biographical research. The Exners left behind a wealth of memoirs and letters alongside their vast publications. Such collections are always subject to various forms of selective filtering, whether by the authors themselves or those close to them. In no case do archives provide an "objective" representation of an individual's life. In the case of a family like the Exners, intent on preserving their own history, this problem comes into sharper relief. Certain documents exist today only as they were copied and edited by succeeding generations. Rather than lamenting this situation as a barrier to historical objectivity, however, I see it as a resource. The filtering process signals to the historian the need for caution, but it also opens a new interpretive dimension. These materials are tangible evidence of the ways in which the Exner family generated memories and forged continuity. They are evidence of the process of the family's self-fashioning.

Although historians of science have devoted careful attention to the dynamics of research schools, they have paid surprisingly little attention to the institution of the family. Families in nineteenth-century Europe frequently became research schools of their own. The Darwins produced three generations of scientists; the De Broglies shared a home laboratory, as did the Curies; and Central European collaborators included the Ostwalds in Leipzig and the Przibrams in Vienna. Unique to this analysis of the Exner dynasty,

75. Stewart, *Public Science*, p. xxi, emphasis added.

FIGURE 1 Brunnwinkl circa 1906. Source: Ernst v. Frisch, *Chronik von Brunnwinkl* (Vienna: self-published, 1906), frontispiece.

however, is Brunnwinkl's character as a distinctly "public" domestic space, a quality specific to the Exners' milieu and their social status.

ONE approached Brunnwinkl from a path leading from the village of St. Gilgen along the bank of the Wolfgangsee. The cottages clustered close to the shore. Beyond them the land rose toward the tall peak of the Schafberg. With their long balconies and window boxes filled with flowers, the houses were designed to achieve a seamless harmony between architecture and landscape.[76] Nature pressed in from all sides. Windows and doors were rarely closed and animals nested in corners and attics. Summer after summer, flood waters swelled over the banks of the brook. Karl Frisch's earliest specimens repeatedly fell prey to mold. Karl's brother Ernst attested that he

76. Monika Oberhammer, *Sommervillen im Salzkammergut: die spezifische Sommerfrischenarchitektur des Salzkammergutes in der Zeit von 1830 bis 1918* (Salzburg: Galerie Welz, 1983).

had known summers on the Wolfgangsee when "the ships' sterns lay dry in the lake and the Schafberg began to burn from drought; of course, also summers in which the cattle were washed from the pastures to the valley." For the entire region, Ernst explained, such extremes were "possible, but not the rule."[77] The gradations between "possible" and "regular"—the limits of predictability—formed the framework within which the young Exners learned to reason.

The summer retreat was, as Emilie Exner said, an "experiment in uniting so many uniquely gifted people in a single spot." For the experiment to succeed, the individual must be shielded from the "herd existence." The family must not become a "sect"; it must preserve "diversity alongside what is held in common." Brunnwinkl's "unwritten laws" therefore guaranteed individual freedom.[78]

At the center of the Brunnwinkl cottages stood the linden tree. Planted soon after the colony's founding, the linden became its communicative hub and its symbolic center. In a poem set to a well-known melody by Schubert, the linden by the spring (*Brunn*) with words of love carved into its trunk was the voice of nostalgia calling the narrator back from his wanderings. Likewise, the Exners saw the linden as the "voice" that held this community together. Its trunk became a "blackboard" on which family members posted letters from the outside world, notes to each other, poems and riddles. These communicative exercises were the counterweight to the independence of mind the family cultivated. In verses that the Brunnwinklers wrote for the linden's birthdays, the tree spoke of the community's past, present, and future. Even in the summer of 1938, after Hitler's annexation of Austria, the family still heard the tree's promise: "What holds diversity together/ Is benevolent understanding."[79]

At Brunnwinkl the Exners honed their strategies for teaching the young to look, listen, and think for themselves.[80] Through nature study, painting, and music, the young Exners practiced an attitude of intellectual independence. At the back of the "cobbler's cottage" was the music room for Brunnwinkl's

77. Ernst v. Frisch, *Sommer am Abersee* (Vienna: self-published, 1938), p. 37.

78. Emilie Exner, *Der Brunnwinkl* (Vienna: self-published, 1906), p. 7. Life at Brunnwinkl was typically described as "ungebunden und frei." Franz Exner, *Exnerei* (Staltach: self-published, 1944), p. 28.

79. Karl Frisch, *Fünf Häuser am See. Der Brunnwinkl: Werden und Wesen eines Sommersitzes* (Berlin: Springer, 1980), p. 49.

80. Emilie Exner, *Brunnwinkl*, p. 7.

FIGURE 2 The linden tree, 1888. Source: Hans Frisch, *50 Jahre Brunnwinkl* (Vienna: self-published, 1931), p. 22.

young players, with the front stoop providing seating for a small audience. The repertoire included Mozart, Haydn, and Brahms (a family acquaintance). Listening to her nephews play, Emilie Exner judged that they were "innocent of the stirring enchantment of this melody, which was perhaps so effective precisely because the players went about it so happily and artlessly."[81] This "Unbefangenheit," an utter absence of manner and pretense, was the aesthetic ideal of Emilie's generation—one embraced, for instance, by such literary realists as Gottfried Keller and Marie Ebner von Eschenbach, both intimate friends of the Exners and guests at their summer retreat. For the proper way to listen to music, this generation looked to the Viennese philosopher and music critic Eduard Hanslick, a former student of the first Franz Exner. In his famous essay on the "musically beautiful," first published in 1854, Hanslick characterized as "pathological" a mode of listening that was purely passive, unconscious, and visceral, epitomized by the intoxicated "Schwärmerei" of Wagner's fans. "To the contrary," Hanslick wrote, "an aesthetic appreciation of music takes place only when one fully 'notices' it, pays attention to it, and becomes directly conscious of each of its attractions."[82] The essential element of this aesthetic experience was its temporal unfolding: "continuously to follow and outpace the intentions of the composer, to find oneself confirmed here in one's suspicions, there pleasantly deceived."[83] Hanslick's account of the play of attention and anticipation of the music connoisseur is an apt description of the attitude the Exners aimed to foster in their heirs: self-disciplined and focused, anticipating a course of development yet welcoming the unexpected.

The young Exners were acquiring the habitus of the Austrian *Sommerfrische*.[84] This fell "between elegance and folksiness, [between] society's aloofness and summer's friendliness."[85] As the Exners' contemporaries recognized, one could only learn to tread this line between individualism and intimacy through early exposure to the proper domestic environment. The composer Johannes Brahms, a frequent visitor to the Wolfgangsee, confessed his own "lack of social qualities, namely, that freedom, security, and informality [*Ungezwungenheit*] of manner, which must be acquired in youth—in

81. Ibid., p. 15.

82. Eduard Hanslick, *Vom Musikalischen-Schönen: Ein Beitrag zur Revision der Aesthetik der Tonkunst,* 2nd ed. (Leipzig: Weigel, 1858), p. 134.

83. Hanslick, *Vom Musikalischen-Schönen*, p. 133.

84. Kriechbaumer, *Geschmack der Vergänglichkeit*; Haas, "Die Sommerfrische."

85. Rossbacher, *Literatur und Bürgertum*, p. 268.

one's parents' house and its environs—and the subsequent acquisition of which is not easy even with great effort and force of will."[86] Brahms's observation came in a letter to the most prominent of the Exners' friends and neighbors at the Wolfgangsee, physician Theodor Billroth. The qualities he described were precisely those the Exners valued at Brunnwinkl.

Like the Wertheimsteins or Billroths, the Exners helped define the "classical" Austrian *Sommerfrische* and its habitus. Tellingly, Brunnwinkl gave birth in the first decade of the twentieth century to a "daughter colony" nearby on the Attersee, where descendants of family friends like the Billroths put down roots.[87] At the summer retreat, the Exners seized on a unique resource for fashioning themselves as liberals: the natural environment. As one of the Exners' guests at Brunnwinkl remarked, "How truly Austrian [is] this association of outdoor life with the most refined culture of intellect and love."[88] As historians have noted, the history of the modern field sciences is intimately tied to that of bourgeois recreational culture.[89] A particular style of interacting with nature became essential to the self-definition of Austrian liberals—not the subjectivity of the Romantic alpinist, but the cunning of the hunter or the foresight of the farmer, the capacity to outwit a capricious natural environment. Nature mediated the cultivation of a manner flaunting independence to a well-calibrated degree. To these liberals, there was no tension between what Schorske labels the "rational" and the "aesthetic." Learning not just to tolerate but to find beauty in the unpredictable was part of cultivating a liberal character.

THE chapters that follow move across eighty years of history and dozens of individual lives. Each focuses on a juncture at which the Exners' development as a family intersected with transformations in Austrian science, culture, and politics. I do not claim that the meaning of the "probable" was constant or uniform throughout. Probability in the Exner circle was variously an epistemological standard, a tool for quantifying the bounds of physical variability, and a model of mental function. It began as a reflection of human ignorance and became a mirror of the fundamental randomness

86. Rossbacher, *Literatur und Bürgertum*, p. 86.

87. Karl Frisch, *Fünf Häuser am See*, p. 91.

88. A friend of the Exner family in 1931 on the occasion of the fiftieth anniversary of Brunnwinkl's founding, quoted in Karl Frisch, *Fünf Häuser am See*, p. 97.

89. Henrika Kuklick and Robert E. Kohler, introduction to "Science in the Field," *Osiris* 11 (1996): 1–14.

of the physical world. Individual members of the Exner family applied and interpreted probability differently, but they embraced it on the basis of one crucial shared conviction. Whether they dealt with physical, psychological, or social phenomena, they believed that the world was too complex to admit of deterministic laws. In their eyes, those who denied this complexity veered dangerously towards dogmatism. No less dangerous in their view, however, were those who denied that such a world could nonetheless be described objectively, for they raised the specter of anarchy. The probabilistic attitude distinguished liberals from these opponents. It justified the liberals' attempts to explain and govern a world where certainty and predictability were always just beyond reach.

Probabilistic reasoning endured as a mark of a liberal character because of the Exners' strenuous efforts in public and private. The qualified success of the Exners' project depended on the configuration of the social world of the Austrian *Bildungsbürgertum* in the latter half of the nineteenth century, in which academic work and family life occupied the same social sphere. In the 1840s, however, that world was still a dream for the young Franz Exner, an idealistic liberal in a Catholic, absolutist empire. In the next chapter we will see how this thwarted philosopher came to seek a model of a postdogmatic mind.

CHAPTER ONE

The Mind Set Free

Preparing a Liberal Society in the 1840s

Among all the secret societies and clubs that the state often prohibits in critical periods, still it unhesitatingly tolerates the family-clubs, which include as many children as we baptize.[1]

FRANZ EXNER'S childhood passed in the shadow of the Napoleonic wars, a time of fear and insecurity in Austria. He was born in 1802, the last surviving child of Joseph Exner, a customs official in Vienna, and his wife Magdalena, born Supper.[2] Exner was three when Napoleon's troops invaded the Austrian capital. In later years, Austrian liberals would recall the French occupation as the moment of the awakening of Austrian national self-consciousness. Exner's father would be remembered for having shown "courage and reason" at the height of the crisis. Anecdotes insist that little Franz likewise displayed

1. Jean Paul, *Levana*, 2nd ed. (Stuttgart and Tübingen: J. G. Cotta, 1845), preface.

2. "Stammbaum," in Franz Exner Nachlass. On Exner, see Solomon Frankfurter, *Graf Leo Thun-Hohenstein, Franz Exner und Hermann Bonitz. Beiträge zur österreichischen Unterrichtsreform* (Vienna: Hölder, 1893); Robert Zimmermann, "Nekrolog: Dr. Franz Exner," *Akademischen Monatsschrift* 1853 (in Franz Exner Nachlass); Wilhelm von Hartel, *Festrede zur Enthüllung des Thun-Exner-Bonitz Denkmals*, (Vienna: Verein deutscher Philologen und Schulmänner, 1893). Frankfurter's 1893 book is especially valuable because it quotes letters from Exner that have since been lost. Frankfurter himself was later the president of the *Verein der Freunde des humanistischen Gymnasiums*, an association to protect the classical *Gymnasium* curriculum, of which Sigmund Exner, Franz Serafin Exner, and Emilie Exner were members. Hartel was minister of culture and education and a friend of the Exners. Zimmermann was Exner's student and the son of a close friend.

"great courage" in the face of the French soldiers.[3] It is only natural that Exner, a future hero of the Austrian state, would be pictured as a fearless child facing down Napoleon's troops.

Joseph Exner's hopes for his only son were kindled by one of the boy's early teachers. "Your boy has much talent," he was told. "Something good can come of him—just be sure to protect him from bad society." Accordingly, Joseph set strict limits to his son's social life once he entered *Gymnasium*. He restricted Franz's visits to theaters and other "distractions" (*Zerstreuungen*). During school vacations, Franz was sent to a relative in the country, where he was introduced to more wholesome pursuits like fishing and bird-calling.[4] His lifelong friend Josef Mozart believed that Franz's ethical conscience was the fruit of his upbringing. In Joseph and Magdalena's house, "vulgar, hostile expressions, speech or actions were never observed or tolerated . . . [A]ll that occurred was carried out with enlightenment, propriety, and gravity." As Mozart hinted, this house was fertile ground for the growth of a moral philosopher. At age sixteen Franz experienced an "ethical awakening," and from then on avoided those of his peers "with ignoble, crude, or frivolous characters." "Highest of all," Mozart recalled, "he demanded purity of morals."[5] In recounting his friend's childhood as an example of good rearing, Mozart stressed the value of self-discipline and the danger of "distraction." These were pedagogical concerns that would come to preoccupy both men as educators.

Franz's father taught him to read and write at a young age, sent him to a parish elementary school, and in 1812 to Vienna's prestigious Akademisches Gymnasium. The training Franz received there was not designed to excite his curiosity. In this period secondary and higher education in the Habsburg Empire was intended to produce competent and loyal public servants.[6] In

3. Josef Mozart, "Biographisches Fragment," 1–2, Franz Exner Nachlass. Mozart was Exner's friend and a participant in the midcentury education reforms. He was apparently of no relation to the composer.

4. Mozart, "Biographisches Fragment," 6.

5. Ibid., 7.

6. On pre- and post-1848 Austrian education, see Helmut Engelbrecht, *Geschichte des österreichischen Bildungswesens*, vols. 3–4 (Vienna: Österreichische Bundesverlag, 1986); Gary B. Cohen, *Education and Middle-Class Society in Imperial Austria, 1848–1918* (West Lafayette, Indiana: Purdue, 1996); Walter Höflechner, *Die Baumeister des kunftigen Glücks: Fragment einer Geschichte des Hochschulwesens in Oesterreich vom Ausgang des 19. Jahrhunderts bis in das Jahr 1938.* (Graz: Akademische Druck- und Verlagsanstalt, 1988); Hans Lentze, *Die Universitätsreform des Ministers Graf Leo Thun-Hohenstein* (Vienna: Bohlau, 1962); Frankfurter, *Thun-Hohenstein*; Sonja Rinofner,

the eighteenth century, Empress Maria Theresa had established a system of secular primary education prompted by the needs of the modernizing state. She and her son Joseph did little to foster secondary and higher education, however. They did not replace the *Gymnasien* that were forced to close when the state banned the Jesuits in the 1770s.[7] Following the Napoleonic wars, the Catholic church regained much of its control over education. At the universities, lectures were strictly circumscribed by Catholic dogma and official texts. The state-approved philosophy textbook of 1838 presented moral philosophy as a subfield of "rational theology," and metaphysics included "onto-theology" and "cosmo-theology."[8] All candidates for the philosophy degree were required to take a "systematic" course in religion, which adhered meticulously to official dogma.[9] Professors literally read aloud from their readers or "Lesebücher," and so students rarely came to class.

The young Franz Exner was a *Gymnasium* student during the calm and optimistic years after the defeat of Napoleon in 1814. With the end of twenty-five years of war came a new emphasis on domesticity and the natural world, the beginning of the Biedermeier age. Politically, the first few years after the Congress of Vienna brought a relaxation of the repressive measures of wartime. In Prussia, Friedrich Wilhelm III promised his subjects a constitution, and between 1814 and 1819 constitutions came into effect in eleven German states. Throughout the German lands, students formed patriotic societies or *Burschenschaften*, showing their support for a united Germany.

In 1819, this hopeful period came to an abrupt end. Austria's mastermind of reactionary government Clemens von Metternich used the murder of a minor monarchist to win the support of the Prussian and Austrian emperors for a new wave of repression. The imperial government feared in particular the subversive potential of student associations. The Carlsbad Decrees prohibited political associations, including the *Burschenschaften*, and instituted widespread censorship. Throughout the 1820s the state police kept close watch on the universities for signs of illegal political activity.

Franz entered the University of Vienna to study philosophy in 1819. By then, Metternich's policies had brought the life of learning in Austria

---

ed., *Zwischen Orientierung und Krise. Zum Umgang mit Wissen in der Moderne* (Vienna: Böhlau Verlag, 1998).

7. Cohen, *Education and Middle-Class Society*, p. 15.

8. *Lehrbuch der Philosophie*, vol. 1 (Vienna: k.k. Schulbücher-Verschleiß-Administration, 1835).

9. *Systematischer Religions-Unterricht für Candidaten der Philosophie*, 2 vols. (Vienna: k.k, Schulbücher-Verschleißes, 1821).

virtually to a standstill. Franz, however, had the good fortune to encounter a professor of moral philosophy who was not daunted by official decrees. Josef Rembold was popular with students for his lively and effective lectures, which often strayed from the approved texts. It would soon be clear that Rembold was treading a dangerous course.

Given the straitjacket on discussions at the university, much of Franz's education took place outside of the lecture halls. He spent summer holidays with three of his closest friends, with whom he cultivated a passion for literature, philology, and aesthetics. In a rented house outside of Vienna, the friends "devoted themselves alternately to nature, studies, and camaraderie." A few years later a slightly larger group gathered in the autumn in a house overlooking a park south of Vienna, where they discussed aesthetics and politics against a backdrop of gardens, hills, and waterfalls. These idylls, painted in the nostalgic colors of middle-aged reminiscences, would feed Franz's later visions of a utopian community. Friendship, nature, and philosophical exchange were the essential ingredients for such a retreat. Mozart identified the values that this small community embodied for Exner: it "formed the highlight of Exner's adolescence. Whatever lent value to life was gathered here; the need for higher education was met by the intellectual resources, a mutual exchange of ideas and knowledge . . . Added to this were the well-developed self-consciousness and confidence of youthful strength at its height, and the open view into the distant and still veiled future."[10] Another member of this circle confirmed that the most important component of these holidays was unfettered dialogue, the "unreserved exchange of sentiments, thoughts and perspectives on everything that claimed their interest."[11]

Under an education system that meticulously restricted the content of discussion in public settings, such private dialogues were crucial. Historians have stressed the importance for the rise of liberalism in Central Europe of apolitical public societies, including reading circles and gymnastic and singing groups. Yet private groups like that of Exner's student days offered the opportunity for more candid dialogue. For instance, when revolution broke out in France in July 1830, Exner and his old classmates used their customary summer gathering to discuss the consequences freely.[12] Soon after, when disturbances in the German lands prompted a new wave of censorship,

10. Mozart, "Biographisches Fragment," 17.

11. L. Blumfeld, "Biographisches Skizze," pp. 29–30, Franz Exner Nachlass.

12. Blumfeld, "Biographisches Skizze," p. 29; see too Mozart, "Biographisches Fragment," 16.

such private associations became the only secure forum for political discussion. Like the pre-1848 *Burschenschaften*, Exner's student group was founded on male friendship and free intellectual exchange. Unlike the "half-public" student associations,[13] however, Exner and his friends met in seclusion. The bucolic settings of their gatherings became part of Exner's vision of a liberal utopia, ambiguously poised between privacy and publicity.

Despite the intellectual stimulation Exner found in this community, after two years of philosophy study he transferred to the faculty of law. Throughout his life he would struggle to choose between contemplative and active pursuits. Still hesitating between philosophy and law, he left Vienna in the fall of 1824 to complete his third year of legal study at the Habsburg university in Pavia, which had been returned to the Habsburg crown in 1815.[14] Anti-Habsburg agitation in the Italian lands was just then becoming a major concern to Metternich. During Exner's stay, students and soldiers clashed frequently. Ironically, while studying law in Pavia, he shifted allegiances back to philosophy. While there, he took the philosophy exams to sample the character of the discipline in the Italian province. He was unimpressed by the conservative Italian philosophers, and his examiners in turn were shocked by his citations of Kant and Locke. Nonetheless, when he returned to Vienna in 1825 he was awarded a first degree in philosophy on the basis of this exam.[15]

This was not an auspicious moment to begin a career in philosophy in the Habsburg lands. When Exner returned to Vienna from Pavia, his teacher Rembold was under official investigation. Rembold was charged with deviating from the prescribed textbooks, and he refused to confide his lecture notes to the government for inspection. In the end, the authorities prevailed: at the age of forty-five Rembold abandoned philosophy and began a career in medicine.[16] Meanwhile, efforts to defend Rembold would occupy Exner and his liberal-minded friends and colleagues for several years. Behind the accusations against his mentor, Exner believed, were "many secret plots that had been set in motion against him."[17] For many of Exner's former classmates who had planned on a career in philosophy, these events were an overwhelming discouragement. Exner, embittered and wary, nonetheless became more determined than ever to pursue an academic career.[18]

13. Judson, *Wien Brennt!* p. 27.

14. Blumfeld, "Biographisches Skizze," p. 24.

15. Ibid., pp. 25–28.

16. Frankfurter, *Thun-Hohenstein*, p. 156.

17. Mozart, "Biographisches Fragment," 14.

18. Ibid.

In 1827 Rembold's replacement at the University of Vienna was incapacitated, and Exner was given the bittersweet opportunity to take up the duties of his old teacher, along with the teaching of pedagogy. His career advanced at a virtually unprecedented pace. At the age of thirty Exner was offered the chair of philosophy at the University of Prague.

THE PHILOSOPHER IN THE CLASSROOM

On first impression Exner's lectures must have been a disconcerting experience for Prague's students, accustomed as they were to white-haired professors reading from the *Lesebücher* in monotone. Exner did not even bring a notebook with him. Yet his lectures flowed seamlessly, peppered with literary allusions and frank criticisms. At times he fell into discussion with students; at other moments he grew pensive, and students watched as thoughts "took him by storm." When he followed aloud a train of thought, "his listeners did not tire of following him with their thoughts and eyes . . . When he had mounted the podium and leaned over the balustrade, it was an interesting spectacle [*Schauspiel*] to observe how it grew ever quieter, until finally even the lightest whisper died. One could not hear a single breath."[19] Students showed up for his lectures who were not even registered for the course, and when illness kept him from teaching the class was despondent. The music critic Eduard Hanslick, one of Exner's students in those years, recalled that "Exner's beautiful intellectual's forehead appeared to us crowned by a kind of halo of perfection."[20]

The impression Exner made on his students in Prague had much to do with the state of university education in the Habsburg Empire before 1848. "One pictured professors as pedants, ignorant of or even averse to the learning of the day," explained Exner's student Robert Zimmermann, who went on to become one of the leading Herbartian philosophers in Austria. "Exner had nothing of the pedant, physically well trained, a lively social companion, a tasteful and fluent speaker."[21] Indeed the Tuesday evening discussions that Exner hosted gave him something of a reputation as a *salonnier* in Prague. Among the regular attendees were the physicist Christian Doppler and the

19. "Prager Schildereien," *Prager Morgenpost*, 6 May, 7 May, and 8 May 1858, in Franz Exner Nachlass; see too description of his lectures in a page from the diary of Exner's student Hammerschlag, in Franz Exner Nachlass.

20. Hanslick, "Aus meinem Leben," quoted in Frankfurter, *Thun-Hohenstein*, p. 56.

21. Zimmermann, "Nekrolog."

learned aristocrat Graf Leo von Thun, who would later be the minister in charge of implementing the education reforms of 1848–49. Exner's students rightly saw him as a link to a larger world, particularly to the latest developments in German philosophy. "He let on that his lectures afforded a view into the philosophical life of the Germany 'outside' [*draussigen Deutschlands*] and did not confine themselves to the limits of the k. k. ["imperial and royal," official state] textbook . . . It flattered them that they had a professor like 'the ones in Germany.'"[22]

The fresh air that Exner brought to philosophy in Prague blew from Prussia, where Johann Friedrich Herbart was developing his pedagogical psychology in Göttingen and Königsburg. Exner had already begun to delve deeper into Herbart's writings during his time in Pavia. Herbart's ambition of an empirical psychology on the model of physics inspired Exner to expand his knowledge of mathematics and to explore the latest developments in physics, chemistry, mineralogy, and crystallography. "My studies are now very dry and rather boring," he wrote to Mozart in the 1830s. "They deal with mathematics, logic, and physics. That the first two are dry, that you'll take my word for; for physics, first of all, the experiments are missing. It's still a matter of airing out forgotten facts [*Facta*] and collecting new ones. But there are so many of them, and they lack a unifying idea. I had never thought that this science was still to such a degree a disorderly hodge-podge [*Gemengsel*] of individual facts as I now see, and that the best physicists still know so little."[23] Exner's critique of physics guided his aspirations for psychology: an empirical science founded on experiments and ordered into laws.

Exner's classroom antics hinted at his concern with the field of pedagogical psychology, a focus of Herbart's. The professor's challenge, as Exner would later formulate it, was to train students to manage their tendency towards distraction. As the experiences of his students attest, Exner was perfecting techniques for engaging a class's attention by making his lectures a form of spectacle. His tactics were part of what historian Jonathan Crary has described broadly as the construction of a subjectivity of attentiveness in the mid-nineteenth century, accompanied by the rise of anxiety over the human propensity for distraction.[24] To Exner and other progressive

22. Ibid.

23. Franz Exner to Josef Mozart, 17 Nov. 1837, quoted in Frankfurter, *Thun-Hohenstein*, p. 68.

24. Jonathan Crary, *Suspensions of Perception: Attention, Spectacle, and Modern Culture* (Cambridge, Mass.: MIT Press, 1999), pp. 4–10.

philosophers of the day, the capacity to direct one's attention offered intuitive evidence for mental freedom.[25]

The meaning that Exner attached to "freedom" diverged from that of Western European liberal thinkers like Rousseau or Mill in that it did not imply a list of powers to be withheld from the state. Rather, in the tradition of Kant, Exner understood freedom as the condition in which humans have the potential to become self-legislating beings. In contrast to French or British liberal thought, freedom in this sense was compatible with a hierarchical and tradition-bound society. According to Kant, an enlightened state demanded obedience in action but permitted dissent in the form of rational debate.[26]

Kant's effort to define freedom led into the realm of metaphysics, a field into which Exner followed. In order to reconcile free will with a view of nature founded on mechanical causality, Kant had posited two separate realms: a world of physical objects governed by deterministic laws and a mental realm in which freedom reigned supreme.[27] Departing from Kant, Hegel had attempted to construct a theory of mind on the view of mental activity as pure spontaneity. To Herbart, Kant's and Hegel's solutions to the problem of free will were equally anathema. Herbart was a realist philosopher afloat in the Prussian sea of Hegelians, liberal-minded but no radical. His highest ambition was a mathematical psychology founded on empirical laws, a vision that became central to Austrian contributions to psychology in the nineteenth century.[28] A specific interpretation of Herbart's philosophy shaped the Austrian philosophical tradition in the nineteenth century, and Exner helped forge that reading. Herbart's project of determining the laws of thought, the goal that Exner believed would make psychology relevant to "moral theory, politics, and pedagogy," was clearly incompatible with the view that mental processes were purely spontaneous. For this goal,

25. Bolzano, *Adversarien* (1806–12), p. 34, quoted in Eduard Winter, *Religion und Offenbahrung in der Religionsphilosophie B. Bolzanos* (Breslau: Müller und Seiffert, 1932), p. 487. On Bolzano, see also William M. Johnston, *The Austrian Mind: An Intellectual and Social History, 1848–1938* (Berkeley: University of California Press, 1972), pp. 274–78.

26. Kant, *Foundations of the Metaphysics of Morals* and *What Is Enlightenment?* trans. Lewis White Beck (New York: Macmillan, 1990).

27. See Henry Allison, *Kant's Transcendental Idealism* (New Haven: Yale University Press, 1983).

28. Though Exner never met Herbart in person, he sent a student to Göttingen to transcribe Herbart's lectures. (Frankfurter, *Thun- Hohenstein*, p. 72.) On Herbart's life, see U. P. Lattmann and P. Metz, *Bilden und Erziehen* (Aarau, 1995), pp. 142–153. On Herbart's influence on Freud in particular, see Sulloway, *Biologist of the Mind*, p. 67; Henri F. Ellenberger, *The Discovery of the Unconscious* (New York: Basic Books, 1970), p. 312; and Johnston, *Austrian Mind*, p. 282.

however, determinism was an equally problematic assumption. To be sure, "determinism" had more than one implication in this period. On one hand, it corresponded to Hegel's view of history as the necessary course of the realization of the world spirit. On the other hand, Hegel's belief that the world spirit was actualized through the lives of individuals suggested a second understanding of determinism, one that traced an individual's actions with the force of necessity to his or her "character" or "passion." Other nineteenth-century thinkers formulated determinism alternatively, for instance as a physiological or environmental theory.[29] Herbart rejected determinism in all these forms. The very possibility of education depended on the unbounded potential of young minds. Nonetheless, Herbart's pedagogical theories required a degree of causal influence between teacher and pupil. In order to arrive at an ethical pedagogy, one that accounted for the existence of psychological laws yet retained the concept of freedom of choice, Herbart needed a new understanding of causality.

He needed, in other words, to redefine the form of necessity that characterized Kantian self-legislating morality. The laws described by Kantian ethics could not rest on the causality that characterized mechanical processes, since the dictates of the ethical self were not "musts" but "oughts." Herbart therefore posited a third form of causality, which he termed "aesthetic necessity." In *Über die ästhetische Erziehung des Menschen* of 1793, Friedrich Schiller had assigned to art the role of portraying new possibilities of human interaction as a first step in the creation of an enlightened society. Schiller celebrated the potential of art to free viewers of class-based conventions and prejudices. Herbart's invocation of aesthetic necessity in the psychological realm was an attempt to capture the complexity of such an interactive and transformational process. It was the fruit of his effort to derive an active model of learning appropriate to a liberal Central European understanding of freedom.[30]

The conflict between causality and free will took on particular urgency in post-Enlightenment Catholic Austria. No one felt this tension more strongly than Bernard Bolzano, a priest and professor of mathematics in Prague, who shaped an enduring tradition of empiricist and reform-minded philosophy in Austria. Bolzano believed that he had found a way to make causality compatible with an ethical theory that gave individuals full responsibility

29. On theories of environmental determinism in the Enlightenment, see James Fleming, *Historical Perspectives on Climate Change* (New York: Oxford, 1998), pp. 11–20.

30. Benner, *Interpretationen*, pp. 17–19.

for their actions. Throughout the 1830s, Bolzano discussed such issues with Franz Exner, his colleague in Prague. Bolzano's solution to the free will problem, with which Exner concurred, rested on viewing "judgments" and "desires" as commensurable psychic forces. Actions could thus in principle be accounted for by the relative strengths of ideas. However, Bolzano posited, many of the grounds for action that formed the links in the causal chains of the mental realm were unconscious. In practice, causality could thus coexist with the absence of determinism. Bolzano stressed that young students must be guarded at all costs from deterministic theories of mind, lest their sense of moral responsibility be stunted. In a letter of 1834, Exner congratulated Bolzano on this solution to the problem of free will. "Upon a closer reading I found myself most pleasantly surprised by the complete confirmation of my views by yours. The point is so important that such an experience brings more than a little pleasure and encouragement."[31]

Herbart's psychology, like Bolzano's, offered Exner a model of the mind that seemed particularly suited to the project of liberal education. Herbart presupposed the "plasticity" (*Plastizität*) or "educability" (*Bildsamkeit*) of the pupil and stressed the active, spontaneous participation of the pupil in the learning process. As Exner optimistically expressed this principle, "In [children] lies yet unbounded the entire infinite range of the human capacity for education [*Bildungsfähigkeit*]; no path is closed to them that ever a human foot, seeking its fate, has wandered."[32] Here, then, was the philosophical context behind Exner's strategies as a teacher. He sought to foster in his students a form of self-consciousness that duly recognized the constraints posed by laws—physical, moral, and political—but which strenuously avoided a deterministic fatalism.

A RESTLESS TRAVELER IN AN UNEASY CONTINENT

Exner's embrace of the latest German philosophy only made his life in Prague more difficult. In 1832 and again in 1837 he wrote to friends that he felt his position at the university to be "provisional."[33] He considered writing a textbook, but expected that anything he would write would be rejected by the

31. Exner refers to the section on free will in Bolzano's *Athanasia*. Franz Exner to Bernard Bolzano, 11 Nov. 1834, in Winter, *Briefwechsel*, p. 61.

32. Franz Exner to Maria von Rosthorn, 22 Jan. 1838, Franz Exner Nachlass.

33. Franz Exner to Josef Mozart, 17 Nov. 1837, Exner to Zitkovsky, 12 March 1832, quoted in Frankfurter, *Thun-Hohenstein*, p. 57.

religious authorities.[34] From time to time he was seized with "a hard-to-overcome disgust" and longed to make "a vigorous leap forward" to find a larger field of action for his "strengths."[35] He doubted that philosophy was in fact his calling. The treatise on logic that Franz long hoped to publish would linger incomplete until his death, ever a source of frustration to him. As much as his colleagues admired him for his ethical nature and his skill as a teacher, it was true that they did not rate him equally highly as a philosopher. Bolzano described Exner to a colleague in 1839: "One may doubt his acuity or the depth of his thought, but one cannot deny the integrity of his character and his rectitude."[36] In short, Exner sensed that he had yet to find his true calling.

Prague in the mid-1830s showed symptoms of a larger breakdown in the European system established by the Congress of Vienna. Czech nationalism could not be contained within the old system, and Exner felt the tensions it engendered. The anti-German sentiment seemed not to disturb his professional position, he told a friend, "doubtless however our social relations."[37]

Living as a bachelor in a foreign city, Exner's sense of dissatisfaction was bound up with a longing for the lost camaraderie of his university days. In 1833 he received news of the death of one of that band of friends, bringing on an outbreak of nostalgia: "So must we gradually learn to expect that the joys of youth vanish and the gravity of life draws ever nearer to us—separation and death."[38] In his solitude he brooded over lost opportunities: "The best forces of our soul are like early sprouting plants, which soon draw in their leaves, wilt, and die. May life only be otherwise. But one becomes a ruin before one's time."[39] The more the present disappointed him, the more Exner idealized the passion and vitality of youth.

During the summer break of 1838 he made an effort to revive his spirits. He traveled through the Italian lands, savoring artistic treasures and visiting friends from his student days in Pavia. In Florence he relished the monuments of the Renaissance. "I wander in the midst of eminent ghosts as in

34. Franz Exner to Josef Mozart, 26 July 1833, quoted in Frankfurter, *Thun-Hohenstein,* p. 67.

35. Franz Exner to Josef Mozart, 17 Nov. 1837, quoted in Frankfurter, *Thun-Hohenstein,* p. 57.

36. Quoted in Eduard Winter, ed., *Der Briefwechsel B. Bolzano's mit F. Exner* (Prague: Böhmische Gesellschaft der Wissenschaften, 1935), p. xiv.

37. Quoted in Frankfurter, *Thun-Hohenstein,* p. 61.

38. Quoted in Frankfurter, *Thun-Hohenstein,* p. 53.

39. Franz Exner to Josef Mozart, 17 Nov. 1837, quoted in Frankfurter, *Thun-Hohenstein,* p. 57–58.

Elysium," Exner wrote to his friend Maria von Rosthorn, "and out of awe hardly trust myself to step outside." He eulogized the heroes of the Renaissance, "who returned art and science to the impoverished race, and brought freedom to the enslaved... Once they were revered as gods like Prometheus, Apollo, and Pallas Athena. And thank God that here one still erects statues to them. Our poor Germany is hardly there yet. But we're sensitive to it."[40] As he communed with the ghosts of Michelangelo, Leonardo, and their contemporaries, Franz's hopes swelled for a humanist rebirth of intellectual life in Austria.

Yet this Italian vacation could not shake him free of his depression. Back in Prague at the start of the new semester, he struck his most despondent note. "I live like a fish in tepid water," he wrote to Rosthorn. "My career is not mine. I am a mediocre teacher, because I don't strive to be a better one; I don't strive because I see that in my situation I can't be a good one." By his "situation" Franz likely meant both the official restrictions on his academic freedom and his current melancholy state. "I am a mediocre scholar," he continued, "because my field, abstract speculation, is suited neither to my tastes nor to the nature of my intellectual powers."[41] He was a harsh judge of his own accomplishments, but he was honest about his inclinations. He missed Viennese society and longed to test his powers against more practical challenges.

Pushing him in a more practical direction was the political climate of the day. Along with the stirrings of nationalist sentiment at the University of Prague, Exner's political conscience was growing more acute. Von Rosthorn, who asked him in 1838 for an overview of the state of philosophy in the German lands, received more of a lesson in politics than she had perhaps expected. Exner described how rationalism was forced into a narrow corner in the Catholic German lands. Bolzano, Rembold, and the Bavarian Jakob Salat had each lost a teaching post for offending the Church. The Catholic capitals were dominated by philosophical schools that were sanctioned by the secular and religious authorities and that were little more than a buttress to tyranny and religious dogmatism. In Vienna, these apologists "with little erudition, some wit and a good deal of insolence, guard the holy sheep-pen." But the situation in the Protestant German lands was, if anything, worse. There such apologists wore the cloak of Hegelian idealism, a philosophy that Exner roundly condemned. In his view, the Hegelians

40. Franz Exner to Maria von Rosthorn, 1 Sept. 1839, Franz Exner Nachlass.

41. Franz Exner to Maria von Rosthorn 23 Oct. 1838, Franz Exner Nachlass.

were no more than apologists for an authoritarian status quo. They had made philosophy "an unexpected but highly potent aid to [legal] positivism." They had, in other words, "stamped the existing church and state as rational."

> To the government this doctrine was quite welcome, and it was licensed as the court and state philosophy.... What Hegel began, his students have completed; they have transplanted Christian dogmas right into the middle of philosophy, and under their protection they flourish now in heads into which they would otherwise hardly have found entry.[42]

Philosophical training was thus in itself no guarantee of open-mindedness. Philosophers, Exner found, could be as dogmatic as clergymen, and their dogmas were surprisingly seductive. Exner was convinced that the Hegelians' reign was coming to an end in Prussia. Yet even the prospect of preparing their fall and liberating their victims could not rouse Exner's flagging energy.

LOVE AND DUTY

It was Charlotte Dusensy who finally pulled Exner from this gloomy state. The daughter of a Jewish tradesman, Charlotte met Franz at her parents' home in Prague in 1840. She was by all accounts a "finely educated" woman, twenty-four years old and a Catholic convert.[43] The couple wed on the thirteenth of May, 1840, and within three months Charlotte was pregnant with their first child, Adolf. Among their children and grandchildren, Charlotte's Jewish origins would be suppressed by collective silence.

The character of the couple's early relationship emerges in a series of letters that Franz wrote to his wife from Marienbad, where his failing health drove him just three months after their wedding. Unfortunately, Charlotte's replies have not survived. She was remembered by her children as youthful and light-spirited, and Franz often took a paternal tone with her.[44] When Charlotte complained of missing him, Franz advised her to find something worthwhile to raise her spirits—a good book, music, ideas: "Beschäftigung!" He responded to her listlessness with all the pent-up passion of a reformer whose lofty ambitions had fallen on deaf ears. He pressed her to rise above

42. Franz Exner to Maria von Rosthorn, 10 Sept. 1837, Franz Exner Nachlass.

43. Her family had apparently changed its name from the recognizably Jewish "Ducheles." Verbal communication, Dr. P. C. Dijkgraaf, 6 March 2004.

44. Franz Exner, *Einiges über die Exnerei* (Staltach: self-published, 1944), p. 5.

petty concerns and shared with her his own aspirations, wavering characteristically between idealism and pragmatism:

> Be the dew of heaven, and enliven and nourish what may blossom in me and around me. Not only shall we avoid iniquity, we shall not loiter on the dusty highway of common life. Let us seek the heights, where we overlook the earth more freely and stand closer to heaven. To arrive there, however, we must not shy from the path through the valley. Conscientious fulfillment of plain but indispensable duties, which the day and the hour present unannounced, is the foundation on which alone there rises a solid structure into the kingdom of perfection.[45]

We can imagine Charlotte responding to Franz's earnest counsel with good-humored indulgence before turning the dialogue to more immediate concerns. She complained, for instance, of the discomforts of her first pregnancy. Her husband in turn advised her to remember that she suffered for the sake of the fatherland. By this he meant a unified Germany, the liberal dream of the 1840s. His response was not quite the comfort she had sought:

> So then, think: you live for the power and glory of Germany: your sons will likely one day have to fight against the presumptions of the French and Russians . . . For that reason we want to raise our boys as future soldiers. They must be hard, bear gladly every deprivation, and above all be self-sufficient in everything, be able to help themselves. Inwardly they shall have pure souls and inspiring ideas. Our girls, though, shall be able to become nothing but mothers who in turn have such sons. [Mothers] will be held responsible, whether the next generation is one of cowards, good for nothing but tending to their own amusements, or of nobly striving men.[46]

Perhaps to Charlotte's dismay, Franz would not discuss domestic cares without reference to the future of Europe. His visions of family life were inextricably bound to his political aspirations. It was clear to him that such an upset to the balance of European power as the creation of a unified Germany would not come to pass without war, and he hoped to see his sons fighting for the new nation. In his view, it was a woman's duty to raise the citizens of the future to be patriotic as well as ethical. Exner clung to these lofty principles as his children grew. Even at home he remained "the trained philosopher," the "teacher" and "caretaker of youth."[47] One of Exner's sons, likely Adolf, confirmed that his father's style of child-rearing tended towards

45. Franz Exner to Charlotte Dusensy Exner, 8 Aug. 1840, Franz Exner Nachlass.

46. Franz Exner to Charlotte Dusensy Exner, 1 Aug. 1841, quoted in Frankfurter, *Thun-Hohenstein*, p. 65.

47. "Prager Schilderein," note 19 above.

the Spartan: "My father was of the opinion that *Erziehung* ended at seven years, and had few rules of *Erziehung*: honesty isn't always easy, and do the hard part first, those were the lessons I remember."[48] With his young daughter Marie, Franz was more gentle but no less compelling. Reluctant, perhaps, to submit her to the influence of the clergy, he took her religious instruction into his own hands. She recalled him gazing deep into her eyes and speaking captivatingly of the majesty of Christ.[49] Unlike her brothers, Marie kept something of her childhood faith. Adolf was more typical of the family in retaining only an admiration for Catholicism's aesthetic achievements, compared to which Protestantism seemed to him impoverished. For Catholic dogma he had no patience. As he wrote to a friend after visiting the Straßburg Cathedral on Whitsunday, "It was certainly not thanks to its dogmas that [Catholicism] conquered the world. For my part I don't believe in them and expect little of them for now, but I love this religion of Germanic, earnest art."[50]

From the letters of these newlyweds and from their children's recollections we can begin to envision Franz and Charlotte—one charismatic, strong-willed and idealistic, the other light-hearted, devoted, and fragile. But their correspondence can also be read for the light it sheds on the convergence of personal and political aspirations among Austrian liberals. Austria's educated liberal elite placed the highest value on the institution of marriage in its secular, post-Reformation incarnation, which reclaimed the value of "earthly happiness." From the perspective of contemporary legal theory, marriage was the final stage of individual development, "an institution for the development of the higher spiritual side of man."[51] As Exner made clear, this process of self-cultivation was ultimately in the service of a greater good. Through the spiritual support he received from his wife, he believed himself better able to serve his nation.

PRUSSIAN VIRTUES AND VICES

In 1841 Exner planned his first visit to the elite centers of learning in Prussia and Saxony. For a man gripped by the dream of a unified German state,

48. Blumfeld, "Biographisches Skizze."

49. Marie Exner von Frisch, "Aus meinen Jugendtagen" (1870), p. 2, typescript, Collection of Dr. Herwig Frisch, Vienna.

50. Quoted in Franz Exner, *Exnerei*, p. 7.

51. Josef Unger, *Die Ehe in ihrer Welthistorischen Entwicklung: Ein Beitrag zur philosophie der Gechsichte* (Vienna, 1850), p. 155.

this was a political as well intellectual pilgrimage. At long last, he looked forward to meeting Herbart, the inspiration for his attacks on the tyranny of Hegelianism. As fate would have it, word reached him as he was setting out of Herbart's death. Still, Exner's journey proved fruitful. He established connections with philosophers such as Moritz Wilhelm Drobisch and Friedrich Adolf Trendelenburg, and with Hermann Bonitz, his future collaborator on the reform of the imperial education system.

At the Prussian universities Exner also witnessed the use of seminars and laboratories to foster original research alongside teaching, making knowledge into a collaborative and open-ended process. This system was less of an innovation than is often thought. It had existed at Göttingen and Halle for roughly four decades before Exner's visit and came to be associated with the ideas of Wilhelm von Humboldt only in the early twentieth century.[52] For Exner, though, the Prussian example made the need for change in Austria all the more apparent. It was time to join Austria to the world of learning in the wider German-speaking world.[53] This was a lesson that would bear fruit soon after Exner's return to Prague.

In addition to the educational model he witnessed and the contacts he made, Exner's time in Prussia was decisive in another way. It showed him that the Hegelian philosophers were still holding fast to their chairs at Prussian universities. He soon decided to make his critiques of their teachings public. In Leipzig he found a publisher for a polemic, "On the Psychology of the Hegelian School," directed against three of the leading Hegelians: Karl Rosenkranz at Königsberg, Karl-Ludwig Michelet at Berlin, and Johannes Eduard Erdmann at Halle. This volume, and the counter-defense that he published after the Hegelians responded, became the best known of Exner's philosophical writings.

Through his critique of the Hegelian psychologists, Exner developed his counter-vision of an empirical, scientific psychology. His attack centered on the inconsistency of the Hegelians' claims alternately for and against mental and historical determinism. The question of psychological causation, he asserted, was "so important that the content and form and value of psychology depend on it."[54] The Hegelians claimed, on one hand, that mental activity

52. Rainer C. Schwinges, ed., *Humboldt International. Der Export des deutschen Universitätsmodells im 19. und 20. Jahrhundert* (Basel: Schwabe, 2001), especially the essays by Rüdiger vom Bruch and Sylvia Paletschek.

53. Frankfurter, *Thun-Hohenstein.*

54. Franz Serafin Exner, *Die Psychologie der Hegelschen Schule* (Leipzig: Friedrich Fleischer, 1842–44), vol. 2, p. 93.

was purely spontaneous, that the mind possessed, as Exner put it, "countless freedoms." Exner quoted statements to this effect from Rosenkranz, such as "Association has no laws"; "Dreams are the existence of the mind as chaos." Yet Rosenkranz also maintained that ideas in dreams are related "to each other only mechanically and chemically; their combination is random." Exner responded: "Here then the absence of cause and law reigns again, and what is more, mechanism and chance, law and lawlessness go amiably hand in hand."[55] In his view, the duality of chance and determinism led the Hegelians into illogic and obscurity.

Exner went on to demonstrate the reactionary implications of Hegelian psychology. Turning to the Hegelian dialectic of master and slave, Exner asked, "Does the author seriously believe that slaves have no self-consciousness in a literal sense? Does he really believe that the North American or other slave owners take their slaves for apes, for nonhumans, for beings without self-consciousness? Then they would probably not make a crime of their education."[56] Exner located a passage in Rosenkranz's work that asserted that slavery was justified in cases where a race remained in a state of "barbarism," such as the "Negroes." Was this anything more than the claim that might makes right? Ultimately, Exner charged, an insidious historical determinism lay behind the Hegelians' apologetics for slavery. The violence of the master-slave dialectic was "necessary" only insofar as one believed in historical necessity. In short, determinism was at the root of the threat the Hegelians posed to liberalism.

Exner ultimately dismissed the Hegelians' discussions of necessity and freedom as "a rather clumsy sleight of hand."[57] "With determinism and indeterminism it is a matter of whether the will actually residing in each of us is necessary due to sufficient causes, or causeless and thus random. . . . The poor men! They have gone to such trouble and not noticed that they are dealing with a trick problem [*Vexierproblem*], and that the will is neither solely determined nor solely indetermined, but rather in truth both at the same time, or what is the same thing, neither of the two!"[58] Exner's understanding of mental causality, like Herbart's, could not be reduced to either determinism or indeterminism. Exner was moving away from the abstractions of pure philosophy, toward the application of psychology to education.

Against the background of the Hegelians' failings, Exner laid out his own vision for psychology. He insisted that the most pressing task facing philoso-

55. Ibid., vol. 2, pp. 92–93.

56. Ibid., vol. 1, pp. 31–32.

57. Ibid., vol. 1, p. 55.

58. Ibid., vol. 1, p. 105.

phers of the day was a precise description of mental states, from which to infer empirical laws of thought. By making psychology scientific in this sense, philosophers could offer the world something of "practical" value. They could provide answers to "the moralist, the politician, the pedagogue, and anyone else who requires basic insight into the human soul."[59] To fulfill its obligations to the politician and the pedagogue, Exner argued, psychology should become to the interior world what natural science was to the exterior: "so shall it teach us to recognize those internal variations that one calls mental states."[60] By contrast, the Hegelian psychologists aimed to preserve the mind as the province of absolute freedom, in which the will reigned supreme. Exner found this picture unworkable from both a pedagogical and a political perspective.

The Hegelians were unable to account, for instance, for the phenomenon of attention, "likely one of the most important psychological matters." "We demand it of children and of adults; thought, feeling, will, action, the person as a whole, all follow its orientation; the pedagogue and the statesman must know for which things they can expect it, for which they can demand it, how it is to be stimulated, how highly it may be excited without its ending in exhaustion and aversion."[61] Recognizing spectacle as a key mode of politics and education (including in his own classroom), Exner saw that an empirical account of attention could inform its use. Rosenkranz maintained instead that "[f]ree subjectivity controls its contents with absolute sovereignty." But observation proved otherwise, Exner argued. "Or does not every schoolboy know the opposite when he observes that he has forgotten something he had learned, even after he already knew it quite well?"[62] The Hegelians simply ignored the significance of those phenomena, such as distraction, that subverted the will's control over the contents of thought. Exner, on the other hand, considered it paramount to teach students to recognize these unconscious tendencies so that they might discipline themselves as best they could.

Given its denial of the existence of mental laws, Exner argued, "[f]or the Hegelian philosophy there is no pedagogy, nor politics, despite many political lectures . . . The Hegelian philosophy permits no science of *Erziehung* at all, neither of the individual, nor of a *Volk*. For the countless freedoms with which, they claim, the human mind teems, utterly exclude each operation of

59. Ibid., vol. 2, p. 2.

60. Ibid., vol. 2, p. 3.

61. Ibid., vol. 1, p. 36.

62. Ibid., vol. 2, p. 70.

the causal law."[63] Without empirical laws of psychology, politicians would have no tools for engaging the interest of the populace, and tyranny would be their only recourse. Nor could pedagogues discover the most effective ways of fostering autonomous, critical thought without the insight of an empirical science of psychology. If there existed laws of the natural world, then Exner believed there must exist laws of psychology as well. But he refused to allow this question to float in the realm of metaphysics, grounding it instead in pedagogical psychology: "That people can be educated [*erzogen*], this alone would suffice" to decide the question of the existence of causal relations in the realm of the psyche.[64]

Exner called on psychology to strive not for "absolute" knowledge as the idealists sought, but rather for empirical laws that had application in the real world. The Hegelians charged that such a scientific psychology would never arrive at laws that were better than *approximations*. Exner fully conceded this point.[65] But approximate empirical laws were, as he saw it, a proper goal for an empirical science. "Is it not possible, too, that one might proceed with the description of all transitions of the mind as the mathematicians do, who, in seeking a line, often determine at first only certain of its special points, such as turning-points or averages?"[66] Exner's suggestion was likely drawn from solutions to the problem of complex phenomena in other contemporary empirical sciences, such as physical geography and meteorology, which relied heavily on approximative methods. It hints at the significance that probabilistic reasoning might acquire in the framework of a liberal-empiricist curriculum.

Exner's conclusion was that the question of determinism versus indeterminism was an illusory problem (*Scheinproblem*) of idealist philosophy. *Approximate* laws would be the focus of an empirical psychology with applications to politics and pedagogy.

A UTOPIA OF TRANSPARENT COMMUNICATION

Exner's 1842 critique characterized the Hegelians' psychology as conservative, even reactionary, juxtaposing it with an empirical psychology that would aid social reform by seeking applications to politics and pedagogy. Already Exner's writing evinced a utopian strain, one that would become more

63. Ibid., vol. 2, p. 91.
64. Ibid., vol. 2, p. 4.
65. Ibid., vol. 2, p. 91.
66. Ibid., vol. 1, p. 4.

pronounced in the next installation of his assault on the Hegelians, his 1843 essay *Leibniz's Universal Philosophy*.[67] Utopianism is a useful genre in which to consider Exner's essays because of its prevalence among liberal and socialist writers of the 1830s and 1840s.[68] In 1845, for instance, Robert von Mohl, the Bavarian legal theorist who popularized the term *Rechtstaat* (constitutional state) laid out an attempt to bring utopian literature to bear on political philosophy.[69] Exner similarly sought to demonstrate the relevance of Leibniz's philosophical utopia for the reform of philosophy in his own day.

Leibniz's plan for the mathematization of philosophy had been heralded by Schleiermacher and other German Romantics for its ideal of a universal language, a notion that was particularly important to the Romantics' literary criticism.[70] Exner challenged the Romantics' reading of Leibniz, insisting that Leibniz's central concern was with clarity. The ultimate purpose of Leibniz's mathematization of philosophy had been to achieve transparent communication, not to create a universal language for its own sake. The "ideal" toward which Leibniz worked was "nothing other than the perfect clarity of all concepts and their relationships, for which the calculus was only a special means of representation. Genuine science and unity among thinkers follow alone from the clarity of thoughts."[71] Exner thus saw Leibniz's ideal of transparent communication, to be achieved through mathematical symbols, as one facet of a utopian vision of a unified community of philosophers. Was this dream merely the stuff of fantasy? Against this interpretation, Exner noted that the natural sciences had recently made great strides toward mathematization, particularly in chemistry and crystallography. Mathematization may have been a distant ideal for philosophy, but Exner declared emphatically that it was not an impossibility.

In fact, a quest for clarity would be the perfect antidote to the sorry state of German philosophy as Exner saw it. "Modern" philosophy—by which

67. Franz Serafin Exner, *Leibniz Universal-Wissenschaft* (Prague: Borrosch & André, 1843).

68. On utopias, see Frank Manuel, *Utopian Thinking in the West* (Cambridge, Mass.: Harvard University Press, 1979) and Frank Manuel, ed., *Utopias and Utopian Thought* (Beacon: Boston, 1965).

69. On Mohl and the *Rechtstaat*, see James Q. Whitman, *The Legacy of Roman Law in the German Romantic Era* (Princeton: Princeton University Press, 1990), pp. 95, 148; on Mohl and utopias, see Frank Manuel, introduction to *Utopias and Utopian Thought*, p. ix.

70. On the significance of Leibniz's idea of a mathematical language for Romantic literary theory, see P. Lacoue-Labarthe and J.-L. Nancy, *The Literary Absolute*, trans. Bernard and Lester (Albany: State University of New York Press, 1988), 111–19.

71. Exner, *Leibniz*, p. 40.

he undoubtedly meant Idealism—had "favored a hazy vagueness over conceptual clarity, audacious leaps over steady development; because Logic protested, she was banished." He accused the Idealists of having made of philosophy "a phantasmagoria,"

> which arose to applause at the beginning of the century in predominantly libertine guise, after a long metamorphosis is now mystical-pietistic—and reemerges to the same applause, while the fair opponent who stands to be suppressed by it first, who was long praised as irresistible, has in her entire arsenal no weapon to stop it; she and Philosophy must endure it. For all have equal claim to *sovereignty that founds itself on arbitrariness.*[72]

Exner's attack on the arbitrary sovereignty and mystical pietism of the Hegelians echoed liberal denunciations of the abuses of secular and religious powers in the 1840s. The link he drew between philosophy and politics was more than metaphorical. As he had written to von Rosthorn in 1838, he judged the Hegelians to be complicit in tyranny in a very literal sense. As much as their philosophy had abetted the suppression of freedom, just as strong in his judgment was the capacity of the new empirical philosophy to foster it. In the place of these tyrants, Exner hoped to see a new philosophical community, one in which language would transparently reflect the empirical world, communication would be unambiguous and consensus attainable.

THE LONG ROAD TO REFORM

With the relaxation of police surveillance in the 1840s, the formation of student societies or *Vereine* at the Habsburg universities opened a new discourse on education reform.[73] Despite Exner's youth, his popularity with students and growing scholarly reputation prompted the local authorities in Prague to consult with him in these years on the selection of new faculty and textbooks. Perhaps they sensed that the volatile student community could best be managed with Exner as intermediary. Exner soon drafted a plan for a general reorganization of Austrian education, above all of philosophical studies, and sent it to the director of the imperial schools commission. This won him the attention of the authorities in Vienna, who in 1844 called on Exner to advise on the reform of *Gymnasien* throughout the empire. Several

72. Ibid., p. 40, emphasis added.

73. On these associations, see Judson, *Exclusive Revolutionaries*.

of his manuscripts on education reform from this period survive, and they show strong continuity with the plans published after 1848.[74]

Exner assigned the state the task of preparing young men for the responsibilities of freedom. Nineteenth-century Austrian liberals considered education a prerequisite for citizenship. A typical pamphlet of 1848 proclaimed: "People are not free in one blow; they become free, they develop and educate themselves to freedom, and freedom exists precisely in carrying out this process."[75] Exner set out to design a system in which students would educate themselves to freedom.

In announcing the lessons of psychology to educators and statesmen, Exner faced the challenge of explaining the mind's relation to laws, both mechanical and political. The problem was to produce citizens who would be law-abiding without being fatalistic, virtuous without being dogmatic. Phrased differently, his obstacle was the apparent paradox of *teaching* people how to be free in a Kantian sense. As in the teaching of psychology in French schools in this period, Exner's challenge was to present abstract philosophy in a way that would shape students' experiences of the workings of their own minds.[76] Like Cousinian psychology in France, Herbartian psychology in Austria was part of a postrevolutionary effort to steer students clear of political extremes.

Exner's work in Vienna was abruptly interrupted by his nomination for a teaching post in Bonn. He quickly found himself caught up in his most unpleasant encounter yet with the forces of Catholic reaction. Apparently in an effort to block Exner's nomination, the bishop of Leitmeritz (near Prague) accused Exner of teaching the eternity of matter. Exner was forced to respond with a contorted attempt to reconcile his physicalism with the orthodox account of Creation. In the end, he did not receive the offer from Bonn. The confrontation undoubtedly strengthened his impatience to see academic freedoms safeguarded. To his disappointment, however, the reform commission in Vienna was dissolved in 1847 without any concrete accomplishments, and Exner's visions of reform began once again to look more idealistic than practical.

74. These pre-1848 manuscripts are in the Manuscript Collection of the Library of the University of Vienna and are discussed in Frankfurter, *Thun-Hohenstein*; and Hartel, *Festrede*.

75. Johann Berger, "Die Pressefreiheit und das Pressegesetz," (Vienna, 1848), quoted in Judson, *Exclusive Revolutionaries*, on p. 51.

76. Jan Goldstein, "Foucault and the Post-Revolutionary Self," in *Foucault and the Writing of History*, ed. Jan Goldstein (Oxford: Blackwell, 1994).

A PRIVATE UTOPIA

In late 1846, as visions of reform were finding freer expression in Austria, Exner let himself imagine another ideal community. This time his audience was a festive gathering of friends and family, including his four young children. He told them he had dreamed of the future: It was the year 1876. At the age of seventy-four, his deepest desires had been fulfilled. His school reforms had been implemented, his treatise on logic published, and his own family and those of his friends were blossoming. Success took political, academic, financial, and domestic forms:

> A new education plan had come into effect, the spirit of science rose from its grave and walked again alive into our schools among our youths. Fresh air began to blow through the land . . . On its currents came free speech and I could speak and teach as spirit and heart inspired me, I had grateful students everywhere and I had written many books that were honored. . . . [Exner's friend Serafin] made trade and industry flourish inside and out and multiplied Austria's honor and power. Each of us had a wife and sons.

Exner reserved special praise for these sons: "powerfully free and noble were their thoughts and words and deeds." They "proudly called themselves sons of Austria" and "brought glory to their fathers' names." With their lives' work in such capable hands, the fathers spent their savings on an estate, a "retirement castle" on the city outskirts, "in which they lived jointly with their wives and younger children." In this philosophers' retreat, "The newest books, newspapers, the best writings of ancient and modern times stand in neat cabinets. The sons, now men, come and go, the young wives with the children bring life and cheerfulness to the house . . . ; old friends and interesting travelers come to visit the well-known men."[77]

Exner's dream was an intricate fusion of liberal and nationalist aspirations, bourgeois assumptions, and aristocratic pretensions. A triumphant battle for the freedom of public expression would end, ironically, with the heroes retreating to a fortress in the countryside. Exner's retirement castle was an imagined reconstruction of the holidays he had shared with friends during university. The aristocratic trappings of this fantasy are relevant less as

77. Franz Exner, Sr., "Toast," 25 Nov. 1846, Sig. 14.627, Manuscript Collection, Österreichische Nationalbibliothek, Vienna. On the architectural trend in this period towards villas beyond the city boundaries, see Geza Hajos, "Die 'Verhüttelung' der Landschaft—Beiträge zum Problem Villa und Einfamilienhaus seit dem 18. Jahrhundert," in *Landhaus und Villa in Niederösterreich 1840–1914*, Österreichische Gesellscahft für Denkmal- und Orstbildpflege (Vienna: Böhlaus, 1982), pp. 9–56.

a sign of a bourgeois "aping of the aristocracy" than as an indication of the significance of inheritance for *Bildungsburger* families. Alongside the material aspects of middle-class inheritance, which have been well studied by social and economic historians, lay the value of intellectual inheritance or what Bourdieu has called "cultural capital." Material and intellectual legacies could not be disentangled; without the stability and continuity created by the former, the latter would be impossible. The novelist Adalbert Stifter, who collaborated with Franz Exner in the education ministry in the early 1850s, made the theme of inheritance central to his most famous work, *Der Nachsommer*. This novel brought to life the model of moral and intellectual development behind Exner's education program, particularly in its fusion of empiricism and humanism. For Stifter's characters, the country home served as a refuge from the volatile worlds of politics and fashion, a haven where a former statesman could dedicate his later years to art and science. In keeping with this atmosphere of seclusion, the novel had no precise historical setting; time was measured only by the passing seasons. In Stifter's fictional world, objects—a castle, a sculpture, a jewel—took on momentous significance when they became heirlooms, and the plot built toward the revelation of the chain of inheritance of two summer homes. Stifter implied that intergenerational continuity was what gave meaning to the productions of art and science: "Art, science, human progress, the state, all rest on the family. If marriage does not lead to a happy family life, then your highest achievements in art and science are in vain, you pass them to a race in moral decay, to whom your gift is ultimately useless."[78]

The utopia of the *Bildungsburger* country house, as it appeared at midcentury in Stifter's novel and in Exner's dream, presented the family as the site of the production of intellectual and social continuity. Even among such passionate liberals, the significance of individual achievements paled in comparison to the inheritance of intellectual traditions and material goods. This tension was reflected in certain mysterious properties of these fictional country homes. They seemed to exist temporally outside of the modernizing city, as an almost timeless space. And they were repositories fit for the print explosion of the nineteenth century, with room to spare for ancient texts. The philosophers' retreat contained within its walls the contingencies and inexhaustible variability of individual and collective lives without being itself subject to such vagaries. To Exner, the dream of the country house held

78. Adalbert Stifter, *Der Nachsommer* (Frankfurt: Insel Taschenbuch, 1982), p. 773.

the promise of continuity, even as he saw his world on the brink of radical change. In the years before 1848, twin utopias crystallized side by side in Exner's imagination: two settings for free intellectual exchange, one private and one public.

MARCH 1848

Exner believed that his visions of a reformed society were not just idle dreams. His appointment as an imperial adviser in 1844 had been a first sign that the government, weakened by a financial crisis, felt pressed to compromise. The emperor's ministers could not ignore the growing signs of discontent throughout the Habsburg lands, most visibly the revolts of Silesian weavers in 1844 and of Galician peasants in 1846. In 1847 Exner declined the nomination of his colleagues to serve a one-year term as rector of the university, citing concern for his health as grounds for the decision. Quite possibly he was also reluctant to compromise his loyalties. As the tribune of the Prague academic legion, Exner's only obligation was to his students. Zimmermann recalled him in his legionary uniform, "the philosopher with white arm-bands and polished saber in hand."[79]

Word reached Austria in February of the revolution in France. In the following weeks civic societies and student associations began to draw up petitions for reform. As a spokesperson for Prague's students, Exner had the ear of the authorities. On the eleventh of March, Exner and colleagues sent Vienna a manifesto demanding the freedoms of teaching and learning, the right to attend foreign universities, and reform of the exam system. Following up on an ambiguous response from the government, on the twenty-eighth of March Exner negotiated with the governor Graf Rudolf von Stadion, who agreed to grant these rights provisionally. Two days later the Bohemian authorities officially conceded the demands of the university's petition. Exner became a hero to the academic legion, which went on to ally with Prague's workers to fight for more radical reforms. The following week these students bade Exner tearful goodbyes as he left Prague for Vienna, where he had been urgently called to serve once again as an adviser to the education ministry.[80]

79. Zimmermann, "Nekrolog."

80. On student activity in the revolutions in Vienna, see Thomas Maisel, *Alma Mater auf den Barrikaden* (Vienna: Universitätsverlag, 1998); on Exner's activities, see Frankfurter, *Thun-Hohenstein*, pp. 89–90.

As his students recalled, Exner embarked "unarmed" on his journey to Vienna; or rather, he was armed only with the education reform plans he had drawn up between 1844 and 1847. With this head start, in just four months Exner was able to hand the ministry his "Outline of the Essential Features of Public Instruction in Austria." The following year, aided by Bonitz, Exner completed a proposal for the reform of the *Gymnasien* and *Realschulen*.[81] While Exner's reform plans of 1848–49 were to be no more than guidelines, they remained the standard, the ideal, against which Austrians judged the success or failure of the empire's schools for over half a century.

Exner's real innovations lay in his program for the *Gymnasium*, since his program for higher education resembled what he had seen at the Prussian universities.[82] For the first time in the history of these elite institutions in the Habsburg Empire, classes would now be taught in local languages. Never before had German literature been deemed a worthy topic of study at *Gymnasium*, of equal value for *Erziehung* as the classics of Greece and Rome. Now there would be classes taught in languages from Magyar to Slovenian. It was not by straitjacketing the empire into a single language of instruction that Exner intended to foster unity.

The ultimate goal of the *Gymnasium* now became not the mastery of particulars but the cultivation of a "noble character." The schools remained responsible for religious instruction, but Exner laid out a firmly secular, humanist basis for ethical education. Exner acknowledged the concern, expressed by Herbart and others, that public schools were not adequate to the task of *Erziehung*. In a Kantian spirit, however, he argued that a liberal state in particular could not neglect this duty: "however much greater the political freedom is, which the young man awaits upon entrance into life, that much more necessary is it, that he, so long as he is not of age, learns to master himself and to obey the laws, both internal and external."[83] Exner here married a Kantian ideal of freedom to a practical plan of pedagogy, showing how a psychologically astute style of teaching could help students develop into self-legislating agents. Exner thus insisted that ethical education was possible in a classroom if teachers heeded the lessons of psychology.

81. Franz Exner, *Entwurf der Organisation der Gymnasien und Realschulen in Oesterreich* (1849), Universitätsbibliothek III 992, M; reprinted by k.k. Schulbücher Verlag, Vienna, 1875. There was lingering debate over the authorship of the *Entwurf*, which Frankfurter laid to rest in 1893 (see Frankfurter, *Thun-Hohenstein*, p. 91).

82. See Engelbrecht, *Geschichte*, for discussion of reforms to the *Realschule*.

83. Exner, *Entwurf*, p. 9.

Throughout his outline of the new curriculum, Exner stressed that mental faculties developed gradually, experientially, through practice. A capacity for geometric visualization (*mathematische Phantasie*), for instance, was "not an exclusive gift of nature, but is instead capable of methodical *Bildung*, through exercises." By focusing on the active participation of the student in the learning process, Exner leaned on Herbart's argument that it was in fact possible to *teach* students to be autonomous. The key was to help them master the faculty of attention. Through their classroom experiences, students would develop a disciplined attention, the basis for their future capacity to profit from political freedom.[84]

"Students should be permitted as little time as possible for mere passive listening, which at any moment can turn into absent-mindedness [*Gedankenlosigkeit*] or distraction [*Zerstreutheit*]." Using Herbart's language to describe the vying of ideas in the mind, Exner recommended that instructors make their presentations "vivid" and "lively." The goal was not "merely a passive receptiveness on the part of the student to studying the required information," but rather "the complete assimilation of this information through the student's own activity."[85]

Exner maintained that the study of natural science lent itself well to this active style of learning. He created a prominent new role for science and mathematics in the *Gymnasium* curriculum, emphasizing the compatibility of these subjects with the goals of humanistic education. In a phrase that would be quoted for decades to come, he wrote that it was the instructor's task to draw out the "humanistic elements, which are plentiful in the natural sciences as well."[86] Thus natural history classes could build on children's "interest" for their natural surroundings and "develop acuity and precision of observation out of the joy of contemplation."[87] Such lessons in observing nature continued even in the study of the German language, for which the youngest *Gymnasium* students used the new reader prepared by Exner's friend Josef Mozart. Among the exemplary "descriptions and depictions" that Mozart selected were natural scientific accounts of meteorological and astronomical phenomena, including one of a solar eclipse by D. J. F. Arago, the president of the French Academy of Sciences and a liberal member of

84. Ibid., p. 50.

85. Ibid., p. 100.

86. Ibid., p. 8.

87. Ibid., p. 171.

parliament.[88] Exner recommended against limiting the study of nature to familiar objects, suggesting that natural history should instead guide "the gaze up and beyond" the realm of everyday nature. In this sense, science could match art in its capacity to open students' minds to new possibilities. But these lessons, Exner emphasized, must "categorically" demand from students "keenness of attention."[89]

These lessons in self-discipline were preparation for the freedom students would encounter when they completed *Gymnasium* and became full citizens. The final stage in this process was the philosophy curriculum that Exner designed for the last year of *Gymnasium*, removing this "propadeutic" (preparatory) task from the universities. Unlike the Prussian *Gymnasien*, which taught only logic in the final year, the Austrian *Gymnasien* would offer logic, empirical psychology, and a further course that Exner called "Introduction to Philosophy."[90] The last of these courses was, as Exner noted, the most novel. He cautioned that such a difficult subject could not be taught without an appropriate textbook, and in 1849 no such work existed. But Exner vividly described the experience that an "Introduction to Philosophy" should give the student:

> The ground of experience, on which one believed oneself able to stand unshakably, becomes shaken by doubts, which threaten to destroy all certainty, indeed all possibility of experience. The highest concepts, which are indispensable in all sciences of nature—for instance the concept of change, of active and passive, of force, of spatial and temporal continuum, of personality, and many more—far from being a lamp that could illuminate the entire field which they govern, are themselves obscured by the greatest difficulties. Ethical judgment with its unproven claims to unconditional validity strikes against the contradiction of opposing equally worthy perspectives of the present as in the historical development of the species.[91]

The point of this exercise was not, of course, to turn students into radical skeptics; it was instead to give them a first taste of the freedom they would enjoy as citizens and, in so doing, to immunize them against the plague of dogmatism.

88. Josef Mozart, *Deutsches Lesebuch für die unteren Klassen der Gymnasien*, Bd. 4, 3rd ed. (Vienna: Carl Gerold & Sohn, 1859).

89. Exner, *Entwurf*, p. 172.

90. In 1855, the Austrian education ministry extended the philosophy course to two years.

91. Ibid., p. 179.

The education ministry commissioned the first textbook for the Philosophical Propaedeutic from Exner's student Robert Zimmermann in 1851, but not until 1860 was Zimmermann able to add a suitable "Introduction to Philosophy." He expressed the hope that this third edition would correspond more fully to Exner's intentions.[92] Indeed, Zimmermann's "Introduction" led students deep into the caverns of skepticism, just as Exner had prescribed.

Zimmermann tried to convince the reader that this uncomfortable experience was necessary. He depicted doubt as a phase through which a young man passes on his way to adulthood. A child knows no doubt until evidence of his own errors begins to shake his confidence in himself. "The foundation of his knowledge sways beneath his feet and he grasps for an anchor to steady himself."[93] This image of a quaking earth, repeated from Exner's 1849 program, would recur in the pedagogical reflections of Austrian liberals well into the twentieth century. This moment of doubt was the precondition of philosophy. Zimmermann proceeded to plant in the reader's head the insidious questions that had plagued René Descartes two centuries earlier, plunging through the list of concepts that Exner had hoped to see "obscured with the greatest difficulties." "He who doubts in order to remain in doubt is a Skeptic," he observed, but "he who does so in order, through doubt, to reach knowledge, is a Philosopher."[94] Assessing the new philosophy textbooks of post-1848 Austria, one educator later wrote that Zimmermann and his fellow authors had fulfilled the task set by Exner's 1849 program, namely, the "stimulation of skepticism."[95]

Zimmermann's *Logic* and *Empirical Psychology* in turn provided the student with tools to help him begin to navigate through this sea of doubt. The text explained that human errors of judgment arise from the finite sphere of experience on which we must base our inferences, which are necessarily, for that reason, "probabilistic." Zimmermann carefully laid out the rules for formulating probabilistic judgments. Yet he explained that even a correct application of the rules could lead one into error. Suppose, for instance, we

92. R. Zimmermann, *Philosophische Propadeutik* (Vienna: Braumüller, 1st ed., 1852, 2nd ed., 1860), p. vi.

93. Zimmermann, *Propadeutik* (2nd ed.), p. 356.

94. Ibid., p. 364.

95. Ludwig Chevalier, "Über den Unterricht in der philosophischen Propädeutik an österreichischen Gymnasien," *Jahresbericht des k.k. Staats-Untergymnasiums in Prag-Neustadt* 4 (1885): 3–39, quotation on p. 8.

judge that from an urn containing ninety-nine white balls and one black ball a white ball will be drawn, and instead out comes a black one. Then we have erred "not in that we judged the drawing of a white ball probable, but rather in that we judged it certain, and we err for that matter not arbitrarily, but necessarily, in that for a probability of 99/100 we must expect the drawing of a white ball to be exceedingly more likely."[96] Zimmermann made clear that error and uncertainty were the common lot of mankind, but he did not let the reader despair. In the *Empirical Psychology* he offered a list of techniques for avoiding error, including the management of attention, and in the *Logic* he explained mathematically how making more observations could increase the certainty of inferences. The reader thus emerged safely from his flirtation with skepticism.

Reading the new curriculum as the culmination of Franz Exner's battle against religious dogmatism and political reaction, it becomes clear that probabilistic reasoning as taught in the new schools was the embodiment of his values. His ideal had always been a mode of thought that surveyed a range of possibilities without losing sight of empirical constraints. Thus he supported Leibniz's ideal of clarity because it was a worthy goal and *not impossible*. Yet he decried the pure speculation of the Hegelians, which appeared to him "like dreams, in whose place one could just as well have dreamed much else."[97] Probability would be the lingua franca of a liberal, multinational society; it would silence dogmatism and enable objective communication.

EXNER had struggled throughout the 1840s to free the debate over mental determinism from the realm of metaphysics and place the issue squarely in the context of pedagogical psychology, where neither determinism nor indeterminism could apply. By 1848 Exner had become convinced that liberal education would have to begin by training students to resist the scourge of determinism unleashed by the Hegelians. In order to avoid infection, students would have to recognize the power of their own wills. Instead of shielding children from the hypothesis that unconscious forces were at work in their minds, pupils would be trained from a young age to recognize and counter such disruptions. This would require their active participation in

96. Zimmermann, *Propadeutik* (1st ed.), p. 104. See too the discussion of probability in Johann von Lichtenfels, *Lehrbuch zur Einleitung in die Philosophie*, vol. 1, *Logic* (Vienna: Wm Braumüller, 1852), sections 80, 100.

97. In Winter, *Briefwechsel*, quotation on p. 37.

every aspect of classroom activity. Empirical science, emphasizing observation and experimentation, would be a new pillar of the curriculum. Above all, students would exercise their powers of attention. Attention was the key to self-discipline, and self-discipline in turn was the prerequisite for gaining political and intellectual freedom. By mastering the less rational tendencies of their own minds, children would learn to think of themselves as malleable, even as they recognized their own agency in the process of education.

In the context of Austria's first experiment in liberal education, probabilistic reasoning had multiple functions. As a framework for pedagogical psychology, the mathematical techniques of approximation showed how the mind could be described as law-like and yet undetermined—in short, as "educable." Individuals who had internalized this psychic model would be fit for incorporation into a liberal state. As a component of the new *Gymnasium* curriculum, exercises in probabilistic reasoning would encourage students to view absolute claims skeptically. Yet probability was also a language in which personal experience (observed frequency) could be reformulated in universal terms (degree of belief warranted for any rational being). For Austria's utopian liberals of the 1840s, probabilistic reasoning was the idiom of a new kind of leader, one capable of undermining appeals to absolute certainty without jeopardizing his own claim to a transcendent perspective.

CHAPTER TWO

In the Stream of the World

Coming of Age in the 1860s

AFTER 1849 Franz Exner's plans for the reform of the empire's education system lay in the hands of others. Although the government had hoped he would serve as minister of education, Exner refused political office. He intended to stay on as an adviser, but a bout of pleurisy prompted his doctor to recommend a warmer climate. In the fall of 1851 he and Charlotte traveled to Italy, leaving their daughter and four sons, ages two to ten, with Exner's friend and colleague Josef Mozart and Exner's sister Antonie in Vienna. Exner died a year and a half later, at age fifty-one, in Padua, where he had been reporting to Vienna from his deathbed on the implementation of school reforms in the Italian provinces. Of the fruits of his labors that had sparkled before his eyes in his dream of 1846, he had lived to see only one: his *Logic* lay incomplete and his sons were still boys, but his education reforms were being realized.

Charlotte returned to Vienna, where she moved with the children to smaller quarters and dismissed most of the servants. Marie Exner, now nine, believed her mother to be on the verge of suicide—"How I came to these thoughts, which are otherwise so foreign to a child, I do not know, but I remember that I often crept behind with a pounding heart when she went outside alone in the evenings." Marie pinned her hopes on her oldest brother Adolf, who, at twelve, would soon be expected to take his father's place: "Mother made him aware that he was her eldest and must soon be her support."[1] Adolf thrived in his new role. The confidence he radiated seemed to Marie to soothe her mother's anxieties and her own. But Karl, just a year younger than Adolf, became the most reclusive and least ambitious of the siblings. The other boys, Sigmund and Franz Serafin (known to family and friends as Serafin), were too young to understand the situation fully. Marie, meanwhile, watched anxiously over her mother, already taking responsibility in her quiet way for holding this family together.

The family also had the support of Exner's friends and colleagues from the education ministry, who visited frequently. These older friends, such as Hermann Bonitz and Josef Mozart, provided Exner's children with a living link to their father, now a figure of national renown. The siblings could hardly have remained immune to these men's excitement as they rebuilt the empire's schools and universities from the ground up. Over the next decade, the education ministry under Count Thun carried out the far-reaching modernization of imperial education that Exner had envisioned in the 1840s. Thun and his collaborators revised the *Gymnasium* curriculum to focus more on natural science and mathematics. They also transformed the universities from professional training schools into research institutes on the Prussian model, with a large measure of self-government and intellectual freedom.

Thun was in many ways emblematic of the contradictions of Austria in the post-1848 era.[2] He was an aristocrat, a conservative, and a pious Catholic, yet he fully supported Exner and his liberal colleagues in their bid to make Austria a center of open intellectual inquiry. On one hand, he backed the 1855 Concordat, which reversed reforms of the Josephist era by giving the Church authority in matters of education and marriage and by restricting the state's power to intervene in the administration of the Church. On the other hand, he was a driving force behind modern intellectual movements at the universities, including the turn to a historical approach to legal theory. This movement's foremost representative in Vienna was Adolf Exner's future mentor Josef Unger—Jewish, liberal, and yet handpicked by Thun to reinvigorate Austrian jurisprudence. It was also under Thun's administration that the University of Vienna opened an institute of physics, charged with training science teachers and fostering original research.[3] The institute's first

1. Marie Exner von Frisch, "Aus meinen Jugendtagen" (1870), p. 6, typescript, Collection of Dr. Herwig Frisch, Vienna.

2. On Thun, see Hans Lentze, "Graf Thun und die voraussetzungslose Wissenschaft," in *Festschrift Karl Eder,* ed. Helmut Mezler-Andelberg (Innsbruck: Universitätsverlag, 1959).

3. On the early history of physics in Vienna, see Lotte Bittner, "Geschichte des Studienfaches Physik an der Wiener Universität in der letzten Hundert Jahren" (Ph.D. diss., University of Vienna, 1949);

director, Christian Doppler, known for his theory of the effect of motion on sound waves, was a friend of Exner's from his time in Prague and another regular guest in his widow's home.

By 1859, as Adolf Exner finished *Gymnasium* and his youngest brother Serafin entered it, the fortunes of their father's liberal friends were again on the rise. Austria's military defeat by the French in Italy that year humiliated a government that was already crippled by financial weakness. Emperor Franz Josef now sought to regain public support by introducing a tentative constitutional system, which would soon guarantee the freedom of research and teaching.

Just as Franz Exner's reforms finally stood to be completed, however, Charlotte Exner's strength gave out. She died in early 1859 with all five children at her bedside.

THE FORMATION OF CHARACTER

For the next few months Marie and her brothers cared for themselves while Mozart and two unmarried aunts decided the children's fates. Finally, the guardians agreed to take the younger boys, Sigmund and Serafin, to live with them just outside Vienna and to send the older two, Karl and Adolf, to stay with distant relatives in the city. Marie was put in the care of a childless friend of her mother's, a daughter of the von Lämel family of bankers in Prague, who struck the young girl as rigid and detached. When Marie looked back on this period from the perspective of a newly married woman in 1870, she thought of the proverb, "Misfortune strengthens the character and kills the spirit." Whether the first was true in her case she could not say, but she had no doubt of the second. She became sullen and lethargic. Most difficult for her to bear was the stifling discipline of her new life: "My dear freedom was entirely taken from me, I had to account for every hour of the day; that was the most painful to me. My whole being revolted against it." A brief reunion with her younger brothers in the country was a tantalizing taste of freedom, to which she responded, "Like a bird set free in the open air after a long imprisonment."[4]

Sigmund ("Schiga" to his family), just a year younger than his sister, was almost as miserable at *Gymnasium* as Marie was in her private lessons: "I

Christa Jungnickel and Russell McCormmach, *The Intellectual Mastery of Nature: Theoretical Physics from Ohm to Einstein,* vol. 1, *The Torch of Mathematics, 1800–1870* (Chicago: University of Chicago Press, 1986), pp. 202–13.

4. Marie Exner v. Frisch, "Jugendtagen," pp. 7–8.

count this *Gymnasium* period among the saddest of my life," Sigmund wrote at the age of forty-five.[5] For his poor performance he blamed the *Gymnasium* itself, which, in his opinion, remained far too focused on grammar and memorization. One day Sigmund's own lectures would be regarded as "dry," delivered slowly and without humor, but he would be praised for the clarity of his presentations.[6] As a boy, Sigmund had developed a love for natural science and began collecting animals, living and dead. He recalled wryly that this hobby met resistance in his aunts' house. Sigmund, like Marie, experienced these years as an arbitrary loss of freedom.[7]

Serafin was lucky enough to remember *Gymnasium* more fondly, since, in his case, an early passion for natural science went along with a facility for languages and an enduring love of classical literature. As an adult Serafin would be known as broadly cultured and unconventional. "Every convention was for him a detested constraint."[8] He too finished *Gymnasium* with a thirst for freedom, which, in his case, took the form of a lasting *Wanderlust*. He later advised that travel should be part of every education.[9]

Already in their youths the Exner siblings were carving out contrasting characters for themselves. Adolf, the eldest, seemed to the others larger than life, a fount of wisdom and fatherly advice. His leadership qualities would blossom into an illustrious academic and political career. Karl, the next born, lived somewhat in his shadow, never as ambitious as his brothers but a dedicated teacher and creative thinker. As a middle-aged man, sitting on the balcony of a summer cottage smoking his pipe, Karl was described by one of his nephews as "contentment personified." Decades later Sigmund's son would sketch the brothers as follows: "Adolf: the cleverest; Schiga: the most industrious; Serafin: the most cultivated; Karl: the most ingenious."[10] Marie was praised for her selflessness, but her wit was equally remarkable.

5. Sigmund Exner, "Auto-biographische Notizen" (1891; no page numbers), Personalakt, Archiv der Akademie der Wissenschaften, Vienna. Sigmund says here and in his curriculum vitae in the Archive of the University of Vienna that he attended Vienna's two premier schools, the Akademisches Gymnasium and the Theresianium, but the records of neither school list any of the Exner brothers among their students, likely because they were "external" students.

6. Karl Frisch, *Erinnerungen*, p. 26.

7. Ibid., and Sigmund Exner, "Auto-biographische Skizze" (1879), Personalakt, Archiv der Akademie der Wissenschaften, Vienna.

8. Franz Exner, *Exnerei*, p. 23.

9. Serafin Exner, "Autobiography" (1917), p. 5, Personalakt, Archiv der Akademie der Wissenschaften, Vienna.

10. Franz Exner, *Exnerei*, p. 30.

The Exner siblings' autobiographical writings imposed narrative order on the chaos of their early lives. Models for these narratives could be found in the *Gymnasium* textbooks and novels of the day. Philosophy textbooks at midcentury sketched the course of individual development quite differently than they had before 1848. The old textbooks had listed, on one hand, the factors that determined development. On the other, they had suggested that man "mastered" these factors "insofar as he conquers temperament and climate, determines diet, defies age, denies race, modifies nationality, outgrows upbringing, regulates social relations, and adopts the true religion."[11] The causal arrows pointed simply and directly from environment to character or from character to environment. There were no effects of feedback and no room for contingency (nor alternatives to "the true religion"). By contrast, the post-1848 textbook described how the formation (*Bildung*) of "ethical character" occurred "in the stream of the world," that is, through experience, as a gradual systematization of empirical rules.[12] *Bildung* in this sense was an interaction between the individual and his environment. The mutual influences were partial and unpredictable. Every step of this process was contingent. In the words of one philosophy teacher in Prague, the teaching of psychology must not lose sight of the "entirely indeterminable, multifarious, and variable ingredients that concrete individuality and the instantaneous current of life add."[13]

The post-1848 *Gymnasium* demonstrated how an individual imposed order on experience by means of the constructive faculties of attention and memory. This psychological model was grounded in Herbart's metaphysics, according to which the self-preservation (*Selbsterhaltung*) of an entity was equivalent to its persistence in the face of an accidental disturbance or negation

11. *Lehrbuch der Philosophie* (Vienna: k.k. Schulbücher-Verschleiß-Administration, 1835), vol. 1, p. 66.

12. R. Zimmermann, *Philosophische Propadeutik*, 1st ed. (Vienna: Braumüller, 1852), vol. 1, pp. 88–95. Among many expositions of the meanings of *Bildung* in the nineteenth century are Reinhart Koselleck, "On the Anthropological and Semantic Structure of *Bildung*," in *The Practice of Conceptual History: Timing History, Spacing Concepts* (Stanford: Stanford University Press, 2002), pp. 170–207; Karl-Ernst Jeismann, "Zur Bedeutung der 'Bildung' im 19. Jahrhundert," in *Handbuch der deutschen Bildungsgeschichte*, ed. Karl-Ernst Jeismann and Peter Lundgreen (Munich: C. H. Beck, 1987), vol. 3, pp. 1–22. See too Walter Benjamin's essay "Schicksal und Charakter," in *Illuminationen: Ausgewählte Schriften* (Frankfurt: Suhrkamp, 1977), pp. 42–49, which emphasizes more specifically several aspects of *Bildung* discussed in this chapter, such as the "Wechselwirkung" between the individual and the external world.

13. Johann Heinrich Löwe, *Ueber den Unterricht in der philosophischen Propädeutik am Gymnasium* (Prague: Verlag A. G. Steinhauser, 1865).

(*Störung*).[14] Here the elements of experience appeared as pure contingencies, not as pieces in a puzzle of divine design. This image of *Bildung* was thus anticlerical, but it was also anti-Kantian. The ego was "a focus for remembered representations," not, as for Kant, a product of innate categories or faculties.[15] In the new model, the synthesis of experience was coextensive with the self. The lesson of Herbart's metaphysics for the *Gymnasium* was thus the definition of "personality" as the "principle of the triumph over chance.[16]

A similar model of character development emerged in the preeminent Austrian coming-of-age novel of the 1850s, *Der Nachsommer,* by Exner's colleague in the education ministry Adalbert Stifter. Stifter's young protagonist learns from art and science to see order and beauty in the tangled chaos of natural landscapes. But these exercises in representing the natural world are simultaneously efforts to mold the psychic interior, as art and science join in a single humanist program: "Insight grew into the beauty of forms, and the delicate and fine was preferred to the heavy and coarse . . . Chemistry and other natural sciences were taken up, and through the reading of beautiful books it was attempted to form the inner nature."[17] The narrator of *Nachsommer* carries his growing portfolio of drawings with him as if this work contains the self he has carved from his experiences. Art and science are his means of fashioning a liberal character. Here, the narrator's imposition of will on contingent experience is a reflection of the novelist's own craft.

German neo-Kantians would articulate a deceptively similar model of *Bildung* in the 1870s, placing their emphasis on the mind's imposition of order on the world. The Heidelberg philosopher Wilhelm Windelband observed

14. Zimmermann offered a mathematical illustration of this principle: let $A = \alpha + \beta + \gamma$ and $B = \alpha + \beta - \gamma$. A and B are each "accidental points of view" (*zufällige Ansichten*), that is, linear combinations of more basic entities. Then γ is the "relationship of the persistence of each against the other." Zimmermann, *Leibnitz und Herbart: Eine Vergleichung ihrer Monadologien* (Vienna: Braumüller, 1849), pp. 42–43.

15. Katharine Arens, *Structures of Knowing: Psychologies of the Nineteenth Century* (Dordrecht and Boston: Kluwer, 1988), p. 95.

16. Franz Heilsberg, "Die Bedeutung des Zufallsbegriffes in der Geschichtswissenschaft," *XI. Programm der Kaiser Franz Josef-Staats-Realschule* (1908–9): 6–26, quotation on p. 6; and discussion of this theme on pp. 22–26.

17. Adalbert Stifter, *Der Nachsommer* (Frankfurt: Insel, 1982), p. 94. Compare p. 205 on the use of drawing in Gustav's natural history lesson. The probability calculus makes an appearance in the novel as a method of reasoning suited to one who is equally naturalist and humanist: "a probability calculation had to be made of the average amount of grain that grew annually on this hill" (ibid., p. 65).

in a treatise on "chance" (*Zufall*) published in 1870 that the *Bildungsroman* introduced accidents (*Zufälle*) to serve as material through which the hero could test his will and shape his character.

> Chance supplies us with varying phenomena, independent of the inner course of our being, and now transported to us, these begin their irresistible operation on us, inhibiting the course of our inner self, disturbing, corrupting. Here chance shows the tragic side of its Janus head, here it proves itself to be the mighty power of the external world, which demands its dictatorial right from our subjectivity: now is the time to assert against [the external world] the triumphant power of character, which, not led astray by accidents independent of it, follows through life the path prescribed by its inner self.[18]

In its characters' vulnerability to the accidents of the outside world lay literature's potential to depict the process of *Bildung*. In the most famous of all *Bildung* novels, *Wilhelm Meister's Lehrjahre*, Goethe had accorded a role to chance, "but such that it must always be steered and guided by the dispositions [*Gesinnungen*] of the characters."[19] Windelband's colleague in Berlin, the neo-Kantian natural scientist Hermann von Helmholtz, explained that the artist sorted experience much as others did, yet with a heightened capacity to "sift out everything accidental and confusing of the doings of the world."[20] Yet this picture "cleansed" of chance had an entirely different significance for the German neo-Kantians than for the Austrians. The neo-Kantians treated this picture as a revelation of "the real";[21] in their view, the process of shearing nature of contingency disclosed truth. Austria's midcentury empiricists, on the other hand, ever wary of dogmatic claims, viewed such pictures as strictly human and conditional creations. To them, the pedagogical value of such exercises in science and art lay in the formative confrontation with contingency itself.

18. Wilhelm Windelband, *Die Lehren vom Zufall* (Berlin: A. W. Schade, 1870), p. 26.

19. Goethe, *Wilhelm Meisters Lehrjahre*, book 5, chap. 7, quoted in Windelband, *Zufall*, p. 26. On the late nineteenth-century aesthetics of chance, see Hacking's discussion of Mallarmé in *Taming of Chance*, p. 10; and Thomas Fechner-Smarsly's analysis of Strindberg's paintings in "Den Zufall kopieren: August Strindbergs (Un-)Ordnungen des Experiments," paper delivered on 7 December 2001 at the Max Planck Institute for History of Science, Berlin.

20. Hermann von Helmholtz, "The Facts in Perception" (1878), in *Science and Culture: Popular and Philosophical Essays*, ed. David Cahan (Chicago: University of Chicago Press, 1995), p. 342–80, quotation on p. 355.

21. Helmholtz: "the artist has beheld the real." Ibid., p. 355; see too "The Relation of Optics to Painting," in *Science and Culture*, pp. 279–308, esp. pp. 307–8.

Through such examples, the Exners may have learned to think of the process of imposing order on experience as part of the cultivation of a strong character. Strength of will was a defining feature of masculinity for the *Bildungsbürgertum*, but its value for young women was more ambiguous.[22] This explains Marie's predicament when she found she could not muster an attraction to one of her admirers because he lacked the strong will she associated with her memories of her father: "A certain feebleness and weakness of will in him had bothered me all along, perhaps precisely because I myself did not feel strong and I recalled the masculine force of my father, on which my mother was able to lean with such confidence."[23] Marie hesitated between seeking strength of will in a male companion and in herself. As her account of the failed romance indicates, sexuality was a pivotal field of experience for the shaping of the *bildungsbürgerlich* self.[24] By the age of twenty-five, when Marie recorded her youth for her new husband, she attested that she had overcome her feminine "weakness." By extricating herself from an unpromising match, she had become one "who, after cruel imprisonment, has finally become free. O freedom!"[25]

FIRST ENCOUNTERS WITH MODERN SCIENCE

As nineteenth-century novels of *Bildung* made clear, the story of a young *Bildungsburger*'s life began with his parents. By adolescence the Exner siblings' memories of their parents were dim. They relied on older friends to supply this part of their story. The celebrated physicist Josef Loschmidt was one friend who offered vivid pictures of their father.

Loschmidt had been Exner's student in Prague in the early 1840s. Coming from a peasant family in Bohemia, Loschmidt supported himself at university by taking a job as a reader for Exner, whose eyesight was failing. Through his reading, Loschmidt plunged far deeper into contemporary philosophy than he otherwise could at a Habsburg university before 1848. He concentrated in particular on Herbart's psychology and metaphysics. In order to pursue Herbart's program of quantifying mental processes, Loschmidt took up the study of mathematics. A passionate student, Loschmidt engaged in lively exchanges with Exner and found in him an eager mentor. When Serafin Exner later recounted this phase of Loschmidt's life, he lingered on a revealing anecdote. After a reading session devoted to Schiller's *Thieves*,

22. Frevert, *Men of Honour*.

23. Marie Exner v. Frisch, "Jugendtagen," p. 12.

24. Koselleck, "Structure of *Bildung*."

25. Marie Exner v. Frisch, "Jugendtagen," p. 17.

Loschmidt had stayed behind a moment alone in Exner's room. Exner returned to find Loschmidt wrapped in his dressing gown, brandishing a paper knife, and declaiming a monologue from the play. "I am fully convinced," Serafin wrote, "that at that moment my father was heartily pleased with this outburst of youthful idealism."[26] Loschmidt's memories gave Serafin a glimpse of the great Franz Exner not as a philosopher and statesman but as a father figure.

In a few years, however, the budding philosopher Loschmidt abandoned his hopes for the Herbartian program, becoming a self-proclaimed "renegade." He decided that his mathematical skills would be more usefully employed in physics and chemistry. This shift was less radical than it might appear. As Serafin would later observe, Herbart's atomistic metaphysics could well have been an inspiration for Loschmidt's early interest in molecular physics. Indeed, in a letter to Exner from 1843 or 1844, Loschmidt recalled having asked him about the apparently unconstrained duration of the "self-preservations" in Herbart's theory. Exner had responded that this was somewhat "opaque," prompting his student to apply "all the force he could muster" to the problem. Loschmidt attempted to calculate the results of "collisions" between "two simple entities" at a given velocity, explicitly treating a question of metaphysics as an exercise in molecular mechanics.[27]

Loschmidt made a smooth transition from his speculations about the mechanics of the psyche to the mechanics of gases. The themes of contingency and self-preservation tied together Herbartian psychology and statistical thermodynamics. But Serafin's account turned this analogy inside out by using these same themes to narrate Loschmidt's life: psyche and world were joined in an endless series of contingent interactions. The biography that Serafin presented was punctuated by encounters between this remarkable individual and a series of fortunate accidents. Serafin struggled to narrate these interactions between will and chance: "With this stroke of luck [*Glücksfall*] the future course of Loschmidt's life was set," Serafin wrote of Loschmidt's entrance to a *Gymnasium*; "if I say stroke of luck, perhaps that is not entirely the correct expression, for the same could have happened to many without their becoming great scholars."[28] Serafin wrote repeatedly of

26. Franz Serafin Exner, "Zur Erinnerung an Josef Loschmidt," *Die Naturwissenschaften* 9 (1921): 177–80, quotation on p. 178.

27. Josef Loschmidt to Franz Serafin Exner, 1843 or 1844, published in P. Schuster and K. Kadletz, "Sechs Briefe Josef Loschmidts an Franz Serafin Exner (1840–1845)," in *Mitteilungen der Österreichische Gesellschaft für Wissenschaftsgeschichte* 14 (1994): 180–93, quotation on p. 188.

28. Franz Serafin Exner, "Loschmidt," pp. 177–78.

the effect of "accidents" on Loschmidt's life, but in each case he stressed that such windfalls were only what the man himself made of them. So, for instance, "It was a stroke of luck [*ein Glück*] that Loschmidt received this infusion of philosophy at an early and impressionable age and not later, in a time when stimuli still have powerful effects, but do not lodge so firmly that they cannot just as well develop in an entirely different direction as is intended."[29] Here Serafin echoed a comment in his own autobiography: he too had benefited from Herbart's philosophy at an age when a student is still untouched by intellectual rigidity.[30]

Loschmidt's studies were soon interrupted by financial difficulties, and he left the university to seek work in chemical factories. After 1848, however, Franz Exner was able to use his influence in the education ministry to find Loschmidt a teaching post in a lower *Realschule*, a position that would allow him to pursue scientific research. When Serafin Exner recounted Loschmidt's trajectory, he stressed that the middle decades of the nineteenth century marked the start of a new era—for Austria and for the eager young scientist from Bohemia. "As the barriers to foreign lands fell, there came as well a new spirit into scientific life." New ideas "streamed in" from abroad, including the work of Clausius and Maxwell on the kinetic theory of gases.[31] Implicit in Exner's account is the fact that atomism had been silenced in Austria before 1848 by clerical opposition to materialism. To those who bridled under the pre-1848 regime, molecular physics therefore stood as a symbol of the fruits of the liberalization of Austria's education system.[32]

Clausius and Maxwell's innovation was the foundation of a model of the behavior of gases based on a purely statistical description of the motions of molecules, entities whose very existence would continue to be debated until the first decade of the twentieth century. The mathematical grounding laid by Clausius and Maxwell led Loschmidt in 1865 to what remains his best-known contribution to molecular physics, his derivation of the approximate diameter of a molecule of an ideal gas.[33] Soon after, this wave of research led Loschmidt's Austrian colleague Ludwig Boltzmann to a statistical formulation

29. Ibid., p. 178.

30. Franz Serafin Exner, "Autobiography," p. 3.

31. Franz Serafin Exner, "Loschmidt," p. 179. See too p. 177: "Das Jahr 1850 brachte nämlich für Österreich . . ."

32. Cf. Rosner, *Chemie in Österreich*, 133. The same was true of evolutionary ideas; see Michler, *Darwinismus und Liberalismus*, p. 42.

33. Josef Loschmidt, "Zur Grösse der Luftmolecüle," *Sitzungsberichte der Kaiserlichen Akademie der Wissenschaften zu Wien II* 52 (1866): 395–407.

of thermodynamics and plunged the two men into debate over the implications of this new mechanics. In its traditional formulation, the second law of thermodynamics was the general rule that heat passes from a warm body to a cold one. Decades of work would bring Boltzmann to the conclusion that this principle was equivalent to the statement that physical systems tend to evolve from less probable to more probable states. States are more probable if they allow for a larger array of possible configurations of their components, so more probable states are the least constrained at an individual level. More probable states are also more stable, because a system forced into an improbable state will revert to a probable one. Thus a vacuum-sealed can, once opened, will be filled with air, because there are more possible configurations of air molecules within the volume of the can plus that of its surroundings than there are solely within the surrounding volume. It is these probable and thus stable states that allow the application of the physicist's usual law-like descriptions. It was on this statistical foundation that Serafin Exner and his physics students would later build.

In 1868, at the age of forty-seven, Loschmidt was finally named professor of physics at the University of Vienna. As Serafin looked back on these developments, Loschmidt and his probabilistic physics became for him a symbol of the renaissance of Austrian science in the age of liberal education. Loschmidt was a link for Serafin and his siblings to their father's values and aspirations, and above all, to the elder Franz Exner's ideal of probabilistic reasoning. Serafin wrote of Loschmidt's approach to science: "it is indeed a proof of proper theoretical as well as experimental thinking, to have foreseen the possible . . . In the unknown, to distinguish the possible from the impossible sets Genius apart."[34] It was, in other words, Loschmidt's virtue to have viewed the world in terms of graded possibilities rather than necessities, to have sought a viewpoint that was neither dogmatic nor merely subjective. Serafin judged Loschmidt according to the rational standard of Austrian liberalism, a standard he saw codified in the probability calculus.

In the 1860s, atomism was a hypothesis that could kindle a young imagination. As Loschmidt was embarking on the calculation of molecular dimensions, Serafin and his brother Sigmund were still *Gymnasium* students. Sigmund had recently had his first glimpse through a microscope of the teeming organic world. He would soon choose as his first research topic at university the phenomenon known since the 1820s as Brownian motion: the zigzag movements of particles suspended in solutions. In 1867 Sigmund

34. Franz Serafin Exner, "Loschmidt," p. 180.

published one of the nineteenth century's only quantitative studies of Brownian motion. Already four years earlier, while still in his second-to-last year of *Gymnasium*, Sigmund had drafted in a quick and eager hand a long, never-published essay with the title "Meditation upon Atoms."[35] This was an exuberant tour of the microscopic world. Sigmund conjured up a universe governed by mechanical principles yet packed with surprises at every scale. The essay's central theme was the relativity of size: the same objects that were unimaginably small to humans might be colossal to smaller life forms. Sigmund's considerations were founded on two principles. First, mechanical forces of attraction and repulsion governed the relationships among all physical bodies, no matter how large or small. Second, given the inexhaustible variety in the sizes of living things and the conditions of life, the general "law" could be formulated: "where something can live, there something does live [*wo es leben kann, da lebt es*]." Sigmund concluded that it was highly "probable" that there were organisms who called atoms home. "It is possible, it has even a large degree of probability, that the atom is not an empty smallest particle, but rather that they serve as homes to living things like ourselves." Sigmund embraced this atomistic worldview as a hypothesis resting on probability, not certainty, much as he and other liberal scientists defended Darwin's theory in this period. Indeed, microscopic civilizations seemed a plausible hypothesis until Loschmidt fixed "a definite [lower] limit" to the size of living things with the calculations he published 1865.[36]

This manuscript helps explain Sigmund's selection of Brownian motion as the subject of his first research project. Sigmund set out to prove that the "particles," as he chose agnostically to call them, were not self-propelled but were instead buffeted by the surrounding fluid. This was a problem poised precariously at the border between the organic and inorganic realms. Did a particle have its own "impulse"? Was its motion the result of "molecular motion" in the solution? Or did the movements arise from external influences like heat and light? Such speculations might have brought charges of materialism a generation earlier in Austria. By siding with the proponents of "molecular motion" in this, his first publication, Sigmund thus laid claim to the legacy of the liberalization of Austrian science after 1850.[37]

35. Sigmund Exner, "Betrachtung über die Atome" (1863), no page numbers. MS 2061, Manuscript Collection, Institut für Geschichte der Medizin, Vienna.

36. Loschmidt, "Zur Grösse der Luftmolecüle," p. 406.

37. Sigmund Exner, "Untersuchungen über Brown's Molecularbewegung," *Sitzungsberichte der Kaiserlichen Akademie der Wissenschaften zu Wien II* 56 (1867): 116–23.

Serafin approached atomism through an alternate path. He followed Josef Grailich, associate professor[38] of mathematical physics at the University of Vienna in the late 1850s. Grailich was convinced that "only the general investigation of crystals can create the foundations of future molecular theory."[39] Serafin published his first study of crystal surfaces in 1873, winning the Vienna Academy of Science's Baumgartner prize for that year. In this work Serafin arrived at a general relationship between the hardness and cleavability of a crystal's faces. He emphasized, however, the limited validity of his own generalizations. He phrased his results qualitatively, noting that the phenomenon "is of far too complicated a nature for it to be possible to say much more in a precise form." His results constituted "laws" only "insofar as this expression can be applied here." They were "empirical laws" derived "directly from a comparison of the collected experimental results; due to their manner of derivation no general validity can at present be accorded to them." Borrowing the language of the new Austrian atomists, Serafin judged that these laws held "with the greatest probability."[40] Serafin emphasized too that much could be learned from "anomalies," from individual differences among crystals.[41] Further research on these "naturally" arising anomalies might demonstrate that physical pressure could deform crystal structures, Serafin suggested. Crystal growth was thus a contingent process, shaped by interactions between internal and external forces, the history of which left each crystal with a visible record of its individual development.

Common to Serafin's and Sigmund's early ventures into molecular theory was a focus on the process through which the course of an individual—whether molecule or crystal—was shaped by its environment, and the potential of that individual in turn to withstand the forces of its surroundings. How did the principles of crystal formation and the forces of the environment interact to produce individual variations? Was Brownian motion an expression of the life force of tiny organisms or the trajectory of inanimate dust given impulse by the fluid that enveloped it? The interaction between will and chance was the reflexive theme of their experiments. This research was an ongoing exchange between the scientist's attempt to infer order and the phenomenon's capacity for disorder. In Vienna, the new molecular theory

38. At the German-language universities, senior professorships were known as "ordinary" and junior as "extraordinary." Typically, there was only one ordinary professor for each field of study.

39. Quoted in Jungnickel and McCormmach, *Intellectual Mastery*, vol. 1, p. 208.

40. Franz Serafin Exner, *Untersuchungen über die Härte an Krystallflächen* (Vienna: k.k. Hof- und Staatsdruckerei, 1873), pp. 106–7.

41. Ibid., pp. 28, 86.

developed through efforts like those of Loschmidt and the young Exners to fashion themselves for a new age according to liberal tropes of *Bildung*.

Yet these scientists could not carve their identities in isolation. The structure Serafin imposed on Loschmidt's life, as on his own, made this clear. Among the *Bildungsbürgertum*, in order to invent oneself, one needed to invent one's family.

"A SILENT REVOLUTION"

In the summer of 1863, the year that Sigmund penned his treatise on atoms, an invitation from friends of their parents made it possible for all five siblings to vacation together on the Wolfgangsee, where Franz and Charlotte had spent happy summers years before. "The collective life in the splendid natural surroundings was almost intoxicating, the mutual excitement was so great."[42] In this idyllic setting, the siblings envisioned the start of a new life together. After mornings spent in "diligent" study, the siblings spent the "too short" days walking together along the lakes or in the hills. More than their Viennese hosts, they took great pleasure in getting to know the woodsman's family they lodged with. Marie was struck by these "endearing, clever people, who have withdrawn from all the world."[43] Her regard for this family was characteristic of the Viennese bourgeoisie's idealization of peasant figures. As we will see, the Exners would take the hunter and the farmer, in particular, as models for their interactions with nature. Nature was far more than a backdrop for the siblings' reunion. The enthusiasm that Karl and his younger brothers brought to the study of nature was essential to the group's experience of this summer as an awakening. As Marie recalled, "Carl had in the meantime become a natural scientist; new worlds sprang up daily before my eyes. The banished dreams of old—a silent revolution—."[44] From their own perspectives, the young Exners had finally found the power to impose their will; they had found freedom through a peaceful revolution. In the following years they began to build a family tradition of summers in rented houses at the Wolfgangsee and other nearby lakes.

The Exners' utopian vision of their new life as a family reflected a "silent revolution" on a much larger scale. The young siblings' vision of freedom, against which they judged the unhappy years after their parents' deaths,

42. Marie Exner von Frisch, "Jugendtagen," p. 9.

43. Marie Exner to Otto Benndorf, 11 Aug. 1863, Manuscript Collection, Austrian National Library, folder 271, letter 32-1.

44. Marie Exner von Frisch, "Jugendtagen," p. 9.

took shape during the renaissance of Austrian liberalism in the early 1860s. Civic associations began to reappear in these years, and, with the dismissal of the chief of state police, discussion became more open and more critical. Vienna's major liberal newspaper, *Die Neue Freie Presse,* began printing in 1864.

Science took on new roles in the framework of this liberal agenda. Historians have traditionally seen science first as a force that drove modernization and industrialization, making possible new tools of bourgeois power from the steam engine to the telegraph, weather prediction, dynamite, and electric street lamps. Science has also been recognized as a definition of rationality, identifying those familiar with it as citizens qualified (by intellect rather than birth) to participate in political life.[45] Austrian liberalism stressed a different though related function for science. As part of a humanistic education, natural science was a means of cultivating a liberal character. This emphasis was part of the liberals' critique of bureaucratic absolutism. The scientists of the liberal era would be a new kind of expert. No longer would they play the narrowly trained pawns of a state machine, conducting surveys of forests and mineral deposits as in the age of enlightened absolutism.[46] They would instead be well-rounded individuals with the capacity and authority to participate in all levels of decision making. The opening of the Austrian Academy of Sciences in May 1848 was a sign of the new role that science would play in the reforming state. One of the academy's first projects was to organize the telegraph network along the monarchy's railroads into a communication system for gathering meteorological data.[47] The academy's resolution carried an implicit message: science would serve the state's technological needs, but imperial resources would also have to serve science.

Only with the beginning of constitutional rule in the 1860s, however, could Austrian liberals make their aims for science explicit. In the 1860s Austrian liberals embraced natural science once again as a means of inoculating the public against clerical dogmatism. Darwinism, for instance, became a central topic of discussion in liberal civic associations, where it was associated with a belief in progress and an opposition to clericalism.[48] In a similar spirit of optimism, the Jewish banker Ignaz Lieben willed a portion of his estate to reward "the general best"—an aim that his son Adolf,

45. Cf. Judson, *Exclusive Revolutionaries.*

46. David Blackbourn, *The Long Nineteenth Century: A History of Germany, 1780–1918* (Oxford: Oxford University Press, 1997), pp. 99–100; Jelavich, *Modern Austria,* p. 26.

47. Christa Hammerl et al., eds., *Die Zentralanstalt für Meteorologie und Geodynamik, 1851–2001* (Graz: Leykam, 2001), pp. 19–20.

48. Judson, *Exclusive Revolutionaries,* p. 157; Michler, *Darwinismus und Literatur,* chapter 1.

a chemist, fulfilled by establishing a prize to be awarded by the Austrian Academy of Sciences for discoveries by Austrian scientists in the fields of physics and chemistry. Fifty-five people, many of them Jews, received the Lieben Prize between 1865 and 1937, among them Sigmund Exner, Karl von Frisch, and several of Serafin Exner's students. The prize brought the winners prestige and a significant sum of money, typically at an early stage of their careers. The Liebens were themselves a minor scientific dynasty, and their patronage of the sciences illustrates another way in which bourgeois family ties contributed to the development of science in a liberalizing Austria.[49]

Even Marie, whose life as a young woman was centered around the home, was caught up in the intellectual transformations of the constitutional era. By 1863, as she recounted, Mozart began to tolerate having all five siblings at home together, since they filled him in on the latest developments in science and politics. "Whatever was new in science or politics, whatever was gathered from lecture halls or laboratories, was sorted through before us in the evenings, to be added to later by friends of my brothers." These evenings long remained Marie's ideal of a "geistreiche Gesellschaft," a witty and clever society.[50]

The pace of reform quickened after 1866, when Austria went to war with Prussia over its exclusion from the German customs union. The defeat forced the Habsburg state to make concessions to domestic critics, leading to the 1867 Compromise with Hungary and the establishment of the Dual Monarchy, with each half of the empire governed by new constitutional laws. Over the next three years, newly powerful liberal politicians extracted the Austrian education system from the papal control to which it had been submitted by the Concordat of 1855. With the provocation of the papal infallibility decree of 1870, Austrian liberals secured the revocation of the Concordat and ensured that secular public education would remain a political priority for the next decade.[51]

In June 1867, a year after Austria's defeat by Prussia at Königgratz, the *Neue Freie Presse* charged itself with stirring popular interest in public education:

> The more decisively the view emerges in the empire that not merely the intellectual training but the material prosperity of the people and the development

49. Rudolf W. Soukup, ed., *Die wissenschaftliche Welt von Gestern: Die Preisträger des Ignaz L. Lieben-Preises 1865–1937 und des Richard Lieben-Preises* (Vienna: Böhlau, 2004).

50. Marie Exner von Frisch, "Jugendtagen," p. 9.

51. On the educational politics of the late 1860s and 1870s, see Cohen, *Education and Middle-Class Society*.

> of military power depend on public education, the more urgent it seems that the press treat this subject comprehensively and candidly . . .[52]

The paper attacked the misconception that the Austrian *Gymnasium* was no more than an import from abroad. In fact, Austria had "clearly and sharply" rejected the educational models of Britain and Germany. The German *Gymnasium* devoted over half of classroom time to studies of the classics, the Austrian *Gymnasium* less than a third. Nearly quoting from Exner's school plan of 1849, the editors explained that the emphasis in Austrian schools fell instead on the "mutual relationship of all subjects of instruction to each other." In other words, Austria's distinct advantage over its European challengers on the battlefield of education lay in its modern curriculum, in the teaching of natural science not merely alongside the classics but as an integral part of a humanistic education. It was with similar ambitions that Adolf Exner expressed his hopes for Austrian scholarship in the preface to his first major legal treatise, also published in 1867:

> In our age of concern for the future it must be of the utmost importance to every *Deutsch-Oesterreicher* who feels in himself the strength for work to demonstrate industriously that we do not want to stand aside from the course of scientific work in Germany . . . [M]ay [this book] confirm once again in its small sphere that the Austrian branch of the great tree of German legal scholarship has not remained barren.[53]

By the late 1860s, Austrian liberals were forging a national identity out of their unique tradition of learning, of which Franz Exner's reforms were a cornerstone.

The moment was ripe for Austrian scholars to fashion themselves as modern, liberal thinkers in the tradition of the heroic reformers of 1848. For the young Exners, this aim was inseparable from their reinvention as a family. The ideal invoked by the editors of the *Neue Freie Presse* was that of "many-sidedness." Many-sidedness had been an overarching goal of the education reforms of the first Franz Exner, inspired by the pedagogical philosophy of Herbart. Unlike alternative responses to modernization at the time, Herbart's solution was neither politically and economically conservative nor revolutionary. The concept of many-sidedness appealed to

52. *Neue Freie Presse*, 6 June 1867, supplement to no. 154, and 19 June, no. 166.

53. Adolf Exner, *Die Lehre vom Rechtserwerb durch Tradition nach österreichischem und gemeinem Recht* (Vienna: Manz, 1867), preface.

nineteenth-century Austrian liberals because it met the challenges of modernization with caution and moderation. Even more importantly, however, the ideal of many-sidedness grounded the liberals' utopian vision of a society whose cohesion would emerge from successful communication, not from imposed uniformity. A many-sided education would ensure that individuals learned to see from a perspective beyond the merely personal.

Herbart had found it impossible to imagine that the state, which required experts, would take responsibility for promoting many-sidedness. Even Franz Exner recommended that the task of *Erziehung* be divided between parents and schools. The Exners' recreation of their family as a *geistreiche Gesellschaft* was thus a transformation in keeping with the times. Then came a summer afternoon in 1868 when the revelry of the siblings and their friends was cut short by a letter for Adolf from abroad. It was an offer of a full professorship at the University of Zurich. To Marie, the imminent separation seemed to mark the end of an era for the siblings: "A break in our life at an unexpected moment."[54]

A POLITICAL EDUCATION IN ZURICH

In fact, Adolf's tenure in Zurich was less of a rupture than Marie expected. Serafin soon followed his oldest brother to Switzerland, and Marie arrived the following summer for the first of two extended visits. For Marie, Zurich's immediate appeal was that it was far from home: she was desperate to escape her implicit engagement to that weak-willed young man for whom she felt so little admiration. Serafin, meanwhile, intended to study experimental physics at Vienna, but its physics institute had never really been equipped for experimentation. In Zurich, he could study at the new Polytechnikum, which had been fully outfitted by its first physics professor, Rudolf Clausius, for research on thermodynamics and physical chemistry.[55] The Exners treated Zurich as a new chapter in their lives, a continuation of the plot of *Bildung*.

The city on the Limmat provided, first, a new context in which the Exners could define themselves as liberals. Since 1830 Zurich had been a magnet

54. Marie Exner von Frisch, "Jugendtagen," p. 15.

55. Clausius had left for Würzburg in 1867, but his replacement, August Kundt, had won esteem for his measurements of the speed of sound in gases. The research Serafin conducted with Kundt, first in Zurich and later in Würzburg, concerned the maximum density of water, an experimental topic in physical chemistry that made use of the high-quality precision instruments in Kundt's laboratory. Franz Serafin Exner, "Autobiography"; Jungnickel and McCormmach, *Intellectual Mastery*, vol. 1, pp. 186–202; vol. 2, pp. 294–97.

FIGURE 3 Group portrait circa 1878. From left to right: Franz Serafin Exner; his first wife, Auguste Bach; the Exners' friend and colleague Ernst Fleischl von Marxow; Marie Exner von Frisch; Sigmund Exner; Sigmund's brother-in-law Alexander von Winiwarter; Marie's husband Anton von Frisch. Source: Irmgard Smidt-Dörrenberg, *Gottfried Keller und Wien* (Vienna: Museumsverein Josefstadt, 1977), p. 41.

for liberals, radicals, and freethinkers. As historian Gordon Craig argues, in the years before 1848 Swiss liberalism served as a model for the rest of Europe.[56] The ideal articulated by the Swiss federation in 1848 was a union based on bonds of friendship and tolerant of diversity. In the 1850s, however, Zurich's republican government became an "oligarchy of wealth and education," dominated by the industrialist Alfred Escher.[57] By the early

56. Gordon A. Craig, *The Triumph of Liberalism: Zürich in the Golden Age, 1830–1969* (New York: Collier, 1988), p. ix.

57. Hans Max Kriesl, *Gottfried Keller als Politiker* (Frauenfeld and Leipzig: Verlag Huber, 1918), pp. 127–29, 164.

1860s, the dominant liberal party faced a growing democratic movement, which attacked Escher and demanded revisions to the Constitution of 1848. When Adolf arrived in Zurich in the fall of 1868, the canton had recently recovered from a cholera epidemic, but its political passions still raged. The elections of 1866 were a victory for the Democrats, who managed to unseat the darling of the liberal party, the novelist Gottfried Keller.[58]

Keller was part of a circle of liberal intellectuals, including the architect Gottfried Semper and the archeologist Karl Dilthey (brother of the philosopher Wilhelm), with whom Adolf quickly became friendly. The great novelist was recovering from a personal tragedy that had dwarfed his political loss. At the age of fifty, Keller had been preparing to marry for the first time. During the elections, he was mercilessly attacked in the press. His fiancée, thirty years his junior, was so horrified by a slanderous article by one of Keller's political opponents that she drowned herself.[59] Over the next three years, while he continued to serve as secretary to the liberal party, Keller moved away from his support for democratic reforms. He questioned whether the public was ready for democracy. For now, the government of the republic seemed safest in the hands of the educated elite. Yet Keller held that elite to a high standard of what Craig calls "civic conscience."[60]

It was during this transitional period that Keller was introduced to Adolf Exner. Adolf had made a striking impression on their mutual acquaintance Otto Benndorf, who saw in him for the first time an example of "the cultured and intellectually cultivated Austrian with the traditional religious indifference, and the desire and ability to enjoy life and not let it be consumed exclusively by work."[61] Although Adolf was half Keller's age, they became fast friends. In their personal histories the two men had much in common. That much Adolf would have known from Keller's autobiographical novel *Der Grüner Heinrich*, which had appeared a decade earlier and went on to become a classic of German literature. Keller's father had been a liberal, civic-minded reformer, a Franz Exner on a far more modest scale, and like Exner, he died in the prime of life. Keller had trained as a painter but had also pursued legal studies, and he read Adolf's scholarly works with enthusiasm.[62] Before long Keller was on intimate terms with the Exner siblings. He visited

58. Ibid., p. 188.

59. Ibid., p. 176.

60. Craig, *Triumph*, pp. 237–263.

61. Otto Benndorf's son Hans, cited in Franz Exner, *Exnerei*, p. 16.

62. On Keller's life, see Jakob Baechtold, *Gottfried Kellers Leben. Seine Briefe und Tagebücher*, 3 vols. (Berlin: Verlag W. Herz, 1897); and Craig, *Triumph*, pp. 237–63. For Keller's comments on

with them at their Zurich apartment and gathered with them over beer at a nearby inn. For Adolf's sister Marie, Keller conceived a shy admiration, to be immortalized in the years of correspondence between these two gifted writers.

In the following years, as Keller fictionalized Swiss politics in the pessimistic second half of his short-story cycle *Die Leute von Seldwyla,* the Exner siblings became a loyal if critical audience for his drafts. Keller painted the dangers of democracy in stories like "Das verlorene Lachen" (The lost laughter) about an unhappy marriage between members of different social classes, which he was writing when he visited the Exners in Vienna in 1874. When Adolf complained that one story had too much political rhetoric, Keller responded that he would leave out such ranting or "Kannegiesserei."[63] When Adolf became tutor to Prince Rudolf, the heir to the Habsburg throne, the young prince became a favorite topic of amusement between the men. "Will we soon be able to name him an honorary member of the Swiss Legal Society," Keller asked in 1875, "provided that he'll spring for a keg of beer?"[64] A few months later, when the emperor awarded Adolf a knighthood for his service, Adolf told Keller not to waste ink on the new title: "since even in Switzerland I didn't really stray so far that I would deliberately let my honest name be fluffed up with a 'Ritter.'"[65] Young Prince Rudolf, who "had taken a great fancy to [Adolf],"[66] eagerly adopted his liberal views and those of other like-minded tutors. Indeed, to Austria's liberals, Rudolf was the promise of a bright future, until his suicide in 1889.

Adolf's years in Zurich did not extinguish his sympathy for democracy entirely. Yet his experiences opened his eyes to the finer nuances of the political spectrum. For a student of politics, he would later write, the Swiss cantons were instructive: they were "especially plastic objects from a political perspective, due to the small space in which the play of governmental forces moves in quick exchanges of effects and counter-effects."[67]

Adolf's scholarship, see their correspondence in *Aus Gottfried Keller's Glücklicher Zeit: der Dichter im Briefwechsel mit Marie und Adolf Exner,* ed. I. Smidt (Zurich: Th. Gut & Co., 1981).

63. Keller to A. Exner, 27 Aug. 1875, in Smidt, *Keller's Glücklicher Zeit,* p. 91.

64. Keller to A. Exner, 27 Aug. 1875, ibid., p. 90.

65. A. Exner to Keller, 28 May 1876, ibid., p. 106. In the same letter to Keller he referred disparagingly to Rudolf as "mein kleines Nepötchen," apparently alluding to the power of the favorite nephew, or "nepot," of Renaissance popes, the etymological origin of the word *nepotism.*

66. Marie Exner von Frisch to Keller, 17 Dec. 1875, ibid., p. 96.

67. A. Exner, *Über politische Bildung,* p. 17.

Swiss liberalism was marked by strong resistance to centralization and bureaucratization. This was an attitude well suited to an Austrian liberal of the late 1860s, when decentralization appeared to be the solution to the ills of the multiethnic empire, with the Compromise with Hungary of 1867 pointing the way. In the fall of 1868, as liberals throughout the German states debated conflicting plans for unification, Adolf used his inaugural speech in Zurich to identify the responsibilities of a faculty of law in an age of constitutional reform and legal codification.[68] Now more than ever, he insisted, the university's goal must be to raise law students "above and beyond the narrow borders" of the laws of a particular time and place. A broad education was necessary to make a graduate of a law faculty more than "a pitiable routine worker, whose entire knowledge and ability is confined to the two-thousand paragraphs of the legal code, and who is a ruined man if this code is done away with overnight." Moreover, Adolf observed, codification had its limits. There was a trade off between a simple, concise style of writing and ease of application. The law student must learn to relate the general principles articulated in the code to idiosyncratic individual cases. He must gain the ability "with a practiced eye to pick out and identify out of the confusing mass of given facts and relationships what is legally essential." This facility for relating general principles and specific instances was in Adolf's view the core of what Adolf alternately called *juristische Bildung* (legal education) or the legal *Kunst* (art). While there might be other means of acquiring this art, Adolf reminded his colleagues and students that the only proven one was the study of Roman law. "It is the total result of a partly conscious and methodical, partly unconscious *labor*, carried out under particularly favorable local and historical conditions." Roman law, in the form in which modern scholars were fortunate to possess it, contained a record of its own development. Visible in its serial incarnations were the methods of the Roman jurists themselves. Such a record of a contingent process of legal evolution provided crucial insight into the relationship of general principles to individual cases, a perspective that Adolf deemed essential in the age of codification.[69]

Adolf's inaugural speech justified the authority of an educated elite against the bureaucratic machinery of a centralized state. Such a critique suited the interests of a scholar of Roman law. The historical school of law

68. A. Exner, *Die praktische Aufgabe der romanistischen Wissenschaft in Staaten mit codificirtem Privatrecht. Akademische Antrittsrede* (Zurich: Meyer & Zeller, 1869).

69. Ibid., pp. 4, 17–18.

had arisen in Austria as a critique of Enlightenment-era natural law, the legal philosophy of the eighteenth-century absolutist state. In Austria, natural law had become associated in particular with the alliance of church and state. Until 1824, for instance, natural law was taught at Austrian universities as a subfield of "rational theology."[70] Legal historicism was thus antiabsolutist and anticlerical. In Austria its roots lay not in Hegelian Romanticism but in the classical humanist ideal of self-realization, the ideal on which the mid-century educational reforms had rested.[71] At the center of this program was the legal expert, whose vision of history would guide the ongoing task of adapting justice to society. To his students in the law faculty Adolf insisted that jurisprudence was not "knowledge of laws" but a method of history.

Perhaps inspired by his younger brothers' scientific pursuits or perhaps by his acquaintance with Helmholtz, whom he had befriended during his studies in Baden in the early 1860s,[72] Adolf had begun to think deeply about the methods of the natural sciences in relation to legal and historical analysis. In the coming years, these reflections would lead him to a natural scientific model for jurisprudence that contrasted strikingly with Helmholtz's ideal of a deterministic science. To his new students in Zurich in 1868 he phrased the comparison thus: "If we carry things so far that you view human rights and political relations with a jurist's eye, that you realize that you are face to face with a phenomenon, as in natural history, then we will have attained what we want."[73]

Crucially, it was the descriptive discipline of natural history and not deterministic physics that Adolf took as his model. Elsewhere in the early 1870s Adolf drew a more elaborate analogy between his empirical approach to jurisprudence and the methods of the empirical sciences. As in the natural world, no two judicial cases were ever exactly alike. The jurist thus needed a taxonomist's gaze: "One cannot say: in identical cases; for identical cases in a literal sense do not exist, there are only similar cases; a case never reoccurs precisely, under the same conditions, a second time. It can, however, be

70. *Lehrbuch der Philosophie* (Vienna: k.k. Schulbücher-Vershleiß-Administration, 1835), vol. 1, p. 10.

71. Werner Ogris, "Die historische Schule der österreichischen Zivilistik, in *Festschrift Hans Lentze*, ed. N. Grauss and W. Ogris (Innsbruck and Munich: Universitätsverlag Wagner, 1969), p. 459ff; Franz Wieacker, *A History of Private Law in Europe (with particular reference to Germany)*, trans. Tony Weir (Oxford: Clarendon Press, 1995), p. 283–347.

72. For brief references to Adolf's friendly relations with Helmholtz, see Franz Exner, *Exnerei*.

73. Quoted in I. Smidt-Dörrenberg, *Gottfried Keller und Wien* (Vienna: Museumsverein Josefstadt, 1977), p. 8.

required that where the difference appears minor, irrelevant, the same decision follows."[74] The jurist had to be as sensitive to contingent variations as the student of natural history. Absolute judicial principles were therefore impossible, a fiction of the bureaucratic state.

In this way Adolf justified the training of a humanistic elite like the circle of scholars into which he had fallen in Zurich, a milieu intimately linked to the new Polytechnikum. This institute had been envisioned in the post-1848 era as a national institute for training the scientists, engineers, and architects who would be responsible for modernizing the Swiss nation.[75] It opened in 1863, with faculty drawn from the cream of the German universities, securing Zurich's status as one of the most cultured and cosmopolitan cities in the German lands.

At the Polytechnikum, the project of training the guardians of the infrastructure of the modern Swiss state developed, characteristically, under the banner of humanism, not technocracy. The institute's building had been designed by Gottfried Semper, professor of architecture at the Polytechnikum, key member of Zurich's liberal elite, and now a friend of the Exners. Semper's architecture expressed in stone the ambition of humanizing the sciences and rationalizing the arts. Sculptures on the façade allegorized the union of art and science, and medallions depicted individual artists and scientists, ancient and modern.[76] Semper exemplified the unity of art and science in his own research. He advocated an "empirical" approach to aesthetics. Against the a priori claims of rationalist aesthetics, Semper viewed art as a thoroughly human product, formed through the observation of nature and free artistic creation. He identified not absolute aesthetic laws but "regularities" common to art and nature, a formulation that left ample room for creative innovation.[77] For Sigmund Exner, who likely visited Zurich from Heidelberg, where he was studying physiological optics with Helmholtz, Semper's empirical aesthetics unlocked new doors. Semper's theories be-

74. A. Exner, "Ueber Recht und Billigkeit," *Juristische Blätter* 3 (1874): 85–87, 105–7, quotation on p. 86.

75. Craig, *Triumph*, pp. 143–47. On the practical orientation of scientific training at the Polytechnikum (later the Eidgenossische Technische Hochschule), see Peter Galison, *Einstein's Clocks, Poincaré's Maps: Empires of Time* (New York: Norton, 2003), pp. 228–29.

76. H. F. Malgrave, *Gottfried Semper, Architect of the Nineteenth Century* (New Haven: Yale University Press, 1996), p. 237.

77. Carrie Asman, "Ornament and Motion: Science and Art in Gottfried Semper's Theory of Adornment," in *Natural History*, ed. Jacques Herzog and Pierre de Meuron (Montreal: Canadian Centre for Architecture, 2002).

came one launching point for Sigmund's investigations of perception. Of his own experience at the Polytechnikum, Serafin Exner wrote that it provided the valuable opportunity "to associate outside of formal studies with a series of men whose effects on me, in the most varied directions, must have been in part very long-lasting."[78] For young scholars like the Exners, then, the community in and around the Polytechnikum was an ideal environment in which to cultivate "many-sidedness" and to fashion themselves as liberal thinkers in the tradition of Keller and Semper. They sought recognition not for the discovery of absolute laws but for the dexterity with which they analyzed contingent processes in nature and society. It was on the basis of this mental agility and absence of dogmatism that they justified the authority of a humanistic elite against that of a bureaucratic state machinery.

FOUR years after his inauguration at Zurich, Adolf Exner received an even more prestigious invitation—to succeed Rudolf Jhering as professor of Roman law at the University of Kiel. "I have too little vanity to let myself be blinded by the situation," he wrote to his sister Marie. Were he to pass up Kiel to return to Vienna, where the study of Roman law lagged far behind, he knew he would encounter "troubles and much frustration." "[B]ut I want to go there [Vienna] because it is ground I know and on which I can be sure of being able to build something sound."[79] To Marie's relief, Adolf did indeed return to Vienna and lived with her and her family in the university quarter until he married.

By the end of their time in Zurich the Exner siblings had formed a new identity. Keller and Dilthey spoke of them with the collective noun "Exnerei," in the sense of a medley of Exners.[80] The Exner siblings' reinvention as a family was an integral part of their self-fashioning as liberals. Keller himself had shown them that the political began with the personal. The revolutions of 1848 had taught him that "there can no longer exist private individuals!"[81] The novel he had drawn from his own life conveyed this lesson. He explained the moral of his *Grüner Heinrich*: "That he who does not succeed in keeping his own relations and those of his family in secure order is also incapable

78. Franz Serafin Exner, "Autobiography," p. 5.

79. Adolf Exner to Marie Exner von Frisch, 6 June, 1872, quoted in Franz Exner, *Exnerei*, p. 8.

80. See, e.g., Dilthey to Marie Exner von Frisch, 2 March 1873, quoted in Smidt-Dörrenberg, *Gottfried Keller und Wien*, p. 8.

81. Quoted in Kriesl, *Gottfried Keller*, p. 101.

of filling an active role in civic life."[82] Keller articulated a key facet of Central European liberalism, which historians have only recently begun to appreciate: that the cultivation of the domestic sphere was not a retreat from politics but a precondition of liberal identity.

82. Ibid., p. 102.

CHAPTER THREE

Memory Images

Models of Reason in the Liberal Age

IN THE decade after their return from Zurich, the Exner siblings took another decisive step toward shaping their identity as a family, namely, marriage. Even without their parents' guidance, their matches bolstered their standing among Vienna's liberal intellectual elite. Marie, having broken her engagement to a suitor of more modest ambitions, enjoyed the attentions of Anton von Frisch, a rising star of the Vienna medical establishment and a guest of her brother Sigmund one summer at the Wolfgangsee. Marie's brothers, meanwhile, courted women who were well-connected in the academic world. Sigmund's wife, Emilie von Winiwarter, came from a family of eminent liberal jurists. She was a niece of Alexander von Bach, the moderate-liberal leader of the 1840s who became a conservative minister of the interior in the 1850s. Adolf's wife, Constanze Grohmann, was the daughter of a banker who had been a close friend of Franz and Charlotte Exner and their host for several summers at the Wolfgangsee.

The intimate atmosphere of the *Sommerfrische* facilitated more than one of these pairings and subsequently helped assimilate the newcomers into the family. As Sigmund's wife Emilie would later write in her history of Brunnwinkl, the marriages of the family's second and third generations were "bound by shared traditions, and the characters [*Musterkarte*] of our sons- and daughters-in-law fuse ever more with that of the whole."[1] These traditions began to crystallize in 1882, when the Exners turned from seasonal renters in the Salzkammergut into property owners.

1. Emilie Exner (Felicie Ewart, pseud.), *Der Brunnwinkl*, p. 30.

At this time, the political atmosphere in Vienna was growing increasingly adverse. The liberals' authority had been weakened by the economic crash of 1873, and their conflicts with the Church sowed further opposition. Anti-Semitism was on the rise, with the medical and scientific faculties of the university becoming targets of hostility towards liberal Jews. The elections of 1879 brought to power a coalition of German and Polish conservatives, clericals, and moderate Czech nationalists under Count Eduard Taaffe, a Catholic conservative and friend of the emperor. Taaffe's administration sought to reverse liberal reforms of the previous decade in part by renewing Church control over the imperial schools, the bastion of liberal power.

Three years into this new administration, the Exners secured a space they envisioned as a liberal utopia on a smaller scale. In the summer of 1882 Marie and Anton von Frisch and their young boys rented the upper level of a house in St. Gilgen, on the Wolfgangsee, an area familiar to them from previous vacations. It was an old farmhouse attached to a corn mill, and the miller and his wife lived on the ground floor. According to Karl Frisch, the miller had a drinking problem and his wife knew nothing about economizing.[2] At summer's end, the wife begged the Frisch family "on her knees" to buy the house from them so they could pay off their debts. Marie and Anton ultimately agreed, and over the next two decades the couple bought up four of the surrounding cottages. The former inhabitants were all small crafts producers plagued by debt and eager for cash. They bought land elsewhere in the region or, in one case, emigrated.

The asking price for the mill house in 1882 was 2,000 gulden, about ten times the cost of a summer rental.[3] "For reckless Viennese like us hardly any money at all," Marie joked to Keller once the purchase was final.[4] She had, however, deliberated carefully over the price. She wrote to Adolf for advice, reporting that the house was in need of repair, but that the surroundings were "enchantingly beautiful." "Yet as a whole I can't be sure if it would always please us." To resell "such a remotely located nest" would be a challenge, as the plight of the current owners showed.[5] In his reply,

2. Karl Frisch, *Fünf Häuser am See*, p. 13.

3. Adolf Exner to Marie Exner v. Frisch, 6 Sept. 1882, reprinted in Karl Frisch, *Fünf Häuser am See*, pp. 17–18.

4. Marie Exner v. Frisch to Gottfried Keller, 22 Dec. 1882, in Smidt, *Keller's Glücklicher Zeit*, p. 111.

5. Marie Exner v. Frisch to Adolf Exner, 4 Sept. 1882, reprinted in Karl Frisch, *Fünf Häuser am See*, pp. 16–17.

FIGURE 4 The Mill House at Brunnwinkl, 1883. Source: Karl Frisch, *Fünf Häuser am See. Der Brunnwinkl: Werden und Wesen eines Sommersitzes* (Berlin: Springer, 1980), fig. 2, p. 14. (c) Springer Verlag, Berlin, 1980. With kind permission of Springer Science and Business Media.

Adolf weighed personal benefits against financial burdens. It was "one of the greatest charms of ownership, that one is certain to find again every year the same dear and familiar trees, dogs and faces, etc., and that gives a splendid feeling of belonging to a piece of earth, which every family should have and which I would particularly like to grant your children."[6]

Adolf's vision of the summer home as a source of intergenerational stability was widely shared in his milieu. We have seen how the country house in Stifter's *Der Nachsommer* appeared insulated from the ravages of time, serving as a stable foundation for intellectual inheritance. In Adolf's imagination, the power of the summer home lay in the visual memories that would become associated with it. Images of "trees, dogs and faces" would create a sense of "belonging," of identity. Sharing such iconic memories was

6. Adolf Exner to Marie Exner v. Frisch, 6 Sept. 1882, reprinted in Karl Frisch, *Fünf Häuser am See*, pp. 17–18.

part of what it meant to be a family, he implied. Through painting, photography, and the written word, the Exners would indeed imprint images of "the familiar" in their minds at Brunnwinkl.

One of the foremost charms of the *Sommerfrische* in the eyes of Austrian liberals was the sense of constancy it provided. Theodor Gomperz, a liberal philosopher and translator of John Stuart Mill, was drawn to the "conservative" function of the summer retreat: "Everything lies in the same place as in earlier years, down to the clothes-brush and train schedule . . . and I understand the idealistic conservative disposition that would leave everything in its place in state and society as well. There really is a great charm in not always having to face novelty."[7] This attraction to a "conservative" environment reminds us of one of the specificities of Austrian liberalism, its resistance to revolutionary tendencies of the political left or right.

For a bourgeois Austrian family, to buy property was to "secure the future," as historian Dietlind Pichler writes.[8] But it also meant taking on risk. Deliberating on Brunnwinkl's purchase, Adolf weighed the value of continuity against the risk of ownership—namely, the possibility "that, in time, something incalculable [*irgend ein Unberechenbares*] could thoroughly spoil the place for you." In the Salzkammergut, where flooding was a frequent problem, the risks were both natural and social. In the early 1880s the area bore no resemblance to the fashionable tourist destination it would become at the turn of the century. After the train from Vienna to Salzburg, it was still a five-hour carriage ride to the lake, and so it would remain until the opening of a rail line in 1893. Wealthier bourgeois families preferred villas on the outskirts of Vienna. Adolf therefore recommended buying the house only if it came with an ample, private garden, all for no more than 4,000 gulden

> since only in this way would the object have a lasting value and be transformed with time by your beautifying hand into a little villa, which will perhaps in the future be sought after by others and bought at a luxury price.

Adolf viewed the property as a source of stability for the family *and* as an investment. Surprisingly, the "conservative" incentive of family life converged with liberal economics in Adolf's calculations. Remarkably, the

7. Quoted in Rossbacher, *Literatur und Bürgertum*, p. 242.

8. Dietlind Pichler, *Bürgertum und Protestantismus: Die Geschichte der Familie Ludwig in Wien und Oberösterreich (1860–1900)*, Bürgertum in der Habsburgermonarchie, 10 (Vienna: Böhlau, 2003), p. 150.

potential for a good return on this investment lay in the wife's hands. In the historiography of the bourgeoisie, the woman's "beautifying [*verschönernde*] hand" and its social value is a familiar theme, but not the *material* value of the woman's aesthetic cultivation of the home. Marie's green thumb and painter's eye received their due. In the Exners' case, it was natural to acknowledge Marie's contribution to the social and financial stability of the family, a role she had assumed since her father's death. By the same token, Brunnwinkl offers historians a rare chance to see how female influence shaped a culture of scientific inquiry.

Almost every summer for the next four decades, Marie rented three of the cottages for a small sum to her brothers Serafin, Sigmund, and Karl, and another cottage was rented out to friends. To Marie's regret, Adolf, whom she admired above all others, was obliged to summer with his wife's family in Tyrol. The week or so he spent at Brunnwinkl each year was greeted with particular joy by Marie and the children. The term "Brunnwinkler" soon arose to designate a member of this extended community.

To the young Exners, Brunnwinkl seemed a haven of security: "in the 'Winkel' we were protected from bad illnesses and serious accidents, even if the famous guardian angel often really had to work like a drudge."[9] Adolf's daughter Nora reflected years after her father's death that "protection may be the best inheritance that parents can leave their children."[10] This sense of security rested on a variety of impressions. Only as adults did the younger generation realize how hard Marie had worked to ensure their comfort. She knew how to "enter into other people's ways of thinking," her son Hans later wrote; "only when you understand someone else's mood can you be fair to him. And so she lovingly indulged all the desires that she got wind of in the Winkel, those of the young and youngest as of the elders."[11] It was she who had ensured "peace in the realm [*der Friede im Reich*]."[12] The invisibility of her labors was her ultimate accomplishment, making the harmony she so carefully composed seem spontaneous. This impression of peace spoke as well of the social and financial stability with which the young Exners associated their summer home; of the widespread view of summer homes as protection against outbreaks of cholera and other diseases in the heat

9. Hans Frisch, *50 Jahre Brunnwinkl*, p. 11.

10. Nora and Hilde Exner to Otto Benndorf, 20 Jan. 1906, folder 640, letter 37-2, Manuscript Collection, Österreichische Nationalbibliothek, Vienna.

11. Hans Frisch, *50 Jahre Brunnwinkl* (Vienna: self-published, 1931), p. 7.

12. Franz Exner, *Exnerei*, p. 29.

and congestion of the cities; and of the new role of the bourgeois family as a source of emotional security, soon to be articulated by Viennese psychoanalysts.[13]

This pervasive sense of stability formed the necessary background for the young generation's intellectual and moral development at Brunnwinkl. In the pedagogical philosophy of the first Franz Exner, skeptical exercises were to be a prophylactic against dogmatism. But they could only perform this function if the pupil began with a strong trust in figures of authority and in the general reliability of his surroundings. In just this way, the summertime explorations of the young Exners rested on their impression of the security of the Brunnwinkl environment. The Exners created at Brunnwinkl a stable structure within which natural contingency could become an object of analysis rather than a source of fear or bewilderment. As Hume observed, inductive reasoning rests on the unprovable assumption that like cases usually have like outcomes. It rests, one might say, on a diffuse trust in the reliability of the environment.

THE HUNTER AS A MODEL OF REASON

It was not uncommon for a lone Brunnwinkler or a small group to disappear entirely from morning till dusk. To an outsider, it would have been hard to say whether they were engaged in research or leisure, since immersion in nature was the precondition for both. Hunting, for instance, was a favorite pastime among the Exner men. It was only after 1848, when legal reforms first allowed common landowners to shoot game on their own property, that the Austrian bourgeoisie had adopted hunting as a pastime. Middle-class hunters decked themselves in the green-gray coats made popular by Archduke Johann, the royal renegade who first linked the identities of hunter, alpinist, scientist, and progressive. In contrast to their aristocratic forerunners, the new bourgeois hunters focused less on the acquisition of trophies and more on the rational management of natural resources, aligning hunting with natural science. Yet aristocratic examples still carried weight. The patriotic iconography of hunting in Austria included abundant portraits of Emperor Franz Joseph with rifle and quarry, surrounded by nature and the

13. In "Symbiosis and Individuation: The Psychological Birth of the Human Infant" (1974), Margaret Mahler recounts that her theory of the infant-mother relationship originated with observations she made in Vienna in the 1920s. In *Selected Papers*, vol. 2, *Separation and Individuation* (New York: Jason Aronson, 1979), pp. 149–65, quotation on p. 149.

simple folk of the Austrian Alps. Hunting was also a favorite diversion of Crown Prince Rudolph, Adolf Exner's former pupil, who was a forceful proponent of science and liberalism in the 1880s.[14] For bourgeois vacationers in the Salzkammergut, hunting was an entryway into the natural world and rural culture. Viennese poet Nikolaus Lenau proudly recounted how he had been introduced to native weapons and medicinal plants by a local hunter.[15] Urban visitors to the Salzkammergut were particularly taken with the natives' profound knowledge of their natural surroundings. The writer Max Mell marveled at a Styrian peasant who knew the Latin name of every plant that grew on the mountains and in farmers' gardens, along with its medicinal properties.[16] Artists and scientists alike identified with the peasant's way of interacting with nature. Peter Altenberg, for instance, a regular visitor to the Salzkammergut, took fishing as his model for the appreciation of nature:

> Being a trout fisher and a nature lover are one and the same! You have to hike from rock to rock. One squats behind each. And this hiking is only satisfying if one truly loves the natural surroundings... The trout fisherman loves the mountain water passionately, it makes him forget his wife and child, often even food. He loses himself in the details of the surroundings, a unique sign of real enjoyment!... He roams along, from rock to rock, and sees everything, everything.[17]

Altenberg undoubtedly saw the affinity between the fisherman's power of observation and the poet's. Artists and scientists alike took their peasant neighbors as models for exploring nature.

For the Exners, hunting was an exercise in independence—of mind, body, and vision. Even a nonscientist like Karl Frisch's brother Ernst learned to value the alpine environment with its diverse flora and fauna because it "offer[ed] attentive senses... always something new." Exploring nature in

14. Krasa, "Erzherzog Johann"; Krasa, "'Mit Büchse und Feder'—Kronprinz Rudolph als Jäger und Ornithologe," in *Jagdzeit: Österreichs Jagdgeschichte. Eine Pirsch* (Vienna: Historisches Museum der Stadt Wien, 1997), 306–7; Frederic Morton, *A Nervous Splendor: Vienna, 1888–1889* (New York: Penguin, 1980). Karen Wonders, "Hunting Narratives of the Age of Empire: A Reading of Their Iconography," *Environment and History* 11 (2005): 269–91; John M. MacKenzie, *The Empire of Nature: Hunting, Conservation and British Imperialism* (Manchester: Manchester University Press, 1988).

15. Kos and Krasny, *Schreibtisch mit Aussicht*, p. 160.

16. Kos and Krasny, *Schreibtisch mit Aussicht*, p. 188.

17. Kos and Krasny, *Schreibtisch mit Aussicht*, p. 164.

such a self-directed way was the sign of knowing, as Ernst put it, how to "use one's freedom well."[18] Sigmund Exner's wife Emilie admired the hunter's independence and sense of freedom. Hunting was an exercise in living, in her words, as "ein freier Mensch in der freien Natur," at liberty in the open country. For the hunter, "no path is too steep, no weather too poor . . . no exertion too large for the pleasure of climbing around in the mountains with a rifle." "Im freien" was a pregnant phrase meaning "in the open," "in nature," and "in freedom." "Nature is the realm of freedom," Alexander von Humboldt had written in 1845, remarking on the aptness of the German phrase "Eintritt in das Freie." (Karl Popper's title *The Open Society* captured a similar resonance in English, where liberalism is often associated with images of openness.) It was not only the hunter's independence that Emilie admired but also his disciplined attention to his natural surroundings. One of her nephews had spent days observing a pair of vultures through a telescope "to get to know their habits." His quarry became prized specimens in the collection of his brother Karl, the budding naturalist.[19]

Outdoor exercise was considered just as healthy for the Exner girls. Sigmund's wife Emilie, who had three brothers, had herself been a "Bubenmädchen," a tomboy.[20] In step with progressive opinion in the United States, Emilie Exner insisted that a young girl "becomes fresher and more impulsive in mind and spirit if one permits her the physical exercises that make for a happy and healthy childhood."[21] Unlike their brothers and male cousins, however, the Exner girls were admired for the grace with which they became part of nature, not for their mastery over it. "In a short peasant dress and a white vest, tanned, with shining eyes and flying pigtails, fearless and gay, wild and yet reliable."[22]

If the boys of the family were hunters, the girls were fawns. When, for instance, young Hilde Exner fell into a swollen brook and was dragged along

18. Ernst v. Frisch, *Sommer am Abersee*, pp. 10–11.

19. Emilie Exner, *Der Brunnwinkl* (Vienna, 1906), p. 14, Alexander von Humboldt, *Kosmos. Entwurf einer physischen Weltbeschreibung*, vol. 1 (Stuttgart: J. G. Cotta, 1874), p. 5. Wonders ("Hunting Narratives") argues that hunting and the popular forms of natural history that accompanied it were a means of masculine self-fashioning for Europeans in the "age of empire" (1875–1914). MacKenzie discusses the moral benefits attributed to hunting in Victorian Britain in *The Empire of Nature*, pp. 42–43.

20. Marie Ebner von Eschenbach, "Exner, Emilie," in "Exner, Emilie (Felicie Ewart)," in *Biographisches Jahrbuch und deutscher Nekrolog*, ed. Anton Bettelheim, vol. 14, *Die Toten des Jahres 1909* (Berlin: Georg Reimer, 1912), pp. 10–18.

21. Emilie Exner (Felicie Ewart, pseud.), *Die Emancipation in der Ehe* (Hamburg: Voß, 1895), p. 70.

22. Emilie Exner, *Der Brunnwinkl*, p. 12.

by the current, it was her cousin Otto's "hunter instinct"—in Emilie's words—which allowed him "immediately to take in the situation at a glance" and to "estimate" where the current would carry her.[23] The hunter, the Exner's symbol of freedom, learned to read the tracks of his prey and so to outwit a capricious natural environment.

As an exercise in the masculine virtues of liberal individualism, hunting served the Exners as one defense against "the herd existence." Yet unmitigated individualism was never Brunnwinkl's social ideal. As Otto's heroic "hunter instinct" showed, the qualities the Exners believed hunting to impart served the community as well as the individual. Among these traits—valor, daring, heroism—were moral qualities that could only be considered valuable in a social context. As Sigmund Exner would argue in 1892, for a Robinson Crusoe there could be no such virtue as courage, only a calculation of the best means of preserving his own life. In this lecture on "morality as a weapon in the struggle for existence," Sigmund presented a naturalistic account of morality in terms of evolutionary pressure and neurophysiology. From nature's perspective, it was the survival of the species and not of the individual that mattered. The "natural" view of morality, which he found embodied in ancient Greek tragedy, was concerned only with the benefit or harm of an action to society, not with the actor's beliefs or intentions.[24] Sigmund's account of the evolution of "courage" suggests that hunting at Brunnwinkl was a lesson in managing the conflicting impulses of individualism and communitarianism. Even as it trained the powers of observation crucial to natural science, the hunt was a form of socialization, an exercise in competition and cooperation.

As a physiologist, Sigmund Exner took professional pride in possessing a hunter's eyes. In an autobiographical sketch composed in 1891 for the Austrian Academy of Sciences, Sigmund recounted how he had held the image of an unidentified bird in his mind for years until, on his first snipe-hunting expedition, he recognized it as nothing other than a snipe. "Thus I had borne in mind for years the manner of taking flight and flying with perfect clarity."[25] Sigmund went so far as to insist that his skills as an observer were all the scientific talent he could claim. He modestly described himself

23. Ibid., p. 27.

24. Sigmund Exner, "Die Moral als Waffe im Kampfe ums Dasein," *Almanach der Akademie der Wissenschaften* (1892): 243–73.

25. Sigmund Exner, "Autobiographische Skizze" (1891), Personalakt, Archiv der Österreichischen Akademie der Wissenschaften, Vienna.

as an "exceptionally untalented person," with only one outstanding quality: "that often when observing a natural phenomenon more strikes me than does others, who observe it with me, and who are justly considered far more talented than I." He portrayed himself on the model of the protagonist of Stifter's *Nachsommer* or Keller's *Grüner Heinrich*. "I can watch animals and observe plants for days on end without being bored, and if one would call that a talent, then I would have this little talent in a high degree, namely, the ability to find my way in the animal and plant world. I rarely know a name, but I can say of every plant one shows me where I have seen it before, if it grows in dry or moist soil, and such. Likewise with animals." Perhaps, he surmised, that was why "I am happiest when I am entirely alone in a forest or a field."

This was hardly an accurate description of Sigmund's research methods or of the limits of his "talents." Sigmund portrayed his scientific skills as self-taught. Yet, as sociologists of science alert us, "the activity of seeing and naming objects in the natural environment" depends on an array of social conventions for the organization of experience.[26] Hunting taught such conventions, and Sigmund's training as a hunter may have facilitated his success as a naturalist. Still, his heavily theoretical research hardly reduced to the work of a mere observer. His youthful treatise on atomism, for instance, had celebrated the value of a hypothesis for which no direct evidence existed. He was also a celebrated experimenter. Rather than a frank self-portrait, Sigmund's autobiographical sketch was part of his fashioning of a scientific persona.[27] He modeled himself on the hunter, the decoder of nature's signs.

Such tropes were not unusual in nineteenth-century science. Carlo Ginzburg has shown the affinity of the hermeneutic methods of Sigmund Freud (Sigmund Exner's former student) to the activities of the hunter and the detective.[28] Thomas Henry Huxley compared the study of the fossil record

26. John Law and Michael Lynch, "Lists, Field Guides, and the Descriptive Organization of Seeing: Birdwatching as an Exemplary Observational Activity," in *Representation in Scientific Practice*, ed. Michael Lynch and Steve Woolgar (Cambridge, Mass.: MIT Press, 1990), pp. 267–300.

27. Cf. Daston and Sibum, *Scientific Personae*.

28. Carlo Ginzburg, *Clues, Myths, and the Historical Method*, trans. by J. and A. Tedeschi (Baltimore: Johns Hopkins University Press, 1989). Freud attended Sigmund Exner's lectures "Physiology and Microscopic Anatomy of the Sensory Organs" in the winter of 1876–77, as well as Serafin Exner's lectures on spectral analysis in the summer of 1876 (Nationalien der Medizinischen Fakultät der Universität Wien). Freud also worked in Brücke's laboratory, while Sigmund Exner and Ernst von Fleischl-Marxow were Brücke's assistants. Freud later wrote that these three were men "whom

to the process of reading animal tracks.[29] And Charles Peirce, in his concept of abduction, sought not only to codify the logical process of signreading but also to claim it as a basic human instinct.[30]

On the other hand, in Sigmund's own field, leading experimentalists had come to scorn the practices of the clinician or naturalist. In Johannes Müller's Berlin laboratory and later in Helmholtz's, where Sigmund himself studied, clinical pathologists were chided for basing their theories on isolated instances, when counterexamples could often be found. Nature was a poor experimenter, the argument went.[31] Helmholtz's ideal of the scientist was of an experimenter capable of ascertaining absolute laws, not a naturalist who haphazardly collected and classified "isolated facts."

Sigmund Exner cultivated a scientific persona at odds with Helmholtz's. While Helmholtz portrayed himself in the laboratory, Exner stressed his clinical experience, aiming to replace mere "desk work" with "observation at the hospital bed."[32] Helmholtz was a precision experimentalist, in close touch with industry. Exner longed to be a naturalist, at home in the field. These diverging personae reflected differences in national context. Prussian industry was a mighty force with which Helmholtz strategically aligned himself. German-speaking Austria, on the other hand, lagged in industrial development, but in the Vienna medical faculty it possessed a scientific institution of unsurpassed renown. Even a reluctant medical student like Exner, who had always favored research over clinical practice, could profit from the high esteem for medicine in Vienna.

Above all, the figure of the naturalist carried associations in Exner's milieu that were absent in Helmholtz's. Faced with challenges from clerical conservatives, nationalists, and socialists, Austrian liberals required a model of reason with which to discredit dogmatism without falling into the trap of radical relativism. Exner's self-fashioning was tuned to an Austrian audience, as is clear from the contrast with Helmholtz's disdain for the methods

I could respect and take as my models." Ernest Jones, *The Life and Work of Sigmund Freud* (New York: Basic, 1953), vol. 1, p. 39.

29. T. H. Huxley, "On the Method of Zadig," *Nineteenth Century* 7 (1880): 929–40.

30. See, for instance, Arthur W. Burks, "Peirce's Theory of Abduction," *Philosophy of Science* 13 (1946): 301–6.

31. Michael Hagner, *Homo cerebralis: Der Wandel vom Seelenorgan zum Gehirn* (Berlin: Berlin Verlag, 1997), pp. 249–52.

32. Sigmund Exner, *Untersuchungen über die Localisation der Functionen in der Grosshirnrinde des Menschen* (Vienna: Braumüller, 1881), pp. v–vi.

of the naturalist. Among the academic elite that wintered in Vienna and summered in the countryside, the figure of the hunter supplied an embodied ideal of liberal reason in action. The hunter could navigate an unpredictable environment by building hypotheses from the clues of accidental evidence. He was a character at home in a world that had forsaken absolute certainty.

SIGMUND Exner's neurological research in the 1880s depended literally as well as symbolically on liberal power in Austria. Materialism had been associated there for decades with the secularizing forces of the French Revolution, then with the failed uprisings of 1848. It was not until the ascension of liberalism in the 1860s that the field was cleared for theories of localized brain function like Exner's.[33] In his 1881 treatise on cerebral localization, Exner's most important innovation was the distinction between "absolute" and "relative" cortical fields.[34] Damage to the "absolute" cortical field *always* resulted in impairment of the associated motor function, while such impairment *frequently*, but not in every case, accompanied damage to the "relative" cortical field. This distinction made it possible to investigate the nature of the connections among the cortical fields themselves. Using an analogy introduced by Helmholtz, Sigmund pictured the nerves as telegraph wires and the brain as the central telegraph office. All the telegraph wires from the outside world joined in one central room of this office, where the incoming messages arrived at the receiver. Elsewhere in the building were rooms where the messages were processed in various ways, such that any given message might require more or less work in any given room. An operator of such a system would want to know not just how a malfunction in the central room would affect the transmission of messages, but also the effects of malfunctions elsewhere in the building. In other words, did the cortical fields have firm boundaries, or was each absolute field surrounded by a region of relative fields?

In Sigmund's investigation of this question, we can see the naturalist at work. Most neuropathologists would have answered with reference to their experiences with a small number of patients. Seeing that the impairment of a particular motor or sensory function frequently accompanied a lesion to a particular area of the cortex, they assumed the damaged region to be the seat of that function. Sigmund pointed out the danger of this approach. Wider experience might reveal cases in which damage to the area in question was *not* accompanied by motor impairment of this kind. True to the

33. Hagner, *Homo cerebralis,* p. 271.

34. Alois Kreidl, "Siegmund Exner zum 70. Geburtstag," *Wiener klinische Wochenschrift* 29 (1916): 426–27.

naturalist's imperative to collect as many samples as possible, Sigmund instead demanded a statistical approach to unraveling the nature of cerebral localization. From the available data it was necessary to distinguish "the lawlike from the accidental," and to this end "the number of cases to be considered can never be large enough."[35] However, since experimenting on human brains was out of the question, there was no systematic way to collect data. The only way to proceed was to construct rules on the basis of which the relevant cases could be distinguished and then worked through to produce a useful, quantitative result. "This result then has a tangible form, one can recognize its dependence on each of the proposed rules, its degree of probability, etc., and can introduce it into each further calculation."[36] On this model, Sigmund introduced what he called "the method of percentual calculation," according to which the cortex was divided arbitrarily into regions, and for each region was determined: 1) in how many of the cases it had been damaged; and 2) in how many of these cases the symptom in question was present. The ratio of the second number to the first was expressed as a percentage, and the cortical field in question was represented on a map of the brain in a shade corresponding to this value. The goal of this approach was clearly not to construct a complete and accurate map of the brain. It was instead to demonstrate the *possibility* of constructing such a map, to prove that functions could in fact be localized in the cortex. To achieve this, it was not necessary to demonstrate an invariable causal connection between region and function; mere statistical correlation was enough to establish, as Sigmund put it, the "probability" of the results.

This emphasis on a hypothesis's probability rather than verifiability was characteristic of Sigmund's research. In 1894, when his career was established enough for him to risk a more speculative work, he published his "Outline of a Physical Explanation of Psychic Phenomena." There he introduced the concept of the neural network and a model of learning analogous to what computer scientists today know as the Hebbian scheme for the strengthening and weakening of neural connections.[37] As Paul Ziche observes, the goal of this treatise was not to prove a specific "explanation" but to demonstrate generally the "explainability" (*Erklärbarkeit*) of psychic

35. Sigmund Exner, *Localisation*, p. v.

36. Ibid., p. vii.

37. Olaf Breidbach, "Neuronale Netze, Bewußtseinstheorie und vergleichende Physiologie," in Sigmund Exner, *Entwurf zu einer physiologischen Erklärung der psychischen Erscheinungen* (1894; Thun and Frankfurt: Verlag Harri Deutsch, 1999), pp. i–xxxviii.

phenomena in physiological terms.[38] Exner's standards for hypotheses therefore had little to do with verifiability, and the brain structures and processes he proposed were drastically simplified. He was interested not in their verisimilitude but in their "possibility" (*Möglichkeit*). In this sense, he resembled British physicists of the day building flywheel models of the ether. But it was the presentation of his work that was characteristic of liberal science in Austria. It was self-consciously speculative but firmly rooted in experience, falling prey neither to dogmatism nor solipsistic fantasy. Sigmund found his prototype for this liberal model of reason in the hunter's ability to construct hypotheses from the clues of accidental evidence. The hunter—and his modern incarnation, the naturalist—provided the patterns for Sigmund's self-fashioning as a scientist and for his standard of liberal rationality.

SIGMUND EXNER ON THE MECHANICS OF MEMORY

Sigmund Exner's study of cerebral localization led him, in the end, to a great mystery of neuropathology: how was it possible that a brain-injured person could see the forms of letters in front of him without being able to read them or recognize a group of objects without being able to count them? Exner's new understanding of relative cortical fields gave him a possible explanation. Certain lesions might leave intact the capacity to sense and become conscious of a given stimulus, while destroying further neural connections to "certain storehouses of psychic attainments."[39] Exner's analysis ended there, but already he had opened a whole new constellation of questions. How did these "storehouses" accumulate, in what form were they stored, and how were they accessed?

It was through his investigations of memory over the following decade that Exner made the hunter's cunning not just a standard of reason but a model of the instinctive operation of the human mind. As we saw in chapter 2, the faculty of memory had become in the nineteenth century a means of understanding the development and persistence of the ego. Over the course of the century, however, philosophical views of memory had shifted significantly. For Hume, memory had been an unproblematic source for

38. Paul Ziche, "Erklärung und Erklärbarkeit. Sigmund Exners Wissenschaftsphilosophie des Physiologie-Psychologie-Verhältnisses," International Workshop Sigmund Exner, 5–6 March 2004, Vienna.

39. Sigmund Exner, *Localisation*, p. 87.

empirical judgments. It was, as Exner put it, a "storehouse" for reasoning. Kant had likewise considered memory to be wholly under the control of volition. Herbart had complicated this picture by introducing the concept of the threshold of consciousness, allowing for the possibility that aspects of memory might lie beyond the domain of the will. The associationist psychology of James Mill showed further how memory could operate fortuitously, linking experiences without logical connection. Building on Mill's theory, Sigmund Exner's Viennese colleague Theodor Meynert was the first to propose in 1867 a cellular conception of the nervous system, in which memories appeared as traces imprinted by the passage of signals. This was the model that set Exner on the path to the novel theory of learning he proposed in 1894.[40]

At the same time, Exner was guided in his investigations by the practices of memory customary of bourgeois Austrian families. Like the expanding portfolio carried by the narrator of Stifter's *Nachsommer*, visual memories in this milieu were the building blocks of identity. At Brunnwinkl the Exners crafted memories with exquisite deliberateness. The colony's anniversary was celebrated with a feast each year on August 28, the birthday of the first Franz Exner and of Goethe, two patron saints of *Bildung*. Meanwhile, the day-to-day work of memory took many forms—recording, preserving, cataloging, arranging images in albums or displaying them on walls, penning them in memoirs or evoking them in toasts; it involved the production of traces, whether with a camera, paintbrush, or pen and ink. A photograph from 1888, for instance, recorded the planting of the linden tree before it had become the family's "blackboard" and the symbol of its communicative ideal. Decades later, reprinted in a memoir on the occasion of Brunnwinkl's fiftieth anniversary, the photograph took on an iconic status, labelled as "a landmark of Brunnwinkl." The written word had similar force. A vivid passage in a letter from Marie to her brother Adolf brought Brunnwinkl to life: "I lay this morning on the cobbler's hill and read Goethe's *Iphigenie*, watched the children romp around on the dock; someone else was lying on the opposite dock and singing Schubert's *Wanderers*; the world was so beautiful to me, I would not have traded places with a king. From the Billroths' one could hear the melancholy sound of scales and three-note exercises." Marie's gaze was at once preternaturally attuned to the present and directed forward to a time when she would look back at that day with nostalgia. And indeed her

40. Cf. Breidbach, "Neuronale Netze."

words would be revived decades later when her son Hans pieced together such scraps of memory into a portrait of Brunnwinkl.[41]

Memorializing was a process of collecting, juxtaposing, and interpreting. In this sense it resembled the work of the naturalist. Indeed, the natural history "museum" established in the donkey stall by the budding young zoologist Karl von Frisch was a memorial to Brunnwinkl as much as a scientific collection. In later years, Frisch saw its specimens as survivals of an epoch in Brunnwinkl's natural history before the advent of mass tourism and as a personal record of his development as a scientist. As we will see in this chapter and the next, these practices of memory reflected and were reflected in new theories of mind in the late nineteenth century. Freud's metaphor of the mind as a magical writing pad, preserving memory traces in its depths while clearing them from its surface, acknowledged the deliberate, visual work of memory in bourgeois households of his day.

Against this background, one aspect of Sigmund Exner's view of memory is remarkable: his treatment of memories as *images*. Memories, in his terms, were not simply *Erinnerungen*, but *Erinnerungsbilder* or *Gedächtnisbilder*—"memory images." For Sigmund, learning was above all a visual process. For this view he was indebted to Gustav Fechner, whose research in Leipzig bridged philosophy, psychology, and physics. In the 1870s Fechner had described the workings of visual memory in his *Vorschule der Aesthetik*. Borrowing from Mill's associationism, Fechner posited an "aesthetic principle of association" according to which the aesthetic impression that an object or scene makes on the viewer is the "resultant" of the visual memories with which that object is unconsciously associated in the viewer's mind. An orange, for instance, is beautiful not by virtue of its color, shape, or texture but because of an image with which the orange is associated, namely, the sun-drenched Italian coast. In this way, Fechner suggestively brought associationist psychology to bear on aesthetics. He hinted further how this model might be given a physical explanation, on the analogy of vector sums ("resultants") of forces in the brain.[42] Still, Fechner's analysis was quite limited in one respect. Was the association of an orange with the "beauty" of the Mediterranean a universal phenomenon? Or did it owe something to the near obsession of Germans and Austrians in this period with the southern climate? Fechner neglected to explain how experiences were translated into memories, and he left unclear the extent to which memory images were culturally and historically contingent.

41. Hans Frisch, *50 Jahre Brunnwinkl*, pp. 22–3.

42. Gustav Theodor Fechner, *Vorschule der Aesthetik* (Hildesheim: G. Olms, 1978), pp. 87, 93.

Sigmund Exner would plumb all these issues while extending Fechner's concept in yet another novel direction: as a theory of aesthetic modernism. (Fechner, who died in 1887, never applied his ideas to fin-de-siècle art.) Inspired, perhaps, by his teacher Ernst Brücke and by his acquaintance with Gottfried Semper in Zurich, Exner made several forays into the history of the arts. He perceived his age as one of rapidly shifting aesthetic standards. He noted for instance that the Eiffel Tower and other iron monuments of the 1880s were greeted with outrage by some, by others as a new, "modern" style. Apparently, what had once struck viewers as ungainly might someday be acclaimed as the height of gracefulness. How could such transformations come about?

As Exner explained, aesthetic standards could be modified both by new material conditions and new theories of the material world. Such transformations hinged on the workings of visual memory. In a lecture at Vienna's Museum for Art and Industry in 1882, Exner argued that humans drew meaning from artistic representations by associating each new perception with stored memory images. The artist profited in this way from a human instinct for probabilistic reasoning: "[W]ithout being conscious of it, we always accept as the cause of a given phenomenon in a specific case the one that has emerged in countless similar cases, according to experience, as the correct one."[43] The viewer thus made an intuitive judgment of the probability that the representation corresponded to a physical reality. Such "unconscious inferences" had been postulated by Sigmund's teacher Helmholtz, following John Stuart Mill, to account for basic perceptual processes like the judgment of an object's location in space or its true color. Perception for Helmholtz was analogous to the scientist's method of induction, except that its premises remained below the threshold of consciousness.[44] Exner used this model of perception to explain the ability of artists to depict "realistically" scenes that were physically impossible.[45] The art world had long

43. Sigmund Exner, *Die Physiologie des Fliegens und Schwebens in den bildenden Künsten* (Vienna: Wm. Braumüller, 1882), p. 35. Reprinted in Olaf Breidbach, ed., *Natur der Ästhetik, Ästhetik der Natur* (Vienna: Springer, 1997).

44. Timothy Lenoir, "The Eye as Mathematician," in *Hermann Helmholtz: Philosopher and Scientist*, ed. David Cahan (Berkeley: University of California Press, 1993), pp. 109–53; Gerd Gigerenzer and David J. Murray, *Cognition as Intuitive Statistics* (Hillsdale, N.J.: Erlbaum, 1987).

45. Sigmund Exner, *Physiologie des Fliegens*. On this lecture, see Peter Geimer, "Das Gewicht der Engel. Eine Physiologie des Unmöglichen," in *Kultur im Experiment*, ed. Henning Schmidgen, Peter Geimer, and Sven Dierig (Berlin: Kadmos Kulturverlag, 2004), pp. 170–90; Olaf Breidbach, "Bemerkungen zu Exners Physiologie des Fliegens und Schwebens," in Breidbach, ed., *Natur der Ästhetik*, pp. 221–23.

abounded, for example, in images of flying angels and floating gods in human form. Sigmund worked through the relevant mechanics and found that such a figure, if real, would have a total body mass of two grams. One would expect these images to be perceived as monstrosities on the basis of a lifetime of memory images of real, gravitationally challenged bodies. Yet they were undeniably beautiful. There must be something in the viewer's storehouse of visual memories to which the figures corresponded. In the end, Sigmund hit on his answer. The attitudes of these flying angels mimicked those of swimmers. The viewer's aesthetic judgment was indeed guided by his visual memories, but by a different store of images than Sigmund had at first assumed.

This dependence of aesthetic judgment on memory implied that aesthetic standards could evolve with time. Exner noted, for instance, that rainbows in Renaissance paintings were often foreshortened. A modern viewer, who learned in *Gymnasium* that rainbows could only appear circular, would find these earlier representations odd, even ridiculous. Theoretical knowledge could thus alter or override physical intuitions, shifting the standards of aesthetic realism. Exner found this to be as true of architecture as of painting. The view of a man-made structure called up visual memories associated with its component materials. If one intuitively judged a structure unstable on the basis of such associations, then it would appear to be an aesthetic failure. Here lay the key to the advent of an "iron style" in architecture. "We find ourselves in an age in which this intuition for the properties of iron seems to be awakening, whereby an 'iron style' for architecture becomes possible."[46]

Exner commented too on the new enthusiasm for abstract, organic ornamentation, a hallmark of the *Jugendstil* or art nouveau movements. He identified the "joy in a beautiful ornament" as "the purest and highest pleasure in the fine arts, since here are absent all the additions foreign to this art, above all the narrative elements of representation." The appreciation of such ornaments rested on the "finest combination of memory images." Among these were memories of the flexible lines of vines and of the graceful motion of plants, "of the functional central position of the center of gravity on a surface, of symmetry, of even distribution."[47]

46. Sigmund Exner, "Ueber allgemeine Denkfehler," *Deutsche Rundschau* 4 (1889): 54–67, quotation on p. 64. On French enthusiasm for the Eiffel Tower as "art nouveau," see Debora Silverman, *Art Nouveau in Fin-de-Siècle France* (Berkeley: University of California Press, 1989), pp. 2–5.

47. Sigmund Exner, *Physiologisches und Pathologisches in den bildenden Künsten* (Vienna: Verein zur Verbreitung naturwissenschaftlicher Kenntnisse in Wien, 1889), p. 8.

By this point, Exner had made Fechner's notion of memory images into a far more powerful explanatory tool. By considering how such memories were rooted in a cultural context, he had begun to develop a theory of aesthetic evolution. Moreover, he was now in a position to explain how it was that aesthetic judgments transcended mere personal taste. As he pointed out in an article for a handbook of physiology in the early 1880s, the effects of stimuli on the sensory organs varied from person to person, since the organs themselves were not identical.[48] Memory was what ensured that the brain recognized a stimulus by virtue of its repeated association with a given set of external conditions. The mind functioned like a scientist, combining data statistically in order to form a judgment based on the widest possible range of available evidence. In the arts as in the sciences, it was this statistical process that overcame the problem of solipsism. It produced judgments that, while falling short of *absolute* validity, well surpassed personal opinion. By the same token, these judgments—scientific as well as aesthetic—remained open to revision, making the history of art as dynamic as the history of science. This was a widely influential perspective. Arthur Schnitzler, for instance, the Vienna medical student turned novelist, echoed Exner's thesis that "[t]he effects of art rest not on illusion but on mental associations."[49]

Still, it remained to uncover the means of the mind's statistical calculation, the unconscious mechanics of memory. Sigmund began to pierce this veil in the 1870s. In Brücke's laboratory he studied the visual effect produced by flashes of light in rapid succession at closely spaced points. If the flashes occurred closely enough in space and time, the viewer perceived not two distinct lights but a moving point. Here the perception of motion, usually understood as a higher-order "perception," had instead the character of a fundamental "sensation." The irreducibility of this phenomenon led to its foundational status in the Gestalt psychology movement, pioneered by Max Wertheimer, who worked briefly in Exner's lab.[50] Exner's own analysis, on the other hand, focused on the role of memory in producing the sensation of motion. To Sigmund, the phenomenon was a special case of a familiar class of experiments, which measured the least perceptible difference between two stimuli. Such experiments produced in the subject (in this case, the

48. Sigmund Exner, "Grosshirnrinde," in *Handbuch der Physiologie Abt IIa*, ed. L. Hermann (Leipzig: F. C. W. Vogel, 1879–83), p. 211.

49. Quoted in Rossbacher, *Literatur und Bürgertum*, p. 351.

50. On this movement, see Mitchell Ash, *Gestalt Psychology in German Culture, 1890–1967: Holism and the Quest for Objectivity* (Cambridge: Cambridge University Press, 1995).

experimenter himself) "the uncomfortable feeling . . . which overcomes one, when one feels oneself to be so little the master of one's sense organs."[51] These experiments, for which Exner received the Lieben Prize from the Austrian Academy of Sciences, demanded that the subject learn to observe his own mental processes just at the margin of consciousness. For lack of a better description, Sigmund explained, he would refer to this state of directed consciousness as "attention." Attention in this sense "fluctuated" in a manner independent of the subject's volition. One stimulus arising at a "favorable moment" would be perceived, but another of equal strength coinciding with an "unfavorable" moment would pass unnoticed. Try as he might to focus his attention in different ways—for instance, concentrating on the perception of the first stimulus, or avoiding focusing on either individually—Exner found the illusion of motion almost always irresistible. "It is as if one were forced to perceive time differentials as motion whenever possible."[52] At the transition between continuous and discontinuous perception, the memory image became irreducible, a composite or average that could not be further analyzed.

This disconcerting phenomenon was Exner's window onto the mechanics of memory. It explained why, for instance, artistic representations of movement were only "approximately correct." The artist was "right"—the image looked real—even if it was not "true." How could this be? Exner took the familiar fin-de-siècle example of stop-action photography. Sequences of photographs of moving horses, for instance, showed positions which, though "correct," would make for poor art—for the simple reason that these positions were *too fleeting* to leave behind visual memories in the viewer. Recall that Fechner had described aesthetic impressions as the geometrical "resultants," the vector sums, of memory images. Exner implied that a memory image was instead a *statistical time average* of a stream of visual experience. As Edward Muybridge would write several years later of his stop-action photography, "The intervals of time between each phase is an average of the intervals of time between all the phases, or an approximation."[53]

The concept of the time average had been developed by the Exners' colleague Ludwig Boltzmann in the 1860s. Through his brother Serafin, Sigmund was well informed of Boltzmann's work. Boltzmann's goal was to

51. Sigmund Exner, "Experimentelle Untersuchungen der einfachsten psychischen Processe," *Archiv f. d. ges. Physiologie* 11 (1875): 403–32, quotation on p. 428.

52. Ibid., p. 431.

53. Quoted in Anson Rabinbach, *The Human Motor: Energy, Fatigue, and the Origins of Modernity* (Berkeley: University of California Press, 1992), p. 103.

give a physical meaning to the statement that a system has the probability *x* of being in a given state. He interpreted *x* as the limiting value of the fraction of time the system would be found in that particular state. Sigmund's explanation suggested that visual perception was a probabilistic inference in precisely Boltzmann's sense of probability. Memory images were quite literally statistical approximations.[54]

Exner's theory of visual memory naturalized a model of reason characteristic of Austrian science and liberalism. Just as Austria's liberal scientists strove to present themselves as resistant equally to the dangers of dogmatism and of solipsistic speculation, Exner showed how the mind was naturally equipped to navigate between these poles. The threat of solipsism arose from the variability of the perceptual apparatus from one individual to the next. What mitigated this problem was the mind's instinct for statistical reasoning. Like a hunter tracking his prey or a naturalist collecting specimens, the mind instinctively gathered accidental evidence to serve in its interpretation of sensations. As Exner explained, "those experiences on which the interpretation of our sensations of the external world rest must be collected in the course of the life of each individual."[55] This process did not follow a neat course. As Exner's laboratory research made all too clear to him, the will had an imperfect grasp on consciousness and attention was an inconstant force. The associative activities of the mind often owed as much to chance as to volition. For Exner, however, this situation was not cause for despair. The discovery of the unconscious was not, as Schorske claimed, a threat to liberal rationality. To the contrary, Exner's model of the reasoning mind fully encompassed the limitations of consciousness. Reasoning meant working within this framework of the accidental and unpredictable. For Exner, the mind was a hunter, and therein lay its power.

THE FARMER AS A STANDARD OF REASON AND JUSTICE

The figure of the hunter suggested to the Exners one model for interpreting and exploiting accidental evidence, but not the only one. The summer retreat provided a second paradigm: the farmer. The farmer became Adolf Exner's

54. Austrian philosopher Alexius Meinong would echo this insight in an 1886 paper on memory in which he stressed that knowledge based on memory had a *probable* character. See David F. Lindenfeld, *The Transformation of Positivism: Alexius Meinong and European Thought, 1880–1920* (Berkeley: University of California Press, 1980), p. 128.

55. Sigmund Exner, *Grosshirnrinde*, p. 215.

model for coping quite literally with an unpredictable environment. In the early 1880s Adolf was engaged with a problem that seemed to him peculiar to the new age of science and industry, of rail travel and steamboat shipping: how should liability be assessed in cases of accidents arising from a nexus of human and natural causes?

The ancient Romans had a legal category for accidents that no human force could have prevented: "vis major." Translated into modern languages as *höhere Gewalt, force majeure,* or acts of God, this concept was regularly invoked in cases of railroad accidents—but it had no precise definition in the Austrian legal code. Adolf asked the obvious question: "In our scientific way of thought, what is an 'act of God?'"[56] The burden of proof lay with the defendant to show that the cause of the accident lay outside his sphere of control. In the case of railway accidents, to decide whether the operator was at fault meant following an intricate and highly technical chain of events backward to some ultimate cause. Judges, however, were ignorant of the new technology, and the investigators hired by the railroads had no incentive to point to a human or technical malfunction. Causal analysis most often came up empty-handed:

> Whoever has attempted to follow meticulously a causal chain such as that which unfolds in the interior of a complex technical organism will have experienced the difficulties and will not estimate very highly the probability of a tidy outcome: it lies above all in the nature of the thing that at many points the connection remains doubtful, that the possibilities branch out, such that the investigation is steered toward entirely different regions, depending on whether one follows one or the other branch at the crossroads. It is natural and human that at such points, where it is really impossible to explain everything down to the last detail, the investigator automatically prefers the alternative that is more favorable to the good reputation of the whole enterprise... which has as a consequence that the entire causal chain empties into the great empire of chance.[57]

For these reasons, judges were increasingly returning the verdict "act of God" and so withholding damages from injured passengers. Adolf sought an "objective" legal standard with which to protect the rights of individuals against large commercial interests. His primary target was the Prussian judge Levin Goldschmidt, a scholar of commercial law, the author of the first

56. Adolf Exner, *Der Begriff der höheren Gewalt (vis major) im römischen und heutigen Verkehrsrecht,* reprint of the 1883 edition (Darmstadt: Scientia Verlag Aalen, 1970), p. 15.

57. Ibid., p. 48.

draft of the German legal code, and dissertation advisor to Max Weber. Goldschmidt was known for his sympathy for tradesmen, and his applications of *vis major* seemed to Adolf to permit far too much judicial bias. Goldschmidt allowed for rulings of *vis major* in cases that involved ordinary "chance" (*Zufall*, or in legal Latin, *casus*) or even negligence.

To right the balance, Adolf supported a return to the Romans' concept of the *bonus et diligens pater familias*. Following recent scholarship by German legal historians, Adolf interpreted the *diligens pater familias* as a definition of the "amount of exertion and caution" that should be expended to avoid damage in any given situation. The *pater familias* was a standard of maturity, a model of someone who "is fully equal [*völlig gewachsen*] to his undertakings." Individuals could be held responsible for damage caused by "chance" events, as long as those events fell within the realm of contingencies against which the diligent *pater familias* was expected to guard.[58]

The question, then, was how to distinguish such minor accidents from "acts of God," for which no one could be held liable. Defining this boundary was the ultimate task of Adolf's study. He argued for two criteria for a judgment of *vis major*. The first was that the cause of the accident must be of "external provenance," that is, it must originate outside of the transportation system itself. If the accident's cause lies within the system, Adolf reasoned, it would be beyond the technical abilities of the judge to decide whether the operator should be charged with negligence. Moreover, in most such cases someone connected to the operating company was to blame. Thus the weight of probability and the safety of the common citizen required that the judge summarily award damages to the plaintiff in cases of internal provenance.

The external origin of the accident's cause was Adolf's first, "qualitative" criterion for *vis major*. His second, "quantitative" criterion addressed the remaining question: "How large—this is the decisive question—must the event be (assuming its external provenance), for it to count as *vis major*, and where is the standard for measuring its size?"[59] Adolf sought an "objective" and universal measure for an "act of God," one that was independent of the strength available to withstand it. This would be a measure not of size or force, he reasoned, but one of probability—in the modern sense of the relative frequency of an event. An event that qualified in the modern world as an act of God would be one that appeared "as something entirely extra-

58. Ibid., pp. 22–31.

59. Ibid., p. 70.

ordinary . . . because it self-evidently and powerfully exceeds the limits of the familiar."[60] Echoing Hume's analysis of the evidence for miracles, Adolf reasoned that such interruptions "of the normal course of human events" would be remembered by numerous witnesses, and in such cases it was "if not certain, still in the highest degree probable" that the event was not the fault of any individual. Adolf cautioned that this criterion of probability could not be applied "mechanically, as if by the reading of a barometer scale." Its application depended instead on local knowledge of the "usual course of events" in the relevant milieu.

The model he chose to illustrate probabilistic reasoning on the basis of local knowledge was the farmer. A farmer could distinguish without hesitation between "accidents" that were part of the normal course of his livelihood—such as bad weather, sick pets, theft—and those that struck like "the hand of fate"—the devastations of war, a major flood or cattle plague, the destruction of his house by lightning.[61] The farmer was a *diligens pater familias* with a special knowledge of the natural world. With his long and detailed experience of the land and climate, he could testify to nature's customary course. It was this capacity that made him so valuable in a modern world bent on subjecting nature to human whims. Behind Adolf's argument was the assumption that human thought is intuitively inductive, that people plan for the future based on perceived regularities in the past. This allowed a limited but significant capacity to deal with the unexpected. Judges might be led astray by causal reasoning, but ordinary farmers seemed to know perfectly well how to deal with "chance."

Adolf's analysis of acts of God was an attempt to resolve the tensions between an ancient legal tradition and modern science and technology. Dazzled by technological innovations and scientific laws, the nineteenth century had made causal explanation its rational ideal. When confronted with the failures of complex systems like the railroad, the modern layman demanded a complete causal account—or else threw up his hands and protested ignorance. Yet even experts could rarely piece together a full causal chain. By these standards it was virtually impossible to hold anyone accountable for large-scale accidents. To Adolf, it seemed that the enthusiasm for causal reasoning had led the courts to abandon a robust sense of moral accountability. Looking around him for a model of upstanding behavior in a highly unpredictable context, Adolf fixed on the figure of the farmer. To Adolf

60. Ibid., p. 71.

61. Ibid., pp. 73–74.

and other Viennese observers of rural life in the 1880s, the farmer was remarkable for his management of an unpredictable natural world, for his resistance to the "incalculables" against which Adolf warned his sister Marie as she contemplated buying land in the Salzkammergut. The farmer exemplified the virtues of a *diligens pater familias*, taking sensible precautions against natural and human contingencies. Like the hunter, the farmer was able to navigate an unpredictable environment by collecting, storing, and reasoning from the accidental clues that nature furnished. This storehouse supplied the raw material for effective probabilistic judgments. In this sense, an individual's most precious resource were memories.

As the Exners themselves became "locals" in the Salzkammergut in the 1880s, they cobbled habits and traditions from rural sources. They drew, on one hand, on the aristocratic culture of hunting, with its values of valor and self-sacrifice; but also on the example of the peasant hunter, with his seemingly encyclopedic knowledge of his natural surroundings. Reckoning with the risks of ownership, they identified with the farmer's duty to cultivate and "beautify" the land. The natives' efforts to stay afloat in a capricious and unforgiving environment struck them as valiant, if not always successful. The country folk displayed a tolerance for uncertainty that liberals could envy. These figures—the hunter and the farmer—were among the Exners' models for interacting with the natural world, as vacationers and as scientists. They guided the Exners' self-fashioning as scholars and their proposed standards of rational conduct in the 1870s and 1880s. The culture of the summer retreat provided the Exners with embodied ideals of liberal morality and reason.

CHAPTER FOUR

The Pigtail of the Nineteenth Century

Determinism in the 1880s

THE clamor of youthful voices at Brunnwinkl in the 1880s earned the hamlet the name "Indian village" among locals. Serafin Exner's daughters Priska and Hilde spent those summers playing in the woods by the Wolfgangsee with their cousin Ilse at games like knights and ladies. The knight would set forth on adventures, while the lady "brilliantly" defended the castle from its enemies. Hilde, always the knight, was the "wildest and most brazen of us all." "We never played with dolls," Priska recalled from the perspective of a newly married woman in 1905; "that was beneath our dignity as 'tomboys.'" That she and her sister were not boys was "naturally the disappointment of our lives; we wanted at least to be as ungirlish as possible, and so we hated all crying and sentimentality and looked down on the less boyish girls like Nora [Exner] and Gretl Conrad."[1]

These games brightened the years that followed the death of Priska and Hilde's mother in 1883. With Serafin devastated by her loss, the girls were left to grieve and pray primarily in the care of their nurse. Their father visited them upstairs only in the evenings, when he would narrate the next episode of their ongoing bedtime story. Not until father and daughters traveled together to Rome in the late 1890s was the rift fully healed. As Priska reassured her father shortly thereafter, "now that it's finally clear that *we* have a dear and good father, and *you* have daughters who are attached to you and love you, our whole life will be much happier and more gratifying than

1. Priska Exner Dijkgraaf, "Aus meinem Leben," in Sven Dijkgraaf, "Biographical Notes on the Dijkgraaf-Exner Family," collection of Dr. Peter Dijkgraaf, Den Helder.

FIGURE 5 The “Exner girls” at Brunnwinkl, 1901. From left to right, Serafin Exner’s daughters Priska and Hilde and Adolf’s daughter Nora. Source: Karl Frisch, *Fünf Häuser am See. Der Brunnwinkl: Werden und Wesen eines Sommersitzes* (Berlin: Springer, 1980), fig. 13, p. 37. © Springer Verlag, Berlin, 1980. With kind permission of Springer Science and Business Media.

before!"[2] Just listening to family conversations, above all the wit and wisdom of their father, seemed to Priska to have done more for her and her sister than any formal education. Through philosophical conversations with other young members of the Exner circle the girls soon lost the faith that had sustained them through their mother's death. They fell into what Priska called a "brazen unbelief." As Priska recalled, the Exner children knew their parents would look indulgently on their "wildness." The common wisdom of Brunnwinkl was that an "uninhibited" atmosphere was equally healthy for young minds and bodies, for girls as well as boys. In retrospect Priska judged it "the principle benefit of our *Erziehung* that we weren't really 'reared' [*erzogen*] at all." "We were raised, so to speak, on the good fertilizer of the Exner tradition."[3]

THE VALUES OF UNCERTAINTY

At Brunnwinkl the Exners fashioned a laboratory for liberal education during the decline of the liberal experiment in the empire at large. In the 1880s the Taaffe administration radically reshaped the politics of education in Austria-Hungary. Taaffe's ministers aimed to solidify an alliance with small farmers, craftsmen, and shopkeepers by opposing the expansion of academic secondary education into populations that, they believed, would be better served by vocational training. The liberals faced both a clerical challenge to secular moral education and rural and petit-bourgeois resistance to academic schooling.[4]

Similar tensions emerged in other modernizing European nations in this period. In Germany, France, and Britain, educators and parents engaged in parallel debates over the merit of a classical education relative to that of training in science and modern languages. In many ways, Austrians used secondary education in similar ways as these other nations to respond to the challenges of modernization.[5] Still, we can see a divergence in the *content* of secondary education in Austria.

Educators in Germany and Britain in this period attached the greatest importance to designing lessons free of speculative content. Only theories that had been definitively proved or universally agreed on were deemed fit for

2. Priska Exner to Franz Serafin Exner, undated, folder 293, letter 50, Manuscript Collection, Österreichische Nationalbibliothek.

3. Priska Exner Dijkgraaf, "Aus meinem Leben."

4. Cohen, *Education and Middle-Class Society*, pp. 97–108.

5. Albisetti, *Secondary School Reform*, chapter 9.

classroom consumption. The proponents of absolute certainty in Germany and Britain included some of Europe's most prominent liberal scientists. Rudolf Virchow famously asserted that scientists must not lose sight of the border between speculative and certain knowledge, above all, when preparing lessons for the younger generation.[6] William Thomson and P. G. Tait's widely used physics textbook earned praise for teaching students to refrain from speculating beyond the basis of physical facts. Helmholtz, who translated this work into German, admired it for similar reasons.[7] Similarly, in the eyes of British schoolteachers, such recent scientific developments as Darwinian theory and non-Euclidean geometry threatened the foundation of the school curriculum—the conception of truth as necessary and provable.[8] Outside of Austria, then, progressive scientists were demanding nothing less than certain knowledge for the classroom.

Faced with a threat from clerical conservatives unmatched in Germany, Britain, or France, Austrian liberals in the 1880s took a very different stance. At stake for them was the liberal project of 1849 to set moral education on a secular and antidogmatic basis. In 1884 the Taaffe administration issued a new set of instructions for the *Gymnasium* curriculum.[9] The revisions affected all disciplines, but most controversial were the recommendations concerning the Philosophical Propaedeutic. The ministry moved to reduce the philosophy course from two years to one and to reverse the order in which it was taught, so that psychology would now precede logic. This demoted psychology to being "entirely in the service of logic," as one critic observed.[10] The ministry insinuated that the field of psychology was in a transitional and uncertain state. And in truth psychology in Austria was sharply divided between the new experimentalists, trained in faculties of philosophy, and the more established, medically trained physiologists. The ministry implied that such inconclusiveness was intolerable in the discipline charged with revealing to students the workings of their own minds.[11]

6. Rudolf Virchow, *Die Freiheit der Wissenschaft im modernen Staat* (Berlin: Wiegandt, 1877).

7. Crosbie Smith, *The Science of Energy* (Chicago: University of Chicago Press, 1998), p. 210.

8. Joan Richards, *Mathematical Visions* (Boston: Academic Press, 1988).

9. *Verordnung des Ministers für Cultus und Unterricht vom 26. Mai, 1884, Z.10.128, betreffend mehrere Abänderungen des Lehrplanes des Gymnasiums und die Hinausgabe von Instructionen für den Unterricht an den Gymnasien, Verordnungsblatt des Ministeriums für Cultus und Unterricht* (1884), Nr. 21, p. 161–224.

10. Wilhelm Jerusalem, "Zur Reform des Unterrichtes in der philosophischen Propädeutik," *Programm des Staats-Gymnasiums in Nikolsburg* 12 (1884–85): 3–32, quotation on p. 3.

11. Alexius Meinong, *Über philosophische Wissenschaft und ihre Propädeutik* (Vienna: A. Hölder, 1885), chapter 3. Alois Höfler quoted a similar sentiment from an official at the education ministry:

In the coming dispute, the pivot on which debate turned was the moral *value* or *danger* of uncertain knowledge. In this context, liberal educators in Austria denied that uncertainty was a liability in the classroom. "Doubt and investigations must not be kept away," attested a *Gymnasium* director in Prague in 1885; "they come on their own in youth."[12] Still, the liberals' goal was to fortify students against dogmatism, not to foster radical skepticism. Liberals agreed that students must be taught to cope with uncertainty by steering between these extremes, but they disagreed over how best to do so.

KNOWLEDGE IN ACTION

Perhaps the most widely cited response to the education ministry came from the eminent Graz psychologist Alexius Meinong. Meinong had been a student of Franz Brentano, whose project for a scientific philosophy set a course for Austrian scholarship well into the twentieth century. Brentano, a former Catholic priest, was called from his home in southern Germany to a post at the University of Vienna in 1874. He was a sympathetic figure to Austria's newly powerful liberals, not least because of his rejection of institutionalized religion in the wake of the papal infallibility decree of 1870.[13] Philosophically, Brentano was interested in observing and classifying mental phenomena as they related not to external objects but to objects existing in the mind. The value of his introspective method, as he saw it, was that it produced knowledge that was transparently, intuitively, true—it had what he called the quality of "evidence."[14] Brentano had no use for merely probable knowledge. Meinong, by contrast, allowed that knowledge that was less than certain might also have a degree of evidence. Meinong found probability a powerful concept and published a treatise on it in 1916. Yet, like Brentano, his goal was to affirm the possibility of certain knowledge.[15]

"It was not necessary to concede to psychology an independent role, since all the psychological theories that can be considered certain and not disputed by any side together hardly have a right to a semester-long course." Höfler, *Die neuen Instructionen für philosophische Propadeutik*, speech delivered on 1 Dec. 1900 in the "Mittelschule" society, Vienna (Linz: Feichtinger, 1901), p. 2.

12. Ludwig Chevalier, "Über den Unterricht in der philosophischen Propädeutik an österreichischen Gymnasien," *Jahresbericht des k.k. Staats-Untergymnasiums in Prag-Neustadt* 4 (1885): 3–39, quotation on p. 17.

13. On Brentano and his influence, see Barry Smith, *Austrian Philosophy: The Legacy of Franz Brentano* (Chicago: Open Court, 1994); Lindenfeld, *Transformation of Positivism*, pp. 42–66.

14. Smith, *Austrian Philosophy*, p. 114.

15. Alexius Meinong, *Möglichkeit und Wahrscheinlichkeit* (Leipzig: Barth, 1915); Lindenfeld, *Transformation of Positivism*, p. 168.

In Meinong's view, then, uncertainty was to be avoided in the finished products of philosophy. But that did not mean that uncertainty should be banished from the classroom. Meinong explicitly rejected the ministry's assumption "that in the middle school, the site of learning not research, only what is completely certain and well-established from each discipline may be offered."[16] All empirical knowledge, all attempts to "take intellectual hold of reality," were fraught with uncertainty, and this was as true of the analysis of historical documents as of psychology. Moreover, proper exposure to uncertain knowledge had intrinsic value for moral education. This was Meinong's more radical argument. "In human life there is no art more important than that of guessing correctly; could it be that in this art alone there can be mastery without practice?" Meinong conceded that humans seem to have a "habit, instinct, or however one wants to call it" that guides our guesses, but this skill nonetheless requires training. Meinong compared human knowledge to a canoe that was tossed about with every wave; "thus it is really a matter of learning to gain a firm footing on the swaying bottom."[17] Here Meinong invoked a trope characteristic of liberal Austrian educators from Franz Exner to Otto Neurath: education was the acquisition not of eternal knowledge but of the ability to find one's footing in an unpredictable world. In the Austrian political context of the 1880s, as in the 1840s or 1920s, this principle represented a liberal strike against clerical dogmatism.

Meinong insisted that psychology was the discipline best suited to "setting the student on his own feet, to making him as independent as possible of the positive knowledge available at the moment."[18] Even if psychology could do no more than tell students what they could *not* know about the thoughts of their neighbors, it would be performing a crucial task. Meinong's younger colleague and friend Alois Höfler likewise responded to the ministry's incursions with a defense of the moral value of the study of psychology. Höfler had studied physics as well as psychology and now taught at the prestigious Akademisches Gymnasium in Vienna. He argued that it was pointless to teach a student how he should think (logic) without showing him how he really did think (psychology). Logic was best approached in the classroom by presenting its *methods* in real-life applications—for instance, in the history

16. Alexius Meinong, *Über philosophische Wissenschaft und ihre Propädeutik* (Vienna: A. Hölder, 1885), p. 30.

17. Meinong, *Propädeutik*, p. 50.

18. Meinong, *Propädeutik*, p. 62.

of science. Höfler stressed that students could best learn "correct thinking" by examining knowledge in its process of production.[19]

Höfler went on to contribute to a new Austrian journal, founded in 1887 and devoted to *Gymnasium* instruction in the physical sciences, Fritz Poske's *Zeitschrift für den Physikalischen und Chemischen Unterricht*. The journal's goal was to explore methods of teaching science not as a set of established facts but as a process of inquiry and so to fulfill "a humanist task."[20] Typical of the antidogmatic approach the journal advocated was the claim that "physics is the least disposed of all sciences to claim its results as 'absolute.'"[21] When the journal's editors noted echoes of their own rhetoric in Germany at the turn of the century, they were quick to point out that this was a mere confirmation of a position that they had "championed since the founding of this journal."[22] In this way, the journal helped define an antidogmatic tradition in Austrian pedagogy.

Also on the journal's editorial board was Ernst Mach, professor of physics at Prague. To liberal-humanists like the Exners, Mach was both ally and threat.[23] He was one of their most prominent Austrian partners in the struggle against dogmatism in the schools, yet he was also the leader of a radical antirealist movement in philosophy. In their effort to dismantle "metaphysics" along with clericalism, Mach and his followers treated the physical world as no more than a composite of sensations, and they dismissed the unified, autonomous subject that was central to liberalism as a metaphysical fiction. In educational politics as well as philosophy, Mach stood to the Exners' left. He took a strong interest in the reform of the *Realschulen*, the secondary schools that focused on science and modern languages in the absence of the classics and that did not confer the privileges of a *Gymnasium* education. Mach sought to provide *Realschule* pupils with the same prophylactics against dogmatism that had previously been reserved for *Gymnasium* students.

19. Alois Höfler, *Zur Propadeutik-Frage* (Vienna: A. Hölder, 1884), pp. 17, 24–29.

20. Fritz Poske, ed., "Ziel und Wege des physikalischen Unterrichts," *Zeitschrift für den Physikalischen und Chemischen Unterricht* 1 (1887): 1–2, quotation on p. 1.

21. "P." [Fritz Poske], review of *Physikalischer Dogmatismus* by Fritz Walther, *Zeitschrift für den Physikalischen und Chemischen Unterricht* 7 (1904): 297–99, quotation on p. 297.

22. "P." [Fritz Poske], "Naturwissenschaftlicher Unterricht und philosophische Propädeutik," *Zeitschrift für den Physikalischen und Chemischen Unterricht* 7 (1904): 365–366, quotation on p. 365.

23. On Mach's influence, see John T. Blackmore, *Ernst Mach: His Work, Life, and Influence* (Berkeley: University of California Press, 1972); and Friedrich Stadler, *Studien zum Wiener Kreis* (Frankfurt: Suhrkamp, 1997).

In this vein, his 1891 science textbook for the *Realschulen* stressed the hypothetical and probabilistic character of physical laws, deconstructed the concept of causality into "conditions" and "changes," and illustrated in detail the calculation of error, all in a pedagogical style that was self-consciously "psychological" and "not dogmatic."[24] Through the activities of reformers like Höfler and Mach, and through Poske's new pedagogical journal, Franz Exner's concept of the "humanistic" value of science study reemerged in the 1880s at the center of a new Austrian pedagogical movement, one that intended to bring science into the classroom with all its unfinished seams in view.

CONFLICT AMONG THE LIBERALS

Meinong and Höfler were reinvigorating the arguments of the midcentury reformers in the face of a new political climate hostile to the goals of humanist education. At the same time, however, they responded to the ministry's 1884 "Instructions" as leaders of a new psychology, a discipline striving to establish itself alongside philosophy and physiology. They charged that Austria had fallen under the "tyranny" of Herbart's philosophy.[25] In particular, they criticized the existing philosophy curriculum for feeding students an overdose of skepticism. As an example of the Herbartian influence on philosophy at the *Gymnasien*, Höfler singled out Zimmermann's "Introduction to Philosophy," the course that Franz Exner had envisioned as an intellectual earthquake. Höfler hoped to replace this barrage of philosophical "problems" with a systematic exposition of philosophy.[26] He denounced as well the Philosophical Propaedeutic's "enthusiasm" for inductive reasoning. Höfler claimed that its authors had falsely depicted the character of many common judgments. He considered an example from Zimmermann's account of probabilistic judgments, a case of the calculation of odds. When rolling two dice, one might reasonably bet one against thirty-six that the total rolled would be two, since there was just one way to roll a sum of two out of thirty-six possible outcomes. Zimmermann explained this belief as the result of an inductive judgment based on past experience of the usual course of events. Höfler, however, took this belief to be "evident," in the sense of a priori

24. Ernst Mach, *Grundriß der Naturlehre für die oberen Classen der Mittelschulen, Ausgabe für Realschulen* (Vienna: F. Tempsky, 1891), preface and introduction.

25. Höfler, *Propadeutik-Frage*, p. 97.

26. Ibid., p. 73.

and intuitive. He charged Zimmermann and other followers of Herbart with misrepresenting to students the nature of their own thought processes.[27]

Höfler followed Meinong in explaining that "evidence" set authentic knowledge apart from apprehension or mere judgment. In his response to the education ministry in 1884, Höfler used the concept of evidence to illustrate his claim that the primary ethical task of *Gymnasium* instruction was to make perfectly clear to the student the wide gap between everyday thinking and logical reasoning. Höfler's opponents were representatives of what he termed the "extreme tendency of particular modern theories." These thinkers treated foundations of rationality such as the principle of causality as no more than empirical "habits of thought." John Stuart Mill was the leading proponent of this view, and the foremost Viennese follower of his psychology was Sigmund Exner. Such theories, in Höfler's view, were a dangerous form of "skepticism," which must not be transplanted to the schools. Höfler insisted that nothing less than the ethical character of *Gymnasium* students hung in the balance. Here stood "serious pedagogical interests at stake." The bedrock of a student's "logical conscience" must lie in the ability to recognize "the absolute distinction between common judgment and thought which, through the mind's labor, achieves evidence." These "solid convictions, won through one's own thinking," stood "in direct correspondence to the ethical interests and so also to the deepest aim of all our *Gymnasium* instruction."[28] It was the school's moral responsibility to teach students to recognize absolute certainty, to feel in their bones the difference between the shaky ground of experience and the conclusiveness of a syllogism.

The "empiricist" psychology that Höfler deplored held that all thought was built up inductively from experience. In the 1880s it stood opposed to the "nativist" program led by Prague professor Ewald Hering, which sought to explain mental functions in terms of innate capacities. Helmholtz's "empiricist" account of spatial perception, for instance, posited that individuals learn to perceive space by associating the muscular sensations of the eye's contractions and torsions with impressions on the retina. Each spatial perception is then an "unconscious judgment," an intuitive induction, based on sensory input and visual memories.[29] Höfler charged that such theories

27. Ibid., pp. 14–19.

28. Höfler, *Propadeutik-Frage*, p. 19.

29. On Helmholtz's theory of spatial perception, see Lenoir, "The Eye as Mathematician," pp. 109–53; Stephen Turner, *In the Eye's Mind: Vision and the Helmholtz-Hering Controversy* (Princeton: Princeton University Press, 1994).

threatened to throw students into an abyss of confusion. Young people knew intuitively that judgments of space, as of color, were a priori.

> Before one sees fit to present principles from the empiricist theory of space, one should carefully put this to a vote in the classroom: how many then truly believe that they do indeed "see" the blackboard's blackness but not its rectangularity; how many will not immediately feel their understanding grind to a halt, if they are given to believe, that they do indeed see colors and brightnesses, but as completely nonspatial, as images without extension and existing nowhere,—that they do not literally "see" the blackboard before them in this spot and in this shape and size.[30]

Höfler presented the a priori status of spatial relations, as of the causal principle, as *intuitively* true—and he insisted that they be taught as such. While students needed to learn how to cope with uncertainty, the classroom was no place to be sowing seeds of doubt over the roots of rational thought.

Where Höfler and Meinong's psychology diverged from Sigmund Exner's was in their insistence on the a priori status of causal reasoning and other mental functions. Exner explained causality as an abstraction from the human (and animal) instinct to attend to change in the environment more than to constancy and to search in each case of change for an associated "changer."[31] For Meinong and Höfler, by contrast, causality was the foremost example of a principle that was a necessary condition of human thought and that, therefore, could not be doubted. When they penned their own version of the Philosophical Propaedeutic in 1890, Meinong and Höfler argued that "the general causal law approaches the evidence of certainty."[32] Meinong later used the concept of evidence in an attempt to "prove" the principle of causality.[33] From their perspective, causality was essential to "right thinking," and so it was the educator's duty to convince students of the

30. Alois Höfler, *Was die gegenwärtige Psychologie unserem Gymnasium sein und werden könnte*, lecture delivered to the pedagogical section of the forty-second meeting of the Deutsche Philologen und Schulmänner (Vienna: Deutsche Philologen und Schulmänner, 1893), p. 205.

31. Sigmund Exner, *Entwurf zu einer physiologischen Erklärung der psychischen Erscheinungen*, pp. 362–71.

32. A. Höfler and A. Meinong, *Philosophische Propadeutik*, vol. 1, *Logik* (Prague and Vienna: F. Tempsky, 1890), p. 189.

33. Alexius Meinong, *Zum Erweise des allgemeinen Kausalgesetzes* (Vienna: A. Hölder, 1918). Barry Smith argues that this group of Austrian philosophers, in anticipation of Husserl, was reconceiving the "a priori" as knowledge that could be gleaned by analyzing relations among objects, rather than, as for Kant, as a framework imposed by the observer. Smith, *Austrian Philosophy*, pp. 305–17.

absolute validity of the causal principle. Their program for teaching students to distinguish degrees of certainty was one tactic in the quest of Austrian liberals to navigate between dogmatism and skepticism. They pursued the pedagogical goals of the first Franz Exner even as they laid waste to his philosophical principles.

After thirty years, Franz Exner's school plan had now fallen under siege from conservatives and even, in parts, from liberals. The Exners could not overlook the critiques of a colleague like Höfler. Karl Exner knew him as a fellow teacher of physics at a Viennese *Gymnasium,* and Serafin and Sigmund Exner later supported his nomination to the Austrian Academy of Sciences, at Meinong's request.[34] The Exners' disagreements with Meinong and Höfler remained friendly; they were, after all, allies in the academic politics of the fin de siècle and beyond. Although Meinong criticized Serafin Exner's indeterministic views in his 1915 treatise on causality, he sent Serafin a complimentary copy. Serafin responded graciously from Brunnwinkl, noting that he had enjoyed this "*measured* and enjoyable holiday reading, which I have been awaiting with impatience."[35] Nonetheless, Höfler's attack on the legacy of Franz Exner's reforms implicated Exner's sons in the debate over *Gymnasium* reforms. They entered the stage of educational politics as Exner's heirs and as proponents of his educational philosophy.

Karl Exner addressed the ministry's "Instructions" as an authority on the *Gymnasium* physics curriculum. Echoing his father's vision for the *Gymnasium,* Karl emphasized the value of empirical learning.[36] He sought ample time for classroom experiments, focusing in particular on the subject of optics. Consistent with the Herbartian program of the original curriculum, optics included phenomena on the borders between physics, psychology, and physiology, such as color theory and stereoscopic vision. Karl complained that the ministry had eliminated from the course "such splendid natural phenomena as the solar and lunar eclipses, twilight, mirages."[37] His partiality

34. Franz Serafin Exner to A. Meinong, 22 May 1916, Forschungsstelle für österreichische Philosophie: "Ihren Vorschlag betreffend A. Höfler werde ich, so weit es mir möglich ist, unterstützen." Sigmund Exner to A. Meinong, 26. May 1916, Forschungsstelle für österreichische Philosophie: "Für Höfler werde ich sehr gerne eintreten. Ich schätze ihn sehr hoch."

35. Franz Serafin Exner to A. Meinong, 26 Aug. 1915, Forschungsstelle für österreichische Philosophie.

36. Karl Exner, "Die neuen Instructionen für den Unterricht in der Physik," in *Stimmen über den österreichischen Gymnasiallehrplan vom 26. Mai 1884,* ed. K. F. Kummer (Vienna: Carl Gerold's Sohn, 1886), pp. 376–82.

37. Ibid., p. 381.

for these phenomena reflected the direction of his own research. Like several members of his family, Karl Exner's research often led to the fuzzy border between the impersonal material world and the realm of perception. He spent one of his first summers at Brunnwinkl studying the scintillation of starlight, taking advantage of an atmosphere free of city dust.[38] At the time, this phenomenon dangled at the contested border between physics and physiology. Karl's analysis of atmospheric optics would set it firmly on the side of physics. As his brother Sigmund would put it, building on Karl's research, the scintillation of starlight could now be seen as an example of a phenomenon at once "subjective" and universally human.[39]

This effort to bridge psychology and physics was characteristic of the "empiricist" program of the midcentury reformers, an example of the Herbartian "tyranny" that Höfler deplored. The significance of such research for the Exners, on the other hand, was tied to its domestic context. At Brunnwinkl, such efforts to translate among subjective experiences were often themselves collaborative. Karl, for instance, corrected the mathematics of Sigmund's analysis of light sensitivity and provided geometrical calculations for Sigmund's Lieben Prize–winning study of insect vision.[40] These collaborative investigations of perception demonstrated the possibility of communication about a shared material world. They were exercises in managing the conflicting desires for independence and intimacy and thus constitutive of the Exners' vision of a liberal utopia.

Meinong and Höfler's program for teaching students to distinguish "evident" knowledge from mere possibility was one answer to the liberals' problem of steering between dogmatism and skepticism. But it was not the only one. The Exners and their empiricist allies weighed the risks differently. To them, there seemed to be more harm than good in teaching students to believe in the certainty of a principle like causality. In the 1880s, psychologist Sigmund Exner and jurist Adolf Exner denounced the false certainty of determinism; they turned to probabilistic reasoning to restore a framework of individual moral accountability.

38. Karl Exner, "Über das Funkeln der Sterne und die Scintillation überhaupt," *Sitzungsberichte der Kaiserlichen Akademie der Wissenschaften zu Wien* 84 (1881): 1038–81, quotation on p. 1057.

39. Sigmund Exner, *Physiologisches und Pathologisches in den bildenden Künsten* (Vienna: Verein zur Verbreitung naturwissenschaftlicher Kenntnisse in Wien, 1889), p. 8.

40. Karl Exner, "Über die Curven des Anklingens und des Abklingens der Lichtempfindungen," *Sitzungsberichte der Kaiserlichen Akademie der Wissenschaften zu Wien* 62 (1870): 197–201; Sigmund Exner, *Die Physiologie der Facettirten Augen von Krebsen und Insecten, Eine Studie* (Leipzig and Vienna: Franz Deuticke, 1891), p. 2.

THE HUMAN WILL IN AN AGE OF SCIENTIFIC DETERMINISM

Several intellectual currents intersected in the 1880s to bring the problem of determinism to the fore of popular discussion. Crimes were said to be "determined" by innate dispositions, threatening (in the words of Adolf Exner) to dissolve criminal law into psychiatry.[41] Methods of suggestion like hypnosis looked to many like direct evidence that free will was a sham. Moreover, various strains of scientific research had converged to cast into doubt assumptions about the nature of human freedom. Among these investigations were case studies of hysteria and other psychosomatic illnesses, laboratory measurements of reflexes and reaction times, and anatomical studies of the localization of mental functions in the brain. Scientists throughout Europe and America strained to reconcile the incontrovertible experience of free will with the growing evidence that the brain was a complex machine.[42] In Vienna, these questions took on a special urgency for liberal scientists. They faced the problem of explaining how it was possible to educate people to be free when freedom seemed no more than a chimera.

In the 1880s, public interest in the question of free will focused on new techniques of accessing and manipulating the unconscious. Hypnosis (translated into German as *Hypnose* or *Suggestion*) had long been treated as little more than a parlor trick by Austrian physicians, even while the eminent French physician Jean-Martin Charcot made it the therapeutic technique of choice at the Paris Salpêtrière in the 1880s. When Viennese psychologists Josef Breuer and Sigmund Freud—the former Sigmund Exner's close friend and the latter his former student—began using hypnosis to treat hysterics in the late 1880s, they faced strident opposition.[43]

In 1891, at the height of the hypnosis controversy in Vienna's medical community, the editor of a literary journal solicited opinions from several leading physiologists and psychiatrists.[44] Sigmund Exner was among the

41. Adolf Exner, *Über politische Bildung* (Vienna: Adolf Holzhausen, 1892), p. 27.

42. Ellenberger, *Unconscious*; Alison Winter, *Mesmerized: Powers of Mind in Victorian Britain* (Chicago: University Chicago Press, 1998); Norton Wise, "How Do Sums Count? On the Cultural Origins of Statistical Causality," in *The Probabilistic Revolution*, ed. Lorenz Krüger et al. (Cambridge, Mass.: MIT Press, 1987), vol. 1, pp. 395–425.

43. They soon became embroiled in a bitter dispute with their more established colleague Theodor Meynert. See Frank Sulloway, *Freud: Biologist of the Mind* (1979; Cambridge, Mass.: Harvard University Press, 1992), pp. 42–50.

44. On scientific popularization in Austria, see Andreas Daum, *Wissenschaftspopularisierung im 19. Jahrhundert: bürgerliche Kultur, naturwissenschaftliche Bildung, und die deutsche Öffentlichkeit,*

luminaries who responded, along with Helmholtz, Du Bois-Reymond, and Meynert. The editor noted that literature had long invoked two contrary convictions, free will and destiny. These beliefs may have been irreconcilable, but poets clung to them until the sciences taught them otherwise. In the nineteenth century, writers had learned "to recognize and heed the constraints that limit or offset free choice: inherited traits, the influence of race, of climate, of *Erziehung*, of custom, of the sum of life's experiences. But they all hold fast to a degree of self-determination, even if only a limited one, and do not believe that we are the playthings of a destiny that governs blindly." By 1891, however, reports arrived ever more often of incidents in which "this capacity of self-determination seems to have been suddenly and completely eliminated from healthy people, indeed through another person's influence—'suggestion'—of his own thoughts, wishes, emotional traits, as he likes or finds useful."[45] In the examples the journal offered, otherwise sane individuals committed inexplicable acts, apparently as a result of being "hypnotized." The journal pointed out that criminal law had already taken account of the limits of free will. What consequences would literature draw?

In Sigmund's view, to ask whether hypnotism was a "worthy" subject for literature was like asking whether the stupor of an alcoholic, or an opium or cocaine addict, was a suitable theme. Responding as a classical humanist, Sigmund implied that literature's task was to portray the human will, not delusions and dreams—a position he would return to as a critique of Viennese modernism. Like Du Bois-Reymond and Helmholtz, Sigmund was skeptical of the hypnotists' claims, and he vehemently denied that the phenomenon should in any way detract from belief in free will. Indeed, it was entirely in an individual's power to let himself be hypnotized or not. To say that a woman had committed adultery while hypnotized was to say no more than that she had acted "in a fit of passion," which, in Sigmund's estimation, was hardly

1848–1914 (Munich: R. Oldenbourg, 1998). Klaus Taschwer, "Wie die Naturwissenschaften populär wurden. Zur Geschichte der Verbreitung naturwissenschaftlicher Kenntnisse in Österreich zwischen 1800 und 1870," *Spurensuche* 8 (1997): 4–31; Michler, *Darwinismus*; Ulrike Felt, "'Öffentliche' Wissenschaft: Zur Beziehung von Naturwissenschaften und Gesellschaft in Wien von der Jahrhundertwende bis zum Ende der Ersten Republik," *Österreichische Zeitschrift für Geschichtswissenschaften* 7 (1996): 45–66; and Mitchell G. Ash, "Wissenschaftspopularisierung und bürgerliche Kultur im 19. Jahrhundert," *Geschichte und Gesellschaft* 28 (2002): 322–34.

45. Karl Emil Franzos et al., "Die Suggestion und die Dichtung," *Deutsche Dichtung* 9 (1890–91): 71–130, quotation on p. 71.

a novel defense.[46] The hypnosis craze looked to Sigmund like a perilous wrong turn in the history of rational thought, akin to medieval witch hunts. In another popular essay published the following year, Sigmund criticized the peculiarly modern morality that judged actions less harshly if they were not performed in full "consciousness."[47]

Instead, Sigmund counseled a return to the ancients' robust conception of moral responsibility. Oedipus, for instance, was a man held responsible for a crime that he had performed in ignorance. Over a decade before his student Freud would transform this myth into a drama of the unconscious, Sigmund Exner made the crucial point: whether the crime was conscious or not was irrelevant. The root of the problem, as Exner saw it, was the modern insistence on causal explanations. In yet another popular article from this period, Exner argued that people too often inferred causal connections where none in fact existed.[48] Evolution may have favored the ability to make quick judgments based on generalizations. Yet to ignore the specifics of the situation at hand and subsume the particular to the general was often a source of error. So, for instance, there endured a widespread belief in the influence of the moon on the weather, despite meteorologists' statistical analyses to the contrary. Or consider the common wisdom that certain people are "luckier" than others. Usually, the association of the "fate and experiences of a person with his qualities" was justified, but not in situations of pure chance, such as gambling.[49] This was another case of a falsely inferred causal connection. As in the model of *Bildung* ensconced in nineteenth-century novels and in the *Gymnasium* curriculum, not causality but chance and will were the appropriate categories of analysis.

These were among the more benign "errors of thought" to which causal reasoning gave rise. More perniciously, Sigmund argued, falsely inferring causal connections could mean pinning responsibility or blame on the wrong head, or eliminating accountability altogether. A particularly vicious "error" had swept through Europe in the Middle Ages: "witches" were held responsible for phenomena with which they had nothing to do. Sigmund presented his prescriptions for correcting errors of thought as gifts of enlightenment.

46. Ibid., pp. 129–30.

47. Sigmund Exner, "Die Moral als Waffe im Kampfe ums Dasein," speech delivered at the Vienna Academy of Sciences, 30 May 1892, *Almanach der Akademie der Wissenschaften* (1892): 243–73, esp. 265–67.

48. Sigmund Exner, "Ueber allgemeine Denkfehler," *Deutsche Rundschau* 4 (1889): 54–67.

49. Ibid., p. 62.

He intended to root out judgments based on the familiar and usual, which, like a contagion, gave rise to the "prejudices" of an age. As a corrective to these lapses of reason, Sigmund set forth the probability calculus. He indicated how probabilistic considerations could correct superstitions about lunar influences or common misconceptions about games of chance. Proper application of the probability calculus could prevent the slide into fatalism.[50]

Sigmund acknowledged that causal reasoning had an instinctive quality that made it seem a priori. And acausal processes were admittedly "unimaginable," at least for now. When it came to analyzing human behavior, the only alternative to the causal principle seemed to be the ill-defined hypothesis of free will. But Sigmund suggested that other possibilities might exist. From the perspective of evolutionary history, causal reasoning in humans might be akin to the swimming instinct in fish: useful only in the environment in which we happen to find ourselves.

> Who can tell us that the central mechanism of causal thought does not fail in certain fields; and does the contradiction we come up against not lead to the thought that we find ourselves in a situation similar to that of a fish on dry land? . . . The property of causal thought really evolved in the first place within the physical phenomena of nature, and we must be prepared to reach a limit in its application in other fields.

Empirical psychology was that border in the history of human evolution. The alternatives of causality and free will were simply inadequate for the study of psychic processes. Sigmund drew an analogy to his experiments on the perception of motion. Observers estimated the speed of a moving object differently if their eyes were moving than if their eyes were at rest. The choice between free will and causality for the empirical psychologist was like that between moving and resting eyes for the measurement of speed: neither was sufficient, both led to illusions.[51] Just as his father had refused to adhere to either determinism or indeterminism, dismissing the choice as an "illusory problem," Sigmund suggested that humans would need to evolve beyond these two alternatives.

For Sigmund, as for his father, the motivation to surpass causal reasoning lay in the need for greater moral accountability and for a satisfying theory of education. Causal reasoning may have been an often useful framework for

50. Ibid., p. 63.

51. Sigmund Exner, *Erklärung*, 372.

scientific analysis, but it was a hindrance in other contexts, even a hazard. "We work mentally with the phenomenon [*Phänomen*] of the causal law and use it as an instrument for the investigation of phenomena [*Erscheinungen*], because in our experience it serves us excellently." Yet even in the natural sciences causal reasoning was at best precarious. Any phenomenon admitted of multiple causal explanations: "a true cause does not exist; there are a boundless number of them. Every member of every chain, and many may yet be attached, is a cause or a reason, and the number of these chains can be multiplied infinitely." In the realm of education—the "*Erziehung* of individuals and nations"—free will was a more appropriate framework than causality. Following in the empiricist tradition of Herbart and of his father, Sigmund made clear that any workable pedagogical philosophy would have to take into account the active role of the pupil. In the realm of moral reasoning more broadly, causal thinking had insidious effects:

> To fall into fatalism on the basis of the law of causality; to ignore people's intentions and aims, because their actions have supposedly been predetermined for millennia by an endless chain of causes; to view the criminal as innocent and therefore his punishment as unjust, and more—these are outgrowths of a misconceived natural scientific worldview. . . . Fatalism must be fought, because ignoring it will cause society's decline, due to a natural scientific misunderstanding, in that a psychic phenomenon is taken to be a natural law governing the external world.[52]

Fatalism, to which determinism could all too easily lead, denied the freedom and responsibility of the individual. In this sense, it undermined the moral structure of Austrian liberalism.

Fatalism was a threat more specifically to the liberal model of masculinity. In studies that were seminal in the field of sexual pathology, Sigmund Exner argued that the exercise of the will was vital to the healthy development of masculinity: "The man desires to love, the woman to be loved; thought that is farther-reaching, less bound to custom, an energetic will distinguishes the psyche of the man from that of the woman and child." Drawing on the literary plot of *Bildung*, Sigmund stressed the critical period of male adolescence, "the period of *Sturm und Drang*, which every youth goes through during his entrance into puberty, the profusion of ideals, of highly ambitious plans, of creative urges, which crowd his young soul." Anticipating Freud's theory of sublimation, Exner stressed the interdependence of sexual and creative

52. This and previous quotations from Sigmund Exner, *Erklärung*, pp. 372–73.

energies." In the case of castrates, he speculated, "everything else that develops and forms [in puberty] must also fall away; thus one understands the lack of bold flights of imagination, of great and idealistic industriousness and creativity. With the productive force of the body vanishes that of the mind as well."[53] Sigmund went on to give a cautionary twist to the familiar model of character development "in the stream of the world," through the interaction of free will and environmental contingencies. Following French psychologist Alfred Binet, Sigmund explained that homosexuality was likely the result of "chance impressions in the period of development"—a momentary lapse of attention could result in the association of a sexual urge with the wrong object. "Everyone knows that in such things *Erziehung* is practically all-powerful, that the formation of associations is therefore in the highest degree dependent on individual experiences."[54] Fatalism, passivity, even the slightest lapse of will, could throw male development irreparably off course.

By the time Sigmund Exner penned this final condemnation of fatalism in 1903, he had been struggling with these issues for two decades. He was hardly alone. As psychologist Wilhelm Jerusalem testified, the publication of the education ministry's "Instructions" in 1884 touched off an intense debate among Austrian philosophers and educators over how best to introduce *Gymnasium* students to the conflict between determinism and free will.[55] Some, like Höfler, portrayed causality as a bedrock of moral education. Sigmund Exner ultimately painted it as a noxious source of moral fatalism. In his view, determinism was not a form of scientific thought but a misunderstanding of the "scientific worldview." Science and society would be healthier without it. Theirs were less two conflicting views of moral education than two perspectives on how best to achieve the same goal: teaching students how to fight dogmatism with reason.

ACADEMIC FREEDOM

This goal was part of the mission of the liberal educator to prepare *Gymnasium* students to manage the freedom they would be accorded at university.

53. Sigmund Exner, "Physiologie der männlichen Geschlechtsfunktionen," in *Handbuck der Urologie,* ed. Anton von Frisch and Otto Zuckerkandl (Vienna: Alfred Hoelder, 1903).

54. Sigmund Exner, "Physiologie der männlichen Geschlechtsfunktionen," p. 226.

55. Wilhelm Jerusalem, "Zur Reform des Unterrichtes in der philosophischen Propädeutik," *Programm des Staats-Gymnasiums in Nikolsburg* 12 (1884–85): 3–32, see pp. 18–19.

That academic freedom, as defined by the midcentury reformers, allowed students to pursue their course of *Bildung* wherever it might lead them and to express their criticisms openly. In the course of the 1880s, however, the value of that freedom itself was cast into doubt as the imperial universities descended into violence. Reflecting the broader political climate, German and Czech nationalist student groups repeatedly clashed at the University of Vienna in the early 1880s. A series of unusually large and violent demonstrations occurred in 1883, after a memorial service for Richard Wagner organized by German nationalist students had erupted into brutal fights. In the Senate chamber representatives debated whether students still merited the privileges they had been accorded in 1849 for the purpose of "following the flight of their ideals with fewer constraints."[56]

Adolf Exner entered this volatile atmosphere when he became dean of Vienna's law faculty for the academic year 1883–84. His popularity with students gave him the upper hand. As Adolf's mentor Josef Unger said of him, "He knew how to combine academic authority with academic freedom."[57] When a large demonstration overtook the university that fall, Adolf successfully appealed to the German nationalist students to let their protest rest.[58] These students likely knew that Adolf sympathized with their anticlerical sentiments and that he valued academic freedom too highly to punish them for demonstrating.

Indeed, Adolf was a liberal educator in his father's mold. In the classroom, he too played the skeptic. "Skepticism," it was said, "was a principal feature of his character."[59] He spurred his students to question all that seemed self-evident in legal systems past and present. His field of Roman law was particularly suited to this enterprise, as it brought to light the genealogy of Austria's modern legal system, making the familiar suddenly look foreign. He insisted that it was the legal scholar's duty "to criticize the current laws and to prepare the way for legislation: his gaze was not directed toward the past alone, but also to the future. . . . [H]is skeptical spirit made him see in every paragraph, whether in the decree of a Byzantine emperor or in any legal clause that he himself had helped bring into effect as a

56. "Studentenvereine und Krawallen," *Vorstadt-Zeitung*, 16 Feb. 1888, p. 1. Wiesinger Newspaper Collection, Stadt- und Landesarchiv, Vienna.

57. Josef Unger, *Adolf Exner* (Vienna: k.k. Hof- & Universität Buchhändler, 1894), p. 4.

58. "Maaßen-Demonstration im neuen Hause," *Neues Wiener Abendblatt*, 29 Oct. 1883, p. 8, Wiesinger Newspaper Collection.

59. "Hofrath Adolf Exner gestorben," *Neues Wiener Tagblatt*, 11 Sept. 1894, pp. 3–4, quotation on p. 4.

member of the Herrenhaus, no more than a historical phase of legal evolution, a momentary truth."[60] Adolf pursued his pedagogical goals through the new framework of Pandectism, a historicist movement that was in part a reaction against the Enlightenment school of natural law. The "Pandect" was the sixth-century Justinian law code meant to furnish a complete system of law. Rather than deducing laws a priori from "human nature," the Pandectists (led in Austria by Unger) aimed to identify the principles that governed Austrian law in its many historical incarnations, beginning with its origins in Roman law. Although it owed much to the historical school founded in Germany by Savigny, Pandectism also broke with the Romantic tendencies of that movement, styling itself as "scientific" or logically rigorous. The Pandectists' ultimate goal was to formalize Austrian law on the basis of a self-contained system of legal doctrines. Adolf's 1867 study was one of the first works to carry out Pandectism's program of systematization.[61] But the Pandectists made clear that their central motivation was pedagogical. Rudolf Jhering, the great historian of Roman law of this period, described the task of Pandectism as primarily heuristic and education-oriented: "To bring out the abstract in its embodiment in the legal case and to give a content that is concrete and easy to grasp and remember to contours which, to the eyes of the legal novice, are still barely visible or indistinct."[62] Adolf himself was careful to avoid giving the impression that the abstract concepts employed by Pandectism were anything more than heuristic tools. Even Adolf's fellow specialists in Roman law, however, disagreed on the pedagogical value of Pandectism. The movement's critics charged that its abstract principles were too far removed from reality. A professor of Roman law at Czernowitz claimed that Pandectists did more harm than good by spoon-feeding knowledge to their students.[63]

From Adolf Exner's perspective as dean in 1884, the controversy over Pandectism was at heart a clash over the same question that had brought *Gymnasium* reforms to an impasse that year: might uncertain knowledge corrupt young minds? Exner took this question seriously. The value of the study

60. Ibid.

61. W. Ogris, "Die Entwicklung der österreichischen Privatrechtswissenschaft im 19. Jahrhundert," in *Handbuch der Quellen und Literatur der neueren europäischen Privatrechtsgeschichte*, ed Helmut Coing (Munich: Beck, 1973), p. 14.

62. Quoted in Herbert Hofmeister, "Jhering in Wien," in *Rudolf von Jhering, Beiträge und Zeugnisse*, ed. Okko Behrends (Göttingen: Wallstein Verlag, 1992).

63. Ernst Hruza, *Der Romanistische Rechtsunterricht in Oesterreich* (Czernowitz, 1886).

of Roman law, as he saw it, lay in the intellectual and moral qualities it fostered in students. Of course, its practical relevance had been doubtful since the codification of Habsburg law at the turn of the nineteenth century. What his discipline could impart to the modern student was a critical attitude toward absolute claims. He insisted that Roman law taught a skill that was essential in the age of codification: the ability to move between abstract principles and particular cases. In a review of a legal textbook published in 1884, Adolf explained that the teacher had a duty to simplify and abstract legal concepts, a task that was necessarily in tension with the infinite variability of real cases to which a practicing jurist might apply these concepts. Adolf put it laconically, "He who lacks the courage to teach something false should not teach at all."[64]

To be clear, Adolf was not advocating that law professors lie. More precisely, he compared jurists to natural scientists—both faced the impossibility of absolute certainty. "What Goethe somewhere says of the products of natural forces," he wrote, "is true too of those of social forces: 'even at their most definite they are always somewhat vague.'"[65] The jurist's skill, like the natural scientist's, rested on the ability to reason under conditions of uncertainty. Behind this surprising analogy lay Adolf's evolving sense that the challenge of modern jurisprudence was to grapple with cases in which natural and human forces were impenetrably entangled. It no longer seemed practicable to apply two different forms of reasoning to the natural and social realms. Just three years earlier, in his analysis of "acts of God," Adolf had rejected causal reasoning because it had failed to support a robust conception of moral accountability. In its place he had promoted a probabilistic empiricism, couched as the intuitive rationality of the common farmer looking out for the interests of his family. In this approach Adolf would soon find a political lesson for his age.

POLITICAL *BILDUNG*

By 1891, with liberal power crumbling and political unrest further battering the Austrian universities, one might well have expected the offer of a coveted professorship in Leipzig to have tempted Adolf Exner. It was, at least, a welcome sign of the reach of his scholarly renown. To the surprise

64. Adolf Exner, review of *Institutionen des römischen Rechts* by Rudolph Sohm, *Grünhuts Zeitschrift* 11 (1884): 622–25, quotation on p. 623.

65. Ibid.

and satisfaction of his Austrian colleagues, however, he turned the offer down. "What holds me here," he explained, "is basically banal patriotism, or whatever else it might be called; I cannot leave my home and country, I cannot think of turning my back on it in the same year in which a memorial is being built to my father as the reformer of the Austrian education system. And I believe too that I, as a good German, can serve the cause of German *Bildung* here better than anywhere else." Adolf's decision thus reflected his sense of the challenges and possibilities that lay in Austria's political future. On one hand, "the cause of German *Bildung*" would have to withstand the separatist demands of the growing nationalist movements as well as the religious control of clerical forces. On the other, due honor was being paid to Austria's tradition of liberal education. Adolf himself had been consulted on the plans for the triple memorial to his father, Graf Thun, and Hermann Bonitz—the architects of the schools system that, despite challenges, continued to shape the citizens of the empire in his own day. As Franz Exner's son, Adolf clearly felt his personal stake in the future of this system.[66]

Adolf's colleagues took his refusal of the Leipzig post as it was intended, as a sign of loyalty to his homeland.[67] Perhaps out of gratitude, they soon bestowed on him their highest honors: he was elected rector of the University of Vienna in 1891 and appointed to the Reichsgericht (Imperial Court) in 1892 and to the Herrenhaus (upper house of parliament) in 1894. He had won the privilege of being able to speak with authority on issues of politics and society. The public thus attended closely to his rectorial address in the fall of 1891. Such events were occasions for an academic elite to shape the public image of their disciplines.[68] The university community expected Adolf to address proposed reforms to the *Rigorosen* examinations for law students, which would shift emphasis away from legal history toward more applied fields. The university's Philosophical Society, under the direction of Höfler, later discussed his speech in this context. But Adolf had a much broader vision to share, a program for "political education," at the heart of which was his evolving critique of causal reasoning.

Adolf opened with his foremost pedagogical principle: "Doubt is the father of all insight." He warned that he intended to make his listeners question their most "self-evident" assumptions. He then called up an image of a

66. Quotation from Franz Exner, *Exnerei*, p. 9. On the memorial, see A. Exner to F. v. Miklosich, 24 May 1890, folder 134, letter 50-3, Manuscript Collection, Österreichische Nationalbibliothek.

67. Josef Unger, "Adolf Exner," p. 4.

68. Daum, *Wissenschaftspopularisierung im 19. Jahrhundert*, p. 436.

student freshly arrived in the capital from his provincial home, sketching the story of a university education that echoed familiar narratives of *Bildung*. "Thousands and thousands may have dreamed since their early boyhoods of completing their general education [*Gesamtbildung*] here; for many, very many, the path was too steep or an adverse fate drove them off course; only a chosen few have reached this gate." In Adolf's image of the new meritocracy of learning, the path to higher education was a struggle between the individual and his environment, between will and chance. He asked the members of his audience to recognize themselves in this account. Of course they had reached the university through their own efforts, but were there larger factors that had made the journey possible? "[I]ndividual strength has led you, certainly, but still only by virtue of its being roused and steeled in the state's workshops of culture [*Culturwerkstätten des Staates*]." The capital city itself offered the student lessons in the workings of the state, but it was above all the university's duty to develop the student's insight into "staatliche Dinge," national matters. This was a lesson the student would bring back to his provincial home "as exquisite provision for a lifetime." In a period dominated by "the new political problems, with which socialism knocks on the door of the twentieth century," the task of integration and education fell above all to the University of Vienna, as "the natural center of gravity for the best of the knowledge-thirsty youth of all the lands and all the peoples of our great Fatherland."[69] Adolf thus drew his audience's attention to the challenges facing the multinational empire and to the increasingly diverse population of its institutions of higher learning.

What the empire needed, Adolf declared, was a new program of political education. He called for historical instruction as a foundation for the "collective memory of a great national past." This demand had several precedents. Count Thun had geared historical instruction in Austria toward civic education, although it had become more narrowly specialized in recent decades. In Prussia, where Wilhelm von Humboldt had once championed a civic-minded "historical *Bildung*," the emperor himself now argued that education, not legislation, was the most effective means to combat socialism.[70] What made Adolf's vision of political education unique was his insistence that the study of the past more closely resembled the investigations of a naturalist than the collections of an antiquarian. The student must learn, Adolf argued, to examine the political structures of the past "as with a microscope"

69. Adolf Exner, *Politische Bildung*, pp. 3–12.

70. Albisetti, *Secondary School Reform*, p. 180.

and to analyze the forces and relations among social facts analogously to "natural scientific *Bildung*." The second side of political *Bildung*, Adolf continued, should in turn direct students' attention toward the future. It should impart a fine sensibility for the possible and impossible outcomes of a given political context.[71]

The university was not merely neglecting this second component of political education, Adolf argued. It was actively stunting students' political growth. The problem was the one-sided emphasis on a particular form of natural science. Trained in the deterministic framework of Newtonian physics, modern man had become incapable of conceiving of possibilities that were not necessities. Things stood differently in the political realm, where the causal nexus was murky ("due to the insensibility of the objects, the impossibility of isolating them through experiment, and the time lag between cause and effect"). This insistence on the heterogeneity of the social realm and the absence of strict laws of society was characteristic of the academic discipline of political statistics in Austria at the time.[72] Social effects, Adolf declared, could be calculated "only approximately and according to probability—a probability calculus on the correct handling of which all practical statecraft rests."[73] As we can now recognize, this "probability calculus" was not a fanciful figure of speech. At least since his study of *vis major*, Adolf had viewed probability as the most objective means of reasoning about moral responsibility when faced with intricate concatenations of human and natural forces. Now Adolf divulged the secret to good statecraft: to recognize within broad historical tendencies a certain latitude for deviations, to develop "Der Takt für das politische Mögliche," a feeling for political potentials, not just realities.[74]

Adolf was arguing that *probabilistic* science—as opposed to Newtonian determinism—was a model of reasoning fit for training the future leaders of a liberal state. "Who does not think of the 'imponderables' of Count Bismarck?" wondered a commentator for a moderate socialist newspaper.[75]

71. Adolf Exner, *Politische Bildung*, pp. 16–19.

72. Theodore M. Porter, *The Rise of Statistical Thinking, 1820–1900* (Princeton: Princeton University Press, 1986), pp. 190–91.

73. "[N]ur annähernd und nach Wahrscheinlichkeit—eine Wahrscheinlichkeitsrechnung, auf deren richtiger Handhabung alle praktische Staatskunst beruht." A. Exner, *Politische Bildung*, p. 20.

74. Ibid., p. 20.

75. "Der Zopf unseres Jahrhunderts," *Oesterreichische Volks-Zeitung*, 23 Oct. 1891, p. 1, Wiesinger Newspaper Collection.

Bismarck, who called his own approach to politics "the art of the possible," was now in retirement at his country estate, where careful observations of his crops convinced him that neither nature nor politics admitted of "scientific laws."[76] Or perhaps Adolf's phrase had more in common with Wilhelm von Humboldt's concept of "historical *Bildung*." This, in Humboldt's view, imparted "the sense for dealing with reality" (*der Sinn für die Behandlung der Wirklichkeit*), a faculty equally essential to the historian and the legislator. "He studies the present direction," Humboldt wrote, "then, according to how he finds it, he furthers it, or he strives against it."[77] Whatever affinities it might have displayed, Adolf's rectorial address reflected above all the years he had spent pondering the relationship between the methods of legal analysis and of natural science. No doubt these reflections were spurred by conversations with his scientist-brothers, as perhaps Humboldt's were by his brother Alexander. To him, the similarity of these disciplines rested not on the deterministic model of Newtonian mechanics but on the approximative approach he associated with Goethe.

Nonetheless, what stuck in the mind of some listeners was Adolf's stick-figure caricature of the natural scientist, deducing from first principles the laws of a frictionless world. The lecture drew quick and impassioned responses, with competing interpretations adding to the blizzard of debate. Some provocatively described the address as "a declaration of war against the natural-scientific orientation of our century."[78] In truth, though, liberal-minded scientists were surprisingly tolerant of Adolf's criticisms. Among Adolf's would-be opponents was Franz Brentano, the Austrian philosopher who had built his career on the claim that the true method of philosophy must be that of the natural sciences. Brentano, as a former priest, had been forbidden to marry but had nonetheless wed the Jewish heiress Ida von Lieben in 1880. His ensuing conflict with the Catholic church led to the loss of his university appointment and his Austrian citizenship. With good reason, then, Brentano shared the Exners' antipathy to dogmatism and their tolerance for uncertainty. His rebuttal to Adolf's speech was in fact more supportive than critical; their apparent disagreements stemmed from

76. Robert W. Whalen, *Bitter Wounds: German Victims of the Great War* (Ithaca: Cornell University Press, 1984), p. 84.

77. Wilhelm von Humboldt, "Über die Aufgabe des Geshichtschreibers" (1821); Herbert Muhlack, "Bildung zwischen Neuhumanismus und Historismus," in *Bildungsbürgertum im 19. Jahrhundert. Teil II: Bildungsgüter und Bildungswissen*, Industrielle Welt, 41, ed. Reinhart Koselleck (Stuttgart, 1990), pp. 80–105.

78. A liberal daily voiced this regret three years later. "Hofrath Adolf Exner gestorben," *Neues Wiener Tagblatt*, 11 Sept. 1894, p. 3, Wiesinger Newspaper Collection.

Adolf's unfortunate caricature of natural science. Where Adolf claimed that the "causal chains" involved in sociological and political analysis were too intricate to follow in detail, Brentano pointed out that the same was true for sciences such as meteorology and mineralogy. Such sciences do not follow a causal, deductive method, Brentano said. Instead they operate empirically, as would someone trying to assess the likelihood of rolling different numbers on a loaded die.[79] There was really no dispute here: the two men were equally convinced empiricists, in agreement that causal reasoning was inadequate for the analysis of nature or society.

Höfler offered a similarly conciliatory critique. He understood that Adolf's conception of political analysis was really an empirical science of society. And so he reformulated the program of political *Bildung*: "Philosophical *Bildung on the basis* of natural scientific, political *Bildung on the basis* of philosophical—and not: the third *instead of* the first two."[80] Another commentator in a liberal daily defended Adolf against accusations that he had slandered the natural sciences. The writer explained that Adolf had only meant to remind his colleagues, in a humanistic spirit, that man had a task "beyond measuring and weighing."[81] Adolf's teacher Josef Unger likewise found the charge of science-bashing "unfounded." In language that accentuated the antidogmatic spirit behind Adolf's speech, Unger explained: "Exner objected only to 'natural scientific chauvinism' and to blind faith in the one holy natural scientific method." Unger was certain that Adolf would not have denied the progress and accomplishments of the natural sciences in the nineteenth century. Moreover, Unger found Adolf true to the principles he espoused in this address: "[H]is mind was always focused on the possible and attainable: he truly possessed political *Bildung*."[82]

Other liberals paid tribute to Adolf by adopting his vocabulary. One was Friedrich Jodl, a progressive philosopher and education reformer. In 1893 Jodl analyzed theories of natural law in Adolf's terms at the Vienna Legal Society, where Adolf would likely have been present. Jodl argued that laws were not properties of human nature but transient products of social

79. Franz Brentano, *Ueber die Zukunft der Philosophie. Mit apologetisch-kritischer Berücksichtigung der Inaugurationsrede von Adolf Exner* (Vienna: Alfred Hölder, 1893).

80. "Über das Verhältnis der politischen zur philosophischen Bildung," manuscript of speech in the Philosophische Gesellschaft on 28 Nov. 1891, emphasis in the original. D.3.5, Österreichische Forschungstelle für Philosophie, Graz.

81. "Das sterbende Jahrhundert," *Wiener Tagblatt*, 23 Oct. 1891, p. 1, Wiesinger Newspaper Collection.

82. Josef Unger, "Adolf Exner," pp. 8, 10.

interaction. "In the moment in which it [a law] is declared and becomes valid it is already out of date." Ideally these changes occurred smoothly, "with the prudent separation of the possible from the impossible and the careful linkage of the work in progress to what already exists." Whether or not such a transition could be achieved depended "on the finer or coarser sensitivity [*grösseren oder geringeren Feingefühligkeit*] with which legislative politics is able to interpret the signs of the times and to bring the voice of ethical conscience into an effective relationship with the law [*die Stimme des sittlichen Bewusstseins für das Recht nutzbar zu machen*]." Jodl's model of legal analysis, like Adolf's, rested on a "fine sense" for politics and the duty to sort "possible from impossible."[83]

Jodl, Unger, Höfler, and Brentano could all sympathize with Adolf's support for humanistic education. So could Theodor Billroth, one of the most stalwart defenders of the classical *Gymnasium* at Vienna's medical faculty. Billroth asked Adolf to send copies of his speech to the music critic Eduard Hanslick, once a devoted student of Adolf's father's, and to the composer Johannes Brahms, a mutual friend of Billroth and the Exners. To Hanslick Billroth confessed, "I don't agree with everything Exner says; but he brings it all off with splendid wit." And he repeated approvingly Adolf's bon mot that the narrow-minded adherence to natural scientific modes of reasoning was the "pigtail of the nineteenth century," as worthy of derision as the wigs of French aristocrats during the Revolution.[84]

Within the academic community, the real critics of Adolf's speech objected not to his philosophy of science but to his politics. From the political left came charges that Adolf's prescription for political education was naïve. The moderate socialist *Volks-Zeitung* asserted that in an age when political fault lines had fractured the university community, one could hardly expect to unite them all with a single version of political education.[85]

It is understandable that these critics questioned the viability of Adolf's bid to unite the university—and the empire—through a program of political education. Adolf's vision of political education was tightly bound to the self-conception of the educated liberal elite. He aimed to foster a mode of

83. Friedrich Jodl, "Über das Wesen des Naturrechts," speech delivered at the Vienna Juristischen Gesellschaft, 1 Feb. 1893, pp. 7, 20, reprinted in *Vom Lebenswege*, vol. 2, p. 79.

84. Billroth to Hanslick, 22 Oct. and 26 Oct. 1891, in *Briefe von Theodor Billroth*, 8th ed. (Hannover and Leipzig: Hahnsche Buchhandlung, 1910), pp. 436–37; also cited in Erna Lesky, *The Vienna Medical School of the Nineteenth Century* (Baltimore: Johns Hopkins University Press, 1976), p. 494.

85. "Der Zopf unseres Jahrhunderts," Oesterreichische Volks-Zeitung, 23 Oct. 1891, 1. Wiesinger Newspaper Collection.

thinking, a social "probability calculus," which reflected the process of self-fashioning implicit in the coming-of-age stories of the liberal era: recognition of the force of external circumstances and evaluation of the scope of individual action: "The individual's impulse is certainly a powerful factor in political as in all human things: but its effect is only proportional to the preexisting conditions, insofar as, having grown out of these, it shapes a future that has already been prepared in the direction of forces that are already present."[86] The goal was to reintroduce a strong sense of moral responsibility into a society saturated with the determinism of a defunct caricature of natural science. Adolf's speech stood for years to come as a defining moment in the history of liberal education in Austria.[87]

Yet Adolf's memorable phrase, "der Takt für das politische Mögliche," was a thinly coded signal to his audience: it designated an ability that depended on experience, not reason. Nineteenth-century physicians likewise spoke of a medical "Takt" that could not be replaced by the rationalizing technologies of instruments or statistics.[88] Adolf implied that the political sense was a skill of the kind that historians of science have called "tacit." Probabilistic reasoning in this sense could not be reduced to an algorithm. Learning to reason probabilistically did not mean merely mastering the machinery of the probability calculus; it meant the proper application of that machinery, a "feeling for the possible," which was ultimately incommunicable. Like the farmer, Adolf's standard of sound reasoning, the statesman had to learn through experience how to estimate the probability of the contingencies against which he was expected to guard.

As Adolf himself indicated, the lessons of political education could not be encapsulated in textbooks. Indeed, there is only so much that historians can learn about liberalism in Austria by studying the public sphere. The diligent

86. Adolf Exner, *Über politische Bildung*, p. 20.

87. It was, for instance, a reference point in the school reform debates of 1908, discussed in chapter 7.

88. Christopher Lawrence, "Incommunicable Knowledge: Science, Technology, and the Clinical Art in Britain, 1850–1914," *Journal of Contemporary History* 20 (1985): 503–20; Thomas Broman, "Rethinking Professionalization: Theory, Practice, and Professional Ideology in Eighteenth-Century German Medicine," *Journal of Modern History* 67 (1995): 835–72. Like Adolf, his predecessor at the University of Vienna Rudolf Jhering also worked closely with analogies between jurisprudence and natural science. Jhering had analyzed "Takt" in an ethical sense in an unpublished section of his major work, *Zweck im Recht*. Jhering emphasized that *Takt* was a product not of intellect alone but also of will and character. See Christian Helfer, "Rudolf von Jhering, 'Der Takt,'" *Nachrichten der Akademie der Wissenschaften in Göttingen* (1968): 73–98.

FIGURE 6 Adolf Exner, Marie Exner von Frisch, and their close friend Gisa Conrad at Brunnwinkl, 1890. Source: Karl Frisch, *Fünf Häuser am See. Der Brunnwinkl: Werden und Wesen eines Sommersitzes* (Berlin: Springer, 1980), fig. 14, p. 38. © Springer Verlag, Berlin, 1980. With kind permission of Springer Science and Business Media.

pater familias, after all, was a legal standard modeled on *domestic* virtue. The "feeling for the possible" was not meant to be a product solely of the urban *Gymnasien* and the imperial university; it was equally the fruit of summers spent exploring nature in the serenity of a summer retreat. Brunnwinkl was the semiprivate complement to the Exners' project of liberal education at the university. So it is suggestive that one of Adolf's most prominent supporters, Theodor Billroth, was the Exners' favorite neighbor at Brunnwinkl, and that another ally, Friedrich Jodl, was vacationing in St. Gilgen during the summer that Adolf composed his appeal for political education.[89] Adolf's probabilistic

89. Margarete Jodl, *Friedrich Jodl: Sein Leben und Wirken* (Stuttgart, Berlin: J. G. Cotta, 1920), p. 142; Georg Gimpl, ed., *Unter uns gesagt: Friedrich Jodls Briefe an Wilhelm Bolin* (Vienna: Löcker, 1990), p. 111.

model of rationality evoked the liberal culture of the *Sommerfrische,* with its celebration of the peasant's experiential knowledge of nature. In the history of liberal education in the Austrian Empire, this little lakeside hamlet, hundreds of kilometers from the imperial capital, would play a surprisingly central role.

CHAPTER FIVE

Afterlife

Inheritance at the Fin de Siècle

IT WAS early September 1894; most of Vienna's academic world was still on summer holiday, and the university was quiet. On his way back from a summer conference, Sigmund Exner disembarked at Franz Josef's Bahnhof and crossed the western edge of the city to the university. As he reached the Ringstrasse he must have seen the mourning flag flying. Perhaps it was the porter at the gate who informed him of the telegram that had arrived an hour earlier. His brother Adolf was dead.[1]

Adolf had spent the summer as usual with his wife's family in Kufstein, Tyrol, but signs of weakness had led him to join in fewer outings. He napped frequently, gave up tea in favor of cocoa, and took his doctor's advice to smoke three Cuban cigars daily.[2] Still, his death was a sudden blow to his family. Like the first Franz Exner, Adolf had passed away at the height of his career, at age fifty-three.

"Deeply shaken," as a newspaper described him, Sigmund left Vienna immediately to see Adolf's widow and children in Kufstein.[3] The Exners had learned early to mask sadness. As Marie explained, it was a family trait—"echt exnerisch"—to put on a cheerful face even in the worst of times.[4] Only

1. "Hofrath Adolf Exner gestorben," *Neues Wiener Tagblatt*, 11 Sept. 1894, pp. 3–4, quotation on p. 4, Wiesinger Newspaper Collection.

2. A letter from Adolf to his physician from 9 Sept. 1894 is quoted in Smidt-Dörrenberg, *Gottfried Keller und Wien*, p. 95.

3. "Hofrath Adolf Exner gestorben," p. 3.

4. Marie Exner von Frisch to Gottfried Keller, 25 March 1884, in I. Smidt, ed., *Gottfried Keller's Glücklicher Zeit*, p. 172: "Dass Ihnen Adolf, als er im Sommer in Zürich war, nichts davon gesagt

a week after Adolf's death, Sigmund, his wife Emilie, and his brother Serafin each managed to send their close friends the Benndorfs warm wishes for the couple's anniversary. These letters offer a glimpse of how the Exners grieved. "We're all indeed not in the spirit to send boisterous congratulations," Emilie wrote. Yet the occasion of the Benndorfs' anniversary brought her some comfort, in the thought of "long years of happy togetherness and of intimate family life," and the "consciousness of what you have in each other."[5] Like his wife, Sigmund found solace in thinking of life's continuities. It comforted him to see how the dead continued to influence the living, an observation he associated with Fechner (himself now seven years gone).

> Fechner's idea of the life of the soul [*Seele*, also "mind"] after death through the effects of every interaction and every word on the survivors comes to mind too in our situation. I would not have the pleasure of knowing you so intimately had Adolf not lived; may this bit of soul [*Seele*] continue to produce its effects! He drew our whole circle together and so in it he lives on, and we would like to hope that in this sense he lingers a while in our midst. I write these lines in a [illegible] and cheerful spirit; hopefully they will affect you only as they were intended.[6]

Serafin too took comfort in thinking of the ways in which Adolf would be remembered:

> In such times it is best to turn one's gaze away from the past toward the future, which may bless us with much joy, whether it be in our profession, in our children, or finally in the spirit of the indeed ongoing world; and if fate is especially good to us then it will give us a death as it gave to our Adolf and a remembrance [*Andenken*] among men as it has secured for him.[7]

What these three ruminations share is the conviction that Adolf would live on in the memories of others. Fechner had offered this consolation in the work that Sigmund cited, the *Little Book of Life after Death*, written for the daughters of a deceased friend. In it Fechner offered comfort in the form of a humanist substitute for the Catholic notion of the afterlife. Fechner's

---

hat, dass ich im Frühling 3 Monate im Bett gelegen bin und am Abkratzen war—ist echt exnerisch; wir sprecher nur gerne von lustigen Sachen."

5. Emilie Exner to Otto Benndorf, 17 Sept. 1894, folder 640, letter 34-1, Manuscript Collection, Österreichische Nationalbibliothek.

6. Sigmund Exner to Otto Benndorf, 17 Sept. 1894, folder 640, letter 39, Manuscript Collection, Österreichische Nationalbibliothek.

7. Franz Serafin Exner to Otto Benndorf, 19 Sept. 1894, folder 640, letter 36, Manuscript Collection, Österreichische Nationalbibliothek.

idea was gracefully simple yet heretical: individuals survived through the memories they inspired in others. Every eulogy spoken, every monument erected, established contact with the dead.[8] Emilie would quote from Fechner's "beautiful words" at the end of a short story that she published in 1899: "Here stands the tree: many individual leaves may fall from it, but its foundation and its structure are solid and fine. It will always sprout new branches and new leaves will always fall; [but] it itself will never fall."[9] This *Stammbaum* or family tree was Fechner's organic image of intergenerational continuity.

The Exners consoled themselves with Fechner's words at a time when they no longer felt assured of a public afterlife, as their control over their own intellectual legacies was beginning to slip. For much of the nineteenth century, Vienna's intellectual elite wielded the authority necessary to construct public and private memory together. They built their legacies simultaneously at the university and at the summer retreat, in the laboratory and in the drawing room. By the time of Adolf's death, the boundaries between the public and private spheres of the *Bildungsbürgertum* had become less fluid. The demographics of higher education in Austria had shifted radically enough to bring a new question to the center of academic politics: were the ideals of liberal education compatible with progressive social goals? Liberal education was meant to be universal, to transcend the boundaries of nation and class (without, of course, laying claim to absolute certainty). By the last decade of the nineteenth century, however, opponents on the right and left challenged the liberals' often repeated claim that higher education as it stood constituted a universal "inheritance," a source of continuity and unity for a fragmented empire. This chapter probes the question of what it meant—in personal, political, legal, psychological, and biological terms—to "inherit" knowledge at the close of the nineteenth century. It is not surprising that Fechner's formulation of the afterlife resonated with the Exners at the moment of Adolf's untimely death. It corresponded to their own practices of memorializing, exemplified in their commemorations at Brunnwinkl. In the Exners' reflections on an intellectual afterlife, we begin to see the mutual construction of a theory of inheritance and the social practices of a bourgeois family.

8. Gustav Theodor Fechner, *Das Büchlein vom Leben nach dem Tode*, 3rd edition (Hamburg: Leopold Voss, 1887).

9. Emilie Exner (Felicie Ewart, pseud.), "Ein Flüchtling," *Velhagen und Klasings Monatshefte* 1: 6 (1898): 602–27, quotation on p. 627.

THE LAW OF INHERITANCE

Adolf had recognized that he was in poor health months before his death.[10] So it was that he had personal reasons in his last months of life to tackle a legal problem at the heart of which lay the relationship between inheritance and immortality. At the time of his death, Adolf had been engaged in drafting a new copyright law for Austria. As we will see, even as the legal theory of inheritance defined familial relationships and duties, the practices of family life also shaped legal theory.

Inheritance as a legal act acquired new meaning in nineteenth-century Austria thanks to the historically minded jurists who gained power after the midcentury educational reforms. Led by Adolf Exner's mentor Josef Unger, the Austrian historical school set its interpretation of inheritance in opposition to the earlier Kantian natural law tradition in jurisprudence. Natural law jurists had made the individual and his subjective will the foundation of all law; they had correspondingly limited the rights of inheritance. By contrast, the historical school insisted that not individuals but "legal powers are what is enduring and essential."[11] Unger quoted Berlin jurist F. J. Stahl, a Hegelian, on the essence of the right of inheritance: "For the entire human race there lies in the family—and in the legal succession formed in its image—the order and continuity that property acquires through the succession of generations, so that property reigns uninterruptedly as the substrate of consciousness and will and preserves the corresponding legal relations of individuals and through them the connection between generations."[12] The passage of material things from one generation to another was the basis for the continuity of "consciousness and will." The rights of inheritance girded the intergenerational structure of families, which in turn grounded a cultural tradition. Like other midcentury reformers in Austria, such as Franz Exner and Adalbert Stifter, Unger equated the particular interests of families with the general interest of "the entire human race."

A generation later, Adolf Exner brought Unger's theory down from its idealist heights to focus on the concrete legal mechanisms of "intellectual

10. An obituary noted that he had been ailing since early 1894. "Hofrath Adolf Exner gestorben," *Neues Wiener Tagblatt*, Nr. 249, 11 Sept. 1894, pp. 3–4, quotation on p. 4, Wiesinger Newspaper Collection.

11. Josef Unger, *Das österreichische Erbrecht systematisch dargestellt* (Leipzig: Breitkopf und Härtel, 1864), p. 1.

12. Ibid; F. J. Stahl, *Die Philosophie des Rechts*, vol. 2, *Rechts- und Staatslehre auf der Grundlage christlichen Weltanschauung*, 2nd ed. (Heidelberg: J. C. B. Mohr, 1845), quotation on p. 384. On Stahl, see Whitman, *Legacy of Roman Law*.

inheritance." In doing so, he threw into question the relationship between particular and universal interests. In the 1890s Adolf was part of a legal commission of the upper house of parliament, advising on a proposed trade treaty with Great Britain. The treaty would effectively give Austria a new law of copyright.[13] During the parliamentary deliberations in the spring of 1894, Adolf explained that the commission's recommendations would set Austria's copyright law apart from Germany's. The difference lay in the Austrians' "sharper emphasis on the significance of individual rights in copyright law." By "individual rights," Adolf meant a profoundly humanistic view of authorship, which he sought to distinguish from the narrowly commercial values with which the concept of copyright was associated in Germany. He argued in particular that financial compensation could not amend the suspension of an author's rights, for "the copyrighted object is not merely the property of the author." "The product of our intellectual labor is more than a piece of our wealth, it is in an intellectual [*geistigem*] sense our child, a piece of ourselves, and there are other, far more tender strings, besides financial interest, which bind us to this child and his fate."[14]

To show that the author had more than a financial interest in his publications, Adolf offered a hypothetical case: an author gives a manuscript to a publisher and receives payment, but instead of printing the book, the publisher destroys it. Adolf imagined himself in the position of this author, who has entrusted the publisher with his "intellectual product, which I intended for the public and from which I expect perhaps my future, perhaps immortality." Although the publisher has not violated the author's rights, "he has however murdered my child, he has fatally intervened in this more delicate, more discreet, sacred relationship."[15] If copyrighted material was an author's intellectual progeny, then it was only natural that the rights to it should pass to the author's children after his death. This claim provoked debate. Legally, the rights of the author's heir would infringe on those of the publisher. The justice minister, speaking for the government, argued for granting greater protection to publishers than the commission's

13. In 1891 Adolf had likewise defended a new patent law that gave more protection to inventors. See "Die Patentgesetzgebung," *Wiener Tagblatt*, 23 Oct. 1891, p. 2, Wiesinger Newspaper Collection. On Austrian patent law, see Paul Ritter Beck v. Mannagetta, "Das österreichische Patentrecht," in *Das oesterreichische Staatswörterbuch*, ed. E. Mischler and J. Ulbrich (Vienna: Alfred Hölder, 1896).

14. Adold Exner, 6 March 1894, *Stenographische Protokolle über die Sitzungen des österreichischen Reichsrathes des Herrenhauses* (Vienna: k. k. Hof- und Staatsdruckerei, 1891–97), p. 48.

15. Ibid.

proposal offered.[16] In support of the heritability of copyrights, Adolf insisted that it was a moral duty to honor the legacy of one's forebears by caring for the products of their creative labors. A quarter century earlier, in his first major legal treatise, Adolf had likewise stressed that the heir must willfully take up his inheritance: "[S]uch an 'acquisition' may not be a thoughtless and passive act, rather it must be taken possession of with consciousness and a will to sovereignty."[17] Now Adolf argued that the heir was entitled to all the rights of the original author, "because he has not inherited merely the monetary value, the commodity, but also the duty to act as deputy for the literary name, the reputation, etc., of the testator."[18] Echoing Stahl and Unger, Adolf was suggesting that material inheritance bound the heir to his forebears through a sense of moral duty, forming the basis for moral order and social stability.

At the same time, Adolf stressed the rights of the individual. In an age when the future of the university was in doubt, it fell to the legal system to guarantee an author intellectual immortality. The legal matter of copyrights roused Adolf to passionate declamations in parliament because it triggered anxiety about the social role of the *Bildungsbürgertum*. By 1894, political challenges had forced Vienna's educated elite to question the value of their labor and the sustainability of their way of life. No longer could academics confidently place their scholarly work in the hands of a tradition-minded culture, sure of reaping some small degree of immortality. In this context, Adolf treated intellectual inheritance not as a process for accruing "universal" knowledge, but as an act that preserved authorship. Any knowledge that could possibly be "inherited" was, by definition, private.

THE BIOLOGY OF INHERITANCE

In Adolf's understanding of copyright, as in Fechner's humanist vision of the afterlife, preserving the memories of the deceased required the willed effort of the heirs. Intellectual inheritance in this sense was an active process on the part of both the individual who inspired memories and those who remembered. This idealistic, nearly spiritual notion of inheritance would seem to have no counterpart in the modern science of physiology in 1890s Vienna.

16. Adolf Exner, "Die Patentgesetzgebung" (see note 13), and discussion in *Stenographische Protokolle* (1891–97), p. 491–92.

17. Adolf Exner, *Die Lehre vom Rechtserwerb durch Tradition nach österreichischem und gemeinem Recht* (Vienna: Manz, 1867).

18. *Stenographische Protokolle* (1891–97), p. 494.

Indeed, a very different view of inheritance had emerged in a lecture by the Prague physiologist Ewald Hering at the Vienna Academy of Sciences in 1870.[19] Hering was the leader of what its opponents called the "nativist" approach to the psychology of the senses. Hering and his collaborators treated perception as a result of innate capacities, while "empiricists" like Hermann von Helmholtz and Sigmund Exner treated it as a learned skill. Hering argued in his 1870 speech for the heritability of psychic traits. What an individual organism learned in its lifetime, he argued, altered its germ cells and could therefore be passed on to its offspring. These traits were inherited in the form of potentials, and they manifested themselves only when the individual encountered an appropriate stimulus. In this sense, memory was a "general property of organized matter."

In Vienna, over the course of the next three decades, scientists reached a broad consensus on Hering's rough theory of psychic inheritance. From Ernst Mach to Sigmund Freud, scientists in a range of fields adopted the heritability of psychic traits as a working assumption.[20] To Mach, Vienna's watchdog against encroaching metaphysics, Hering's theory of inheritance seemed to offer a materialist alternative to the "mysticism of the unconscious," the vague notions often invoked to explain similarities between parents and children.[21] To Freud, psychic inheritance was instead evidence of the workings of the unconscious mind. He interpreted taboos, for instance, as an "inherited psychical endowment." "Who can decide whether such things as 'innate ideas' exist.?"[22] Whether given a materialist or psychoanalytic interpretation, intellectual inheritance on Hering's model was a passive process.

The hardening consensus around psychic inheritance limited the scope of mental function that could be attributed to learning, posing a serious obstacle to adherents of an empiricist theory of mind. Sigmund Exner saw a way out of this dilemma for the empiricists. His solution formed the basis of his most ambitious project, the *Outline of a Physiological Explanation of Psychic Phenomena*, published in 1894, the year of the copyright debate in parlia-

19. Ewald Hering, "Über das Gedächtnis als eine allgemeine Funktion der Organisierten Materie," in *Vier Reden* (Amsterdam: E. J. Bunset, 1969), pp. 5–31.

20. Sulloway, *Freud*, esp. pp. 92–94. On this theme more broadly, see Laura Otis, *Organic Memory: History and the Body in the Late Nineteenth and Early Twentieth Centuries* (Lincoln: University Nebraska Press, 1994).

21. Ernst Mach, *Analyse der Empfindungen*, 2nd edition (Jena: G. Fischer, 1900), p. 59.

22. Sigmund Freud, *Totem and Taboo: Some Points of Agreement between the Mental Lives of Savages and Neurotics*, trans. by James Strachey (New York: Norton, 1950), p. 31.

ment and of Adolf's death. Building on the ideas of his colleague Theodor Meynert, Sigmund argued that inborn and learned mental faculties can be distinguished structurally. The latter sat in the brain's periphery, the cortex (*Hirnrinde*). The cortex consisted of neuronal paths worn by repeated associations among familiar experiences, which produced a topological structure of "facilitations" (*Bahnungen*) and "inhibitions" (*Hemmungen*). On this model, a large portion of mental functioning was learned through experience. Sigmund considered the periphery a uniquely human structure, which "is influenced by the lived experiences of the individual in the most exquisite way" and which "serves . . . the adaptation of the individual to the particular circumstances of his life." Perhaps alluding to Herbart's psychology of education, Sigmund described the periphery as "plastic," a word that Herbart had used as a synonym for "able to be educated." Sigmund suggested that the cortex was the location of personality and the site where education unfolded.

If the cortex was the seat of learning, then the spinal cord and brain stem were the repositories of inherited characteristics. These were the organs that "transfer those adaptive functions that have taken on a generalized form in the course of thousands of generations." Sigmund's theory of inheritance departed from Hering's in assigning relatively greater significance to the plastic as opposed to the inborn portion of the nervous system. A generation earlier, Sigmund's father had spoken for his liberal peers in proclaiming that children possess "the entire infinite range of the human capacity for education [*Bildungsfähigkeit*]." Now evolutionary theory had pushed even the most committed empiricists to regard educational capacity as strictly finite. As Sigmund conceded, the freedom of intellectual development was not absolute.[23]

Sigmund's *Outline* was the target of widespread criticism, particularly from psychologists trained in philosophy rather than medicine, who were still struggling to win recognition from the established field of physiological psychology in the medical faculties. (The first laboratory for experimental psychology in Austria, Meinong's institute at Graz, had just been recognized by the government in 1894, and this would remain the sole center of experimental psychology in Austria outside of the medical faculties until after the First World War.[24]) Participants in the Third International Congress for Psychology in Munich in the summer of 1896 complained repeatedly of the

23. Sigmund Exner, *Erklärung*, 334–39. Franz Exner to M. v. Rosthorn, 22 Jan. 1838, Franz Exner Nachlass. On the history of brain physiology, see Olaf Breidbach, *Die Materialisierung des Ichs. Zur Geschichte der Hirnforschung im 19. und 20. Jahrhundert.* (Frankfurt: Suhrkamp, 1997); and Hagner, *Homo cerebralis*.

24. Lindenfeld, *Transformation of Positivism*, pp. 62, 222.

"hubris" of the "new brain physiology." Since Meynert's death in 1892, Sigmund Exner found himself as the principal "brain physiologist" in Austria. To Emilie Exner, who attended this conference with her husband, the atmosphere was shockingly contentious. She was appalled to see Franz Brentano, the *Doktorvater* of most of Austria's experimental psychologists, speak for half an hour longer than the strict fifteen minute maximum. What was more, she reported, he had been very angry upon being interrupted.[25]

Of particular concern to the new psychologists was the prospect that brain physiology would infiltrate the schools, either as a topic in textbooks or as a basis for pedagogical psychology. Not surprisingly, *Gymnasium* teacher Alois Höfler led this attack. He denied that physiological psychology had any place in the Philosophical Propaedeutic. To include physiology would give the dangerously false impression that one could derive psychology from anatomy. Instructors who lost sight of the distinction between physiology and psychology might misattribute a student's difficulties to his physical constitution and so consider him unteachable.[26] Ironically, while Meynert had left himself open to such accusations, Sigmund Exner had designed his theory of the cortex precisely to eliminate such deterministic implications. Nonetheless, the prominence of theories of psychic inheritance at the close of the century was an obstacle to the liberals' belief that education could make a man universal.

As in the case of theories of legal inheritance, theories of biological inheritance both shaped and responded to the social practices of bourgeois families. Psychological theorists like Mach and Freud regularly cited observations of their own families in support of their ideas, much as Darwin had before them. Mach saw in his children's fear of the dark an evolutionary survival, and Freud found much to interpret in his children's neuroses.[27] Sigmund Exner was no exception to this custom. He demonstrated, for instance, the "influence of emotions on the cortex" by describing the transformation in the behavior of a man or woman upon becoming a parent.[28] Most

25. Entry dated 9 August 1896, Marie von Ebner-Eschenbach, *Tagebücher,* vol. 4 of *Kritische Texte und Deutungen,* ed. K. K. Polheim and N. Gabriel (Tübingen: Max Niemeyer Verlag, 1995), pp. 333–34.

26. Alois Höfler, "Wie soll der psychologische Unterricht an Mittelschulen und wie soll die pädagogische Psychologie zu den Postulaten der modernen Gehirnphysiologie Stellung nehmen?" speech delivered at the Sixth German-Austrian Middle-School Day in Vienna, 13 April 1897, in A. Höfler and S. Witasek, *Physiologische oder experimentelle Psychologie am Gymnasium?* (Vienna: Alfred Hölder, 1898), pp. 5–21.

27. Mach, *Analyse der Empfindungen,* p. 57; Peter Gay, *Freud: A Life for Our Time* (New York: Norton, 1998), pp. 438–46.

28. Sigmund Exner, *Erklärung,* pp. 343–34.

vividly, we can see the mutual shaping of psychological theory and social practice in the Exner family's private efforts to trace continuities from one generation to the next.

NARRATIVES OF INHERITANCE

The Exners structured life at Brunnwinkl so as to pass on the values and ambitions of the older generation to the younger. Marie and her brothers recreated on the Wolfgangsee the holiday life they had known as children, complete with the natural surroundings and the intellectual pursuits their parents had held dear. After the death of their former guardian Josef Mozart in 1892, the five siblings discovered a collection of letters and diaries written by their parents. As Emilie Exner later described, it gave the siblings "bittersweet joy to finally obtain a glimpse into the inner life of the dearly departed." Among these papers they found the first Franz Exner's dream of 1846, in which he had envisioned a "retirement castle" in the countryside. This dream had "become real" at Brunnwinkl, Emilie declared.[29] Indeed, Marie and her brothers created new traditions at Brunnwinkl to embody old aspirations, exemplified by the linden and its promise of transparent communication. The inheritance of the house and land carried with it, in the words of Marie's son Hans, a responsibility for the "cultivation" simultaneously of the physical property and of the values it embodied. After Marie's death, Hans wrote that she, Brunnwinkl's creator, had "left us a legacy that is sacred to us, which is the maintenance and cultivation of her creation in her spirit... And if we succeed in passing on as Brunnwinkl's unshakeable founding principle to the new, almost grown third generation of the Frisch family the spirit of friendly and peaceful togetherness, then we are acting in her spirit and just as she expected of us."[30] Inheritance in this sense was not simply a legal act or a biological fact; it was the transfer of a moral duty. This was a transaction that could only be fulfilled through the conscientious efforts of the heirs.

When the Exners recounted Brunnwinkl's history, however, they explained the continuities between the generations differently. A passive model of inheritance effaced the work of remembrance. Emilie pointed to Franz Exner's dream of 1846 as evidence that children could develop "unconscious reflections" of their parents even long after the parents' deaths. She argued that the Exner siblings had been left with only dim memories of their parents,

29. Emilie Exner, *Der Brunnwinkl*, pp. 9–10.

30. Hans Frisch, *50 Jahre Brunnwinkl*, p. 15.

the younger ones depending on the clearer recollections of Adolf and Marie. Not until reading their parents' papers as adults had members of the second generation realized what they had created at Brunnwinkl:

> An unconscious legacy, taken from father Exner, realized itself after a generation, and the wishes and hopes of the long since vanished generation are reflected today in the way of life of the children, grandchildren, and great-grandchildren . . . There was no ongoing tradition, no direct influence, no adopted way of thought, and yet there was the most surprising similarity in outlook and in the values attached to possessions.[31]

Emilie framed this version of Brunnwinkl's history as the answer to a theoretical question. It is frequently observed, she claimed "that children grow more and more like their parents in the course of their lives, even that the correspondence of thought and behavior only emerges in later periods of life." What causes this mirroring? Should one "attribute the structure of a life to the *willful* impulse of a purposeful character or to the silently working powers of inheritance and tradition? . . . One can wonder whether the power of a repeated example is at work there, or whether the thoughts that give the impetus are independent [of such examples] and enabled to develop solely through inheritance. I would like to claim the latter, and as evidence to tell the history of Brunnwinkl in the last twenty-five years."[32] In this way, Emilie mustered the Exner family history as evidence for a theory of psychic inheritance akin to Hering's or Freud's. The appeal of such a theory to the Austrian *Bildungsbürgertum* at the turn of the twentieth century is no mystery. Haunted by premonitions of socialist or nationalist revolutions, the German-speaking middle class found comfort in the promise of an *inevitable* transfer of memories, values, and aspirations.

Emilie attributed this particular theory of inheritance to Goethe. What Goethe termed "reflections" were, on her account, "psychic phenomena in which the memory of conditions deep in the past awakens the feelings and thoughts that were then active, which then take hold anew."[33] Goethe himself had famously depicted inheritance as a willed act in the first scene of *Faust*: "What you received but as your father's heir / Make it your own to gain possession of it!"[34] Faust delivers this line as he contemplates the scientific instruments he inherited from his father, frustrated at his inability

31. Emilie Exner, *Der Brunnwinkl*, p. 9.

32. Ibid., p. 5.

33. Ibid.

34. "Was du ererbt von deinen Vätern hast/Erwirb es, um es zu besitzen!" *Faust*, trans. by Walter Arndt (New York: Norton, 1976), part 1, act 1, scene 1, line 682.

to put them to good use. It is an assertion of his own responsibility to prove himself a worthy heir. Nonetheless, a Viennese reader at the turn of the century could interpret these lines as evidence of psychic inheritance. In *Totem and Taboo* (1913) Sigmund Freud presented this line as an illustration of his own theory of the unconscious "inheritance of psychical dispositions."[35] Emilie went further, examining the theme of inheritance in Goethe's life alongside his work. In 1899, on the 150th anniversary of the poet's birth, she published a study of the relationship between Goethe and his father, an imperial adviser.[36] Few previous commentators had identified any continuities at all between the two men, yet Emilie boldly argued that the similarities were pronounced and essential to understanding Goethe's moral development. The most precious quality the poet had inherited from his father was his hard-nosed common sense. "The solid, unshakeable foundations of an active, orderly existence, which gave Goethe the immeasurable advantage of a continually productive exchange between practical activity [*Handeln*] and poetry, this he took over from his father and adapted to his needs. The imperial adviser raised him to be a *universal man*, without which the *universal poet* would not have been possible."[37] This is a surprising explanation of the bourgeoisie's claim to universality. How was it possible for economic self-interest, the force behind *Handeln*, to inspire a "universal" art? Implicit in Emilie's account is the Herbartian principle that the acquisition of many-sidedness depends on a dialectic between contemplation and application. The educated bourgeoisie, with one foot in the world of trade and the other in the aeries of high culture, used this ubiquitous account of learning to lay claim to a transcendent point of view. Modern poetry, Emilie suggests, transmuted the mundane experiences of the marketplace into something transcendent. Universality must be understood here as a distinctly bourgeois trait, carried by men, and characterized by the happy coincidence of practical ability with a facility for linguistic self-expression.

Fittingly, on the title page she quoted Theodor Billroth's aphorism, "One enters the aristocracy of intellect, like the aristocracy of caste, only through inheritance." Universality was apparently a heritable quality. Yet "inheritance," as she allowed Goethe himself to characterize it in her conclusion, was nothing if not the transfer of particularity. She closed with the poet's own assessment of his heritage:

35. Freud, *Totem and Taboo*, p. 159.

36. Emilie Exner (Felicie Ewart, pseud.), *Goethe's Vater. Eine Studie* (1899; Schutterwald: Wissenschaftler Verlag, 1999).

37. Ibid., pp. 154–55.

Vom Vater hab' ich die Statur,
Des Lebens ernstes Führen,
Vom Mütterchen die Frohnatur
Und Lust zu fabulieren.

Uhrahnherr war der Schönsten hold,
Das spukt so hin und wieder;
Urahnfrau liebte Schmuck und Gold,
Das zuckt wohl durch die Glieder.

Sind nun die Elemente nicht
Aus dem Komplex zu trennen,
Was ist dann an dem ganzen Wicht
Original zu nennen?

Could the German-speaking world's "universal Genius" have been no more than a patchwork of his parents' traits?

EDUCATION AS A UNIVERSAL LEGACY: THE CASE OF MEDICINE

It was the predicament of Austrian liberals to aspire to "universality," to a perspective transcending the confines of personality, ethnicity, or class, while shunning all claims to absolute certainty. Universality or many-sidedness was supposedly the fruit of proper education and thus the intellectual legacy of one generation to the next. With their own authority in jeopardy in the 1890s, liberal educators invoked the trope of inheritance to suggest that education involved a moral duty to maintain the traditions that bound together young and old, living and dead. It was customary, for instance, for students and professors to use familial metaphors to describe mentoring relationships. Serafin Exner's students spoke of him as "father" and of their research group as a "family." Since male students often married the daughters of their professors, this language had a basis in the social strategies of the *Bildungsbürgertum*. When Brücke retired in 1890 and Sigmund Exner took over his chair in the medical faculty, he likened the aging physiologist to "a father, who with the sacrifice of his own time and productive energy teaches his intellectual sons . . . to guard and multiply the scientific riches produced in a successful lifetime, so that they will someday know how to manage these for the benefit of mankind."[38]

38. Sigmund Exner, "Ernst v. Brücke und die moderne Physiologie," *Wiener klinische Wochenschrift* (1890), Nr. 3, 807–12, quotation on p. 807.

This rhetoric of intellectual stewardship was matched by aesthetic evocations of a universal tradition. Even a new temple to science like the Physiological Institute that opened under Sigmund Exner's direction in 1899 was cloaked in a classical design and crowned with a Latin aphorism. Exner himself explained the impact of such allusions. According to his theory of perception, aesthetic judgments rested on an individual's visual memories, but these memories were in turn shaped by artistic traditions: "The nature of these impressions, however, is conditioned by the conduct of generations who preceded us by centuries, even millennia; in other words: these traditions play the most important role for our impressions of art."[39] The rhetorical power of historicist aesthetics was on full display at the unveiling in May 1893—the fortieth anniversary of the death of Franz Exner, Sr.—of the monument to the architects of the midcentury educational reforms: Thun, Exner, and Bonitz. The commemorative address was delivered by Wilhelm von Hartel, a liberal philologist, friend of the Exners, and future minister of education. Retracing the history of the midcentury education reforms, Hartel claimed that the real monument to these three educators was not the set of marble busts but the reforms the men had instituted, which stood unblemished by the tests of time. Hartel pointed to the decade-old neo-Renaissance palace of the university, symbol of universal learning and of the liberals' rise to power in the 1870s,[40] as confirmation that the nation's ideals remained those of Thun and his collaborators. This was the edifice "that our noble emperor built for scholarship as a dazzling image of the progress that science and art have celebrated under his auspices."[41] As Hartel implied, the university was an embodiment of the universalist aspirations of the empire and of its liberal civil servants.

However, the changing demographics of higher education in the 1890s belied Hartel's optimism. In the eyes of contemporaries, the test case for the future of liberal education in the Austrian empire was the debate over medical school reforms. The question that plagued liberals in this debate was whether education could really make a man universal—and whether universality was even a worthy goal to seek. This was a dispute at once over the character of modern scientific knowledge and the character of modern scientific knowers—male, female, Jewish, Catholic, rich, or poor.

39. Sigmund Exner, "Ueber allgemeine Denkfehler," p. 65.

40. Schorske, *Fin-de-Siècle Vienna*, chapter 2.

41. Wilhelm von Hartel, *Thun-Exner-Bonitz*, p. 32.

Medical faculties at Austrian universities in the 1890s reflected the broader influx into higher education of students from previously underrepresented ethnic, social, and geographic backgrounds. Medical enrollments had reached a first high-water mark in the early 1870s, after which the introduction of more stringent requirements, combined with the economic recession, briefly lowered the number of students. By the mid-1880s the numbers had reached a new high that was not to be surpassed until after 1910. Overcrowding was a persistent problem at the medical faculties from the early 1870s through the turn of the century.[42] Moreover, employment prospects looked dim for this new surplus of the educated. As in Germany in this period, Austrians worried that the creation of a population of unemployed university graduates was akin to throwing a lighted match into the haystack of political unrest.[43]

The rising number of Jewish university students contributed significantly to the mushrooming enrollments. The proportion of medical students who were Jewish rose from 30 percent in 1869 to 61 percent in 1884, nearly twice the frequency of Jewish students in the university as a whole at the later date.[44] In 1895, when Parliament took up the matter of reforming the medical curriculum in order to accommodate the growing numbers of students, anti-Semites made Jews out to be the root of the problem. Karl Lueger, the leader of the populist and anti-Semitic Christian Social party, focused resentment on one of the university's most renowned physicians, Hermann Nothnagel, a non-Jewish member of the Society to Combat Anti-Semitism.[45] To Vienna's vocal anti-Semites, the medical faculty had become a threatening symbol of Jewish power.

Many also viewed female students as intruders. Not until 1900 would women win admission to the medical faculties as degree-earning students. Yet women had been allowed to audit classes in the medical faculty since the early 1880s, and their presence stirred controversy.[46] The standard arguments against enrolling women, as articulated by the university's academic senate in 1873, were that it was not feasible to censor lectures for the sake

42. Cohen, *Education and Middle-Class Society*, esp. pp. 114–17.

43. Konrad Jarausch, "Universität und Hochschule," in *Handbuch der deutschen Bildungsgeschichte*, vol. 4, pp. 313–45.

44. Cohen, *Education and Middle-Class Society*, p. 166.

45. Richard S. Geehr, *Karl Lueger: Mayor of Fin-de-Siècle Vienna* (Detroit: Wayne State University Press, 1990), p. 167.

46. Cohen, *Education and Middle-Class Society*, p. 76.

of young women's modesty; that the fairer sex would be a distraction to male students; that women should not be encouraged to ignore their "natural" duties; and that the university was ultimately responsible for preparing students for professions, including that of physician, that were and would remain *male*.[47] Austria's universities were among the last in Europe to welcome women, with the faculties of philosophy and medicine opening their doors in 1897 and 1900, respectively, and those of law and theology not until after the First World War.[48] Even supporters of women's education in general were skeptical about training women as physicians. As we will see, both Sigmund and Emilie Exner doubted that women were intellectually suited for medical study, and they feared the effects that the higher education of women would have on family life.

The problem facing the medical faculties, then, was not just overpopulation, but a population that was far more heterogeneous than ever before. Beginning in the mid-1890s, proposals for reforming the medical curriculum flooded into the education ministry. For students from lower middle-class backgrounds, there was a financial incentive to complete medical studies as quickly as possible and with adequate preparation to begin practicing immediately. For this reason, many reform proposals, often from untenured teachers at provincial universities, focused on streamlining the curriculum and weighting it more heavily towards clinical experience.[49] Such critics of the current curriculum complained that theory had been divorced from practice by the reforms carried out under Count Thun and the first Franz Exner in the 1850s. In 1873 the dissolution of the university's "Doktorenkollegien" had eliminated the only forum for practicing physicians and jurists to have a voice in academic affairs.[50] At the same time, the university medical faculties had monopolized medical training and certification, shutting down respected institutions like the Josephinum, the imperial training institute for

47. Waltraud Heindl, "Zur Entwicklung des Frauenstudiums in Österreich," in *"Durch Erkenntnis zu Freiheit und Glück . . . ": Frauen an der Universität Wien (ab 1897)*, ed. W. Heindl and Marina Tichy (Vienna: WUV-Universitätsverlag, 1990), pp. 17–26, quotation on p. 19.

48. Ibid., 17. Imperial Germany delayed even longer, with the first women students enrolling in Baden in 1900; M. Tichy, "Die geschlechtliche Un-Ordnung: Facetten des Widerstands gegen das Frauenstudium von 1870 bis zur Jahrhundertwende," in Heindl and Tichy, *"Durch Erkenntnis,"* pp. 27–78, quotation on p. 27.

49. Allgemeines Verwaltungsarchiv Unterricht, doc. 9067 (9 April 1897), on stipends for interns, folder 96 (Medizin, 1896–1906); and doc. 12548 (12 May 1897), on proposal to ease financial strain on students, folder 2607 (Med. Rigorosen), Österreichisches Staatsarchiv, Vienna.

50. On the legal side, Rudolf von Jhering charged that academic jurists in Vienna had shown remarkably little interest in the implementation of legal reforms since the 1867 constitution. For a response to the demands for a more pragmatic legal education, see Hruza, *Romanistische Rechtsunterricht*.

military surgeons.[51] A university degree was now the only path to medical practice, and there was no consensus that it was an efficient one.

In 1895 the education ministry opened an official investigation into medical school reforms. One year earlier the ministry had engaged Sigmund Exner, chair of physiology at the University of Vienna, as an adviser on medical education. In this capacity, Sigmund read and commented on all reform proposals received by the ministry from medical faculties throughout the Austrian half of the empire. These proposals dealt overwhelmingly with reforms to the examination system and with the year of hospital internship that followed the completion of course work. Professors proposed changes in particular to the *Rigorosen*, a daunting series of exams in natural history and other scientific fields administered in the first two years of medical study. Since the first round of *Rigorosen* reforms in 1872, medical students had attended lectures on these topics in the philosophy faculty, and philosophy professors had often acted as examiners for medical students in these fields. Reforms to the medical curriculum would therefore have repercussions beyond the medical faculties themselves. Any changes would also directly implicate the *Gymnasien* as well, since natural science topics no longer taught at university would require more coverage in secondary school. Thus the debate over medical education in Austria repeatedly circled back to the potential value for future physicians of a classical *Gymnasium* education.

According to one contemporary caricature, opinion in Austria divided between "conservatives" (or "humanists") and "progressives."[52] This characterization is useful, if rough. "Progressives" tended to frame the issue of curriculum reform as a social and economic matter. Berthold Hatschek, professor of zoology at Prague, warned that "we live in a time that is desperately poor in time."[53] Hatschek's recommendations were common among progressives: to streamline education by shortening the *Gymnasium* curriculum and creating *Einheitsschulen*, secondary schools that would prepare students to enter either a university or a technical college. In Hatschek's view, the reform of medical education was "a part of the great social questions which form the political problem of our day." Hatschek looked forward to the rise

51. On the closing of the Josephinum and the new process of certification, see Cohen, *Education and Middle-Class Society*, p. 50.

52. Adolf Hromoda, *Briefe über den naturhistorischen Unterricht an der medicinischen Facultät und am Gymnasium* (Vienna: Carl Gerold's Sohn, 1897), p. 5.

53. Berthold Hatschek, *Medicin, Naturwissenschaft und Gymnasialreform* (Prague: Calve, 1896), p. 12.

of a populist movement for education reform, led by "fathers who wish to see their sons trained for a lifetime's profession."[54] Here, Hatschek pointedly reversed the familiar liberal view that the family's interest lay in providing children with a well-rounded education, while the state would naturally focus on narrower professional training. Hatschek implicitly rejected the liberal ideal of "many-sidedness" on which the Austrian education reforms of the 1850s had rested. Universality, he implied, was a luxury of the rich.

Hatschek's "conservative" opponents, on the other hand, were depicted as men who viewed themselves as the "guardians and agents of the final and therefore that much more valuable remainder of humanism and idealism . . . "[55] To their opponents, these humanists naturally seemed backward-looking. Rhetorically, they relied heavily on historicist images and evocations of Austria's tradition of liberal education, with its intellectual ideal of "many-sidedness." Yet these educators defended tradition not for its own sake, but for its value as a source of unity within the academic community and beyond. Julius Wiesner, a former student of Ernst Brücke's and a proponent of women's education, argued on these grounds for continuing to require medical students to obtain a broad and rigorous background in natural science. Wiesner appealed to the "long-standing unity of the university, the solidarity of its members," which it was the duty of the philosophy faculty to promote. "It lies in the nature of the philosophy faculty . . . that it in particular has a supportive and stimulating [*befruchtend*] influence on the intellectual life of the other faculties."[56] Here Wiesner enlisted a metaphor of intellectual cross-fertilization and procreation in order to argue that the disciplines were "related" by the *universally* fertile methods of philosophy.

As Wiesner hinted and other humanists made clear, the source of this unity was the "universal" nature of intellects trained in philosophy. Theodor Hromoda was a teacher of physics at a *Realschule* and a practicing physician. More progressive than Wiesner, Hromoda sought to make medical education more efficient without eliminating classics, philosophy, and natural science. Like Wiesner, however, he portrayed the universal knower as a living ideal. Hromoda characterized philosophy as "the theory of those standards of human thought, which have been valid ever since humans began to think

54. Ibid., p. 14.

55. Hromoda, *Briefe*, p. 5.

56. J. Wiesner, *Die Nothwendigkeit des naturhistorischen Unterrichtes im medicinischen Studium, aus Anlass der bevorstehenden Reform der medicinischen Studien an den österreichischen Universitäten* (Vienna: Alfred Hölder, 1896), p. 1.

at all and will always be valid . . . These standards are removed from the battles of the philosophical schools; they are not scholastic odds and ends but revelations of our human nature."[57] In Hromoda's formulation of the humanist position, the thinker trained in philosophy's "standards of thought" was the source both of continuity and unity at the university. Hromoda's concept of universal knowledge, like Wiesner's, was an intellectual and social ideal, a vision not of homogeneity but of harmony. It recalled an age before the nineteenth-century expansion of the university population and the new diversity of goals it ushered in.

One physician dared to attack these changes directly. He was Theodor Billroth, the Exners' friend and neighbor in St. Gilgen, one of the most strident guards of the old order at Vienna's medical faculty. As early as 1886 Billroth had derided the latest crop of medical students with thinly veiled anti-Semitism. "In Austria the physicians unfortunately come all too rarely from families in which a secure economic position, untroubled family relations, and a fine, benevolent, idealistic civic sense are traditions. To be from 'a good house' is of far more value to a physician than a [*Gymnasium* diploma]."[58] Billroth's diatribe was phrased more baldly than most, but he expressed doubts shared by many of his colleagues. The midcentury reformers had been radical in their conviction that the state *could* and *should* take over a portion of responsibility for *Erziehung*, for educating well-rounded individuals in the place of narrowly trained bureaucrats. The reforms of the 1850s were meant to produce schools and universities capable not merely of imparting a body of knowledge but of cultivating noble personalities on an imperial scale. In his characteristically exclusionary language, Billroth pinpointed the flaw in this program, the disjunction between its purported goals and the prejudices it had failed to root out:

> It is a widespread false belief in our day that one can become an educated [*gebildeter*] person and enter into the intellectual aristocracy by virtue of having learned a good deal. Knowledge is only the raw material for thought; only one who wields his knowledge nimbly and widely belongs to the "Gebildeten." One enters the aristocracy of intellect, as that of birth, only through inheritance. Nature is cruel; she gives much to those who already have much.[59]

57. Hromoda, *Briefe*, p. 48.

58. Theodor Billroth, *Aphorismen zum Lehren und Lernen in der medizinischen Wissenschaften* (Vienna: C. Gerold's Sohn, 1886), p. 53.

59. Ibid., p. 4.

Billroth's attack amounted to a deconstruction of the ideals of "liberal education" in Austria. Since its Herbartian conception, Austrian liberal education had stressed the ability to relate theoretical and practical knowledge, the quality that Herbart called "many-sidedness" and others more loosely called "universality." Universality in this sense had by no means implied uniformity, either of the raw material of education or of its product. *Bildung* was meant to foster not homogeneity but breadth. Yet Billroth wanted to expose the capacity to acquire breadth for what he believed it was: a specifically bourgeois trait, as much style as skill, one that could not be taught, only "inherited." He blurred all familiar oppositions between bourgeois and aristocratic values, between meritocracy and aristocracy. In his attempt to topple fifty years of liberal arguments for public education, he alerts us to the very real contradictions in the liberals' program. How could a process of "inheritance," as the humanists described education, possibly produce "universality" rather than particularity?

In late 1899, four years into the investigation of the medical curriculum, Sigmund Exner's friend Hartel became minister of culture and education. His appointment brought hope to Viennese liberals alarmed by the growing popularity of the Christian Socials on the right and the Social Democrats on the left. Emilie Exner exclaimed that here finally was "a man who understood something of educational matters!" Yet the ministry appeared to her "continuously in danger." "How long will it last?" she wondered.[60] "How long will Hartel stay at the helm? Poor Austria!"[61] Try as he might to fulfill the wishes of his former colleagues, Hartel later lamented to Sigmund that in the early years of his ministry he lost many friends.[62]

Meanwhile, Sigmund threw himself into his work for the ministry on the medical reforms. In 1899 and again in 1902 the ministry reached a provisional agreement with the medical faculties on reform measures, with a final ordinance issued in 1903. By January 1902, as the pressure to conclude the revisions reached its high point, Emilie reported that Sigmund was in an anxious state:

> The stories are more horrid than the onlookers suspect, and when in this case the work of many years and the [undertaking?] of a lifetime are at stake, it

60. Emilie Exner to Marie von Ebner-Eschenbach, 4 April 1899, folder 81082Ja, Stadtarchiv, Vienna.

61. Emilie Exner to Marie von Ebner-Eschenbach, 7 Dec. 1899, folder 81082Ja, Stadtarchiv, Vienna.

62. Sigmund Exner, "Wilhelm Ritt. V. Hartel," *Wiener klinische Wochenschrift* (1907), Nr. 4.

> is a difficult and upsetting matter. Schiga [Sigmund] does not yet know today which position his colleagues will take, how the ministry will weather the storm. He is regarded as the single author of the examination regulations, while his proposal, quite different from the current law, lies printed in his desk and is really fundamentally different. Night and day I preach indifference without feeling it myself. This storm too will pass like many before it.[63]

Sigmund's own proposal, consigned to his desk drawer, remains a mystery. Yet his opinions have been preserved in his notes on the documents he reviewed for the ministry. These show his efforts to strike a compromise, to reconcile the ideals of liberal education with the concerns of socially progressive critics. On one hand, Sigmund, like Wiesner, sought to maintain a substantial course load in natural science for medical students, in order to preserve the ties between the medical and philosophy faculties. After the publication of the first Rigorosen reforms in 1899, the ministry received several questions and complaints regarding the new division of teaching between the medical and philosophical faculties. Sigmund repeatedly advised that medical students be allowed to attend lectures in the philosophy faculty, even if the same subject was being taught by a medical professor. He based this argument first on the principle of the freedom of learning, guaranteed by the Constitution of 1867, and secondly on the content of the 1872 and 1899 ministerial ordinances. In Sigmund's interpretation, these ordinances demanded that "the lectures shall consist not of medical chemistry as an applied science but rather of chemistry with regard to the needs of the future physician." He concluded that these laws therefore recognized "the right of every chemist to give these lectures." Sigmund was similarly in favor of allowing medical students to attend the normal experimental physics lectures and laboratory exercises for natural science students.[64] In short, Exner intended to uphold the broadest possible basis for medical education. This could be seen as a "conservative" or "humanist" move in defense of the goal of producing "universal" thinkers.

Nonetheless, Sigmund tried to make this system of universal education compatible with the goals of the "progressives," particularly their concern with the prospects of economically disadvantaged students. Although he expressed skepticism about some practicalities of clinical learning, he

63. Emilie Exner to Marie von Ebner-Eschenbach, 28 Jan. 1902, folder 81082Ja, Stadtarchiv, Vienna.

64. Allgemeines Verwaltungsarchiv (AVA) Unterricht, folder 2607 (Med. Rigorosen) 15356/1900, Österreichisches Staatsarchiv, Vienna.

supported the effort to give medical students more clinical experience, such as an obligatory hospital internship. He also addressed the problem of how student interns would support themselves, since stipends only covered the period prior to exams. Sigmund acknowledged the difficulty this posed to some students, and he attempted to negotiate with the rector of the University of Vienna to provide stipends to interns.[65] To be fair, however, the university would then be obliged to pay equal stipends to students in the natural sciences, jurisprudence, and pharmacy who had completed their exams and gone on to internships. "It has until now been absolutely out of the question to grant a university scholarship for the period of this practical occupation, or to extend the grant to this period."[66] Revising this policy, he noted, would mean a radical change in the definition of a student, from someone who was acquiring knowledge to someone who had begun to apply that knowledge. Ultimately, he saw no solution to the financial problem. Exner recognized that liberal education must evolve to meet the needs of a changing society but failed to see just how it could do so.

After consigning his own reform plan to a desk drawer and confining himself to confidential remarks to the ministry, Sigmund Exner finally allowed himself one moment of relatively unguarded public expression. The occasion was the opening of the university's newly built Physiological Institute in 1899, at the height of the reform debates. The new institute wore the inscription "Doctrina fundamentum artis," indicating that theory, or theoretical natural science, was the foundation of the "medical art." Playing on the rich ambiguities of this expression, Sigmund titled his speech "Kennen, Können, und Erkenntnis in der ärtzlichen Kunst."[67] ("Kennen" and "Erkenntnis" do not have exact English equivalents; both are usually translated as "knowledge," with *Kennen* having the additional meaning of "acquaintance." *Können* instead denotes practical ability.) He explained that these three concepts "describe three stages of psychic work of ascending value." This epistemology owed debts to Fechner and John Stuart Mill, but it built most immediately on the ideas of Helmholtz, from whom Sigmund acknowledged having learned much "despite his reticence."[68] Reticent he

65. AVA Unterricht 96 (Medizin), 12548/1897, Österreichisches Staatsarchiv, Vienna. Sigmund also discussed the stipend problem in 9067/1897.

66. Ibid.

67. Sigmund Exner, "Kennen, Können, und Erkenntnis in der ärtzlichen Kunst," *Wiener Zeitung*, 19 Oct. 1899, pp. 2–3.

68. Sigmund Exner, "Autobiographische Skizze," Archiv der Akademie der Wissenschaften, Vienna.

may have been, but Helmholtz made clear that the most valuable form of knowledge was that which was "capable of expression in words." The notorious "unconscious inferences" that Helmholtz believed formed the basis of perception served as examples of inarticulable knowledge. In the hierarchy of knowledge, Helmholtz ranked *Wissen* above *Kennen* because it was subject to expression and therefore to synthesis: "[S]peech makes it possible to collect together the experience of millions of individuals and thousands of generations, to preserve them safely, and by continual verification to make them gradually more *certain* and *universal*... [A]ll deliberately combined actions of mankind, and so the greatest part of human power, depend on language."[69] Helmholtz was arguing that the social unity and historical continuity of human efforts of all kinds depended on the possibility of articulating experiential knowledge in a "certain" and "universal" form. Sigmund Exner would adopt the second criterion, though not the first.

Helmholtz himself had participated in discussions of reforms to the medical curriculum in Germany in 1877 and again in 1890, echoes of which can be found in the Austrian debate of 1894.[70] In the 1870s Helmholtz still saw the question of curriculum reforms from the perspective of the liberal struggle for academic freedom of the 1840s, "the fight between learned tradition and the new spirit of natural science, which would have no more of tradition, but wished to depend upon individual experience."[71] In the 1890s, however, when debate erupted anew in Germany, Helmholtz fell on the side of the "conservatives." In 1890, just four years before his death, he pressed colleagues to retain the requirement of *Gymnasium* education for physicians, and he defended the study of the classics as the "most tried and tested" method of intellectual cultivation.[72] Others championed a classical education as essential to the psychological development of the physicians. Only a humanistic education could prepare doctors "to try to understand another person deeply."[73] In this case, liberal educators in Austria echoed their

69. Helmholtz, "The Recent Progress of the Theory of Vision," in *Science and Culture*, pp. 127–203, quotation on p. 200, emphasis added.

70. For comparisons between the Austrian and German debates of the 1890s on medical education, see the senator Robert Pattai, "Das klassische Gymnasium und die Vorbereitung zu unseren Hochschulen," in *Reden und Gedanken*, 2nd edition (Vienna: self-published, 1909), p. 95.

71. Helmholtz, "On Thought in Medicine," in Cahan, *Science and Culture*, pp. 309–27, quotation on p. 309.

72. Quoted in Pattai, "Das klassische Gymnasium," p. 95.

73. Ibid., p. 95.

counterparts in Germany. Where they differed, however, was in the virtues they associated with "universal" knowledge and the classical tradition.

In his 1899 address to Vienna's medical students and faculty, Sigmund modified Helmholtz's terminology. His hierarchy ran upward from *Kennen* to *Können* to *Erkenntnis*. He used *Kennen* in the sense of a capacity to associate a new perception with existing "visual memories." *Können* improved on *Kennen* because it included the ability to articulate the basis for these associations of new experiences with older memories. Thus a physician might be able to identify the disease of a patient based purely on *Kennen*, on an unarticulated association with past cases of that disease. He would only reach the stage of *Können* if he were able to articulate the grounds for his diagnosis, to state the similarities between the patient's symptoms and those of past cases. Such knowledge was more valuable but still imperfect: it "rests on visual memories of immediate sensory impressions and dies with the brain in which the visual memories lie." *Können* was thus knowledge trapped in the form of subjective perceptual experiences. The highest form of knowledge, *Erkenntnis*, was that which could be articulated in a universal form, and which could therefore impart to others the abilities categorized as *Können*. "The scholar, however, can communicate his ability to others, and his discovery is preserved in the intellectual store of mankind." Preserved in a systematic form, *Erkenntnis* became "Doctrina," theory: "These experiences of thousands of people, . . . ordered and recorded in writing or transmitted orally, form that *Doctrina* which is named in the inscription of our institute and on which you, gentlemen, shall build your medical art."[74]

Doctrina could not, however, lay claim to certainty, since causal knowledge could never be complete. Sigmund explained: "Now the cause [*Grund*] of a phenomenon is normally of a complex sort, and each causal factor [*Ursache*] is only a link in the chain of an apparently endless causal chain. For that reason a complete view of all causes is out of the question, our knowledge [*Erkenntnis*] penetrates the causal chain only to a certain extent and is accordingly more profound or superficial." The physician could not see through the tangle of "possible causes and their linkage" on his own. To do so would require "a glimpse that the experiences of *one* lifetime are not able to give." Much as Adolf Exner had argued of railroad accidents, the etiology of disease involved too many factors to make a causal analysis tractable. Thus, while Helmholtz and Exner agreed on the empiricist building blocks

74. Sigmund Exner, "Kennen, Können."

of medicine, they characterized the pinnacle of knowledge differently. In their war on dogmatism, Austrian liberals like Exner continued to stress that their goal was not certainty but probability.

Fortunately for physicians and patients, *Doctrina* rested on a large enough store of collective experience to compensate for the impossibility of analyzing each medical case in full causal detail. Systematically collected over generations, *Doctrina* supplied the physician with the insight he needed to skirt this forest of causal branches. In this way, the physician's *Können*, his practical ability, rested on a foundation of theoretical knowledge, just as the inscription on the new institute proclaimed.

Finally, from the hierarchy of knowledge Sigmund moved on to the hierarchy of knowers. Not everyone was capable of ascending from knowledge that was transient and personal to the permanent and universal. "A man can be very intelligent, but at the same time exceedingly inept in the motivation of his judgment," he had written a decade earlier. "The associations of a clever farmer can be quite plentiful, the judgment of the significance of individual series of ideas can be correct, but since all this unfolds in him without conscious intervention, he can be very far from being able to clothe these processes in words."[75] Standing before his new laboratory, Exner again held up the farmer as his example of common reason: "[W]ere the farmer asked which sensation caused him to identify the seed of a distant field as barley rather than rye, he would not be able to say. Even leafless, he knows the common species of trees and the shape of their branches . . . but could not say what conditions his judgment."[76] Exner praised the farmer's "connoisseurship," drawing a comparison to a certain young physician whose powers of observation allowed him to recall the flight of a species of bird for years until he came upon its name. That young physician was, of course, Sigmund himself. As in Adolf Exner's theory of acts of God, the farmer served Sigmund as a model of intuitive rationality. But Sigmund also noted the farmer's limitations. A farmer could form accurate empirical judgments, but he was unable to articulate the grounds for his judgments and so render them in a universal form. This ability to speak from a perspective beyond the local and personal set the educated bourgeoisie apart from mere peasants.

Privately, Sigmund judged that this faculty was the preserve not simply of the *Bildungsbürgertum* but of its male representatives. So he told Marie von Ebner-Eschenbach, who asked in 1904 for his opinion on medical

75. Sigmund Exner, "Ueber allgemeine Denkfehler," p. 60, footnote 1.

76. Sigmund Exner, "Kennen, Können."

training for women. Sigmund replied that he had never had a problem with teaching men and women together, and his women students always studied diligently. However, the women tended "to focus more on the exam than on utilizing what has been learned." Women showed a pronounced weakness in the practical application of theoretical knowledge: "In any case one must in general be prepared for less dazzling answers when one presents tasks on an exam that can only be solved by the practical application or utilization of what has been learned, than when poses questions whose answers are contained in the textbooks. This difficulty with utilizing and applying what has been learned in cases from everyday life is to be found in male students as well, but it seems to me to be present in a greater proportion of female students."[77] The *man's* ability to relate *Kennen* to *Erkenntnis*, to bridge personal, experiential knowledge and the common fund of theory, was the key to making knowledge universal.

Still, Sigmund was more generous in his evaluation of women than many of his medical colleagues, some of whom insisted that the female anatomy or reproductive functions ruled out higher reasoning.[78] Nor did Sigmund's wife Emilie, an activist for academic education for women, judge women's capacities any more favorably. She had heard it said "hundreds and hundreds of times" that female students were "industrious," able to reproduce the lessons of the textbooks, but that they were largely unable to put that knowledge to use. She had demurred until she saw "how true it is. For in girls' elementary instruction as in their advanced examinations we find the same incongruity between *Wissen* and *Können*."[79] Sigmund added to this critique a biological explanation: this disparity was to be expected, since women's thoughts were naturally more closely tied to habit and thus to instinct.[80] Emilie in turn advised her female readers to acknowledge their own limitations: "Know thyself," she counseled.[81]

What was it that women lacked? John Stuart Mill, whose writings Sigmund and Emilie had studied carefully, had similarly sought an explanation for women's apparent inability to move between experiential knowledge and

77. Sigmund Exner to Marie von Ebner-Eschenbach, 25 Jan. 1904, folder 70086, Stadtarchiv, Vienna.

78. Tichy, "Geschlechtliche Un-Ordnung," pp. 33–35.

79. Emilie Exner (Felicie Ewart, pseud.), *Eine Abrechnung in der Frauenfrage* (Hamburg: L. Voss, 1906), p. 45.

80. Sigmund Exner, "Physiologie der männlichen Geschlechtsfunktionen."

81. Emilie Exner, *Frauenfrage*, p. 45.

theoretical statements. Mill, whose wife contributed to his own publications, did not doubt that women often had important ideas; but "no justice could be done" to them "until some other person, who does possess the previous acquirements, takes it in hand, tests it, gives it a scientific or practical form, and fits it into its place among the existing truths of philosophy or science."[82] Like the Exners, Mill observed that women had not learned to articulate intuitive knowledge in a form in which it could become part of an intellectual tradition. Yet Mill did not see this as a peculiarly female trait, nor did he see the ability to translate knowledge in this way as peculiarly male. Both qualities were merely the result of less or more formal education.

The Exners, on the other hand, spoke of "universality" as a trait unique to a small segment of the population, and not by virtue of schooling alone. There was, for instance, no contradiction in portraying Goethe, the Germans' "universal" genius, as an exemplar of bourgeois values, as Emilie Exner did in her biography. In Vienna, the universal knower had become a specifically male, bourgeois ideal.

WHY would women even desire universality, Emilie Exner asked, when men adored them for their idiosyncrasies? "All our inherited flaws are what charms you, what you love in us, what makes your lives bitter and exciting at the same time." In putting forth her proposals for women's education in step with the medical school reforms of the 1890s, she reassured her readers that she would steer clear of a "homogenization" of the female population.[83]

She admitted, nonetheless, that the diversity of the female population posed an obstacle to her reform goals. When she looked around at the women of her own social circle, she saw a broad spectrum of educational achievements, and she feared that these differences would fragment any movement to improve the lot of women. "Where will there arise a unified perspective, a collective effort, a shared goal? And how little the individual achieves, how weak are the attempts even of the industrious, to make a breach in age-old prejudices and customs?"[84] Disunity was all too evident a problem for Vienna's women's movement circa 1900. Emilie was a member of the Society for the Gainful Employment of Women (*Frauen Erwerb Verein*), which had been founded in 1866 with the goal of promoting the entrance of women

82. J. S. Mill, *The Subjection of Women* (Buffalo: Prometheus, 1986), p. 76.

83. Emilie Exner (Felicie Ewart, pseud.), *Die Emancipation in der Ehe: Briefe an einen Arzt* (Hamburg: Voß, 1895), p. 55.

84. Emilie Exner, *Emancipation*, p. 54.

into "suitable" professions. This moderate group clashed with left-wing women activists like Auguste Fickert, leader of the General Austrian Women's Society, who criticized those fighting merely for equal rights under the law, rather than for broader social change.[85] In calling for unity among women reformers, Emilie elided the key differences that separated her from more progressive activists. Even as she praised female individuality as an enticement to men, social and political differences vanished from her view. In a manner characteristic of liberal reasoning in Austria, the perspective that achieved this disappearance was *statistical.*

In Emilie's prescriptions for women's education, her point of reference was what she called the "average woman," *die Durchschnittsfrau.*[86] Emilie was well-versed in the language of statistics and probability. She too saw inductive reasoning as an evolutionarily ingrained habit of thought.[87] She was proficient enough in statistics to have published a study of the likelihood of a male versus a female birth in a given family, given the genders of the previous children. Using data on births from the Gotha Calendar, she analyzed the relation between the probability of having two male or two female babies in a row and the length of time between the pregnancies. As long as her data were free from systematic error, she argued, her analysis was valid.[88] Her claim was that statistical inferences were insensitive to small fluctuations in individual behavior, and this was likewise the view she applied in her analysis of education for women. Relying on the benchmark of the "average woman," she implied that it was possible to make prescriptions for the gender as a whole, despite the individual variability or "inherited flaws" that gave women their charm.

Emilie also assigned the average woman a normative role. She criticized women who strove too hard to excel, who "raise themselves above the average."[89] This was a motley category comprising overly ambitious intellectuals as well as stereotypes of women of loose sexual morals, such as actresses, singers, dancers, and women who rode animals in circuses. Women who

85. On Vienna's women's movement, see Harriet Anderson, *Utopian Feminism* (New Haven: Yale University Press, 1992).

86. See, for example, Emilie Exner, *Frauenfrage*, p. 6; *Weibliche Pharmaceuten*, speech delivered in the Frauen-Erwerb-Verein, 7 March 1902 (Vienna: Verlag des Frauen-Erwerb-Vereins, 1902), pp. 4, 15.

87. Emilie Exner, *Frauenfrage*, p. 21.

88. Emilie Exner (Felicie Ewart, pseud.), *Zur Kenntnis der Geschlechtsbestimmung beim Menschen* (Bonn, 1908), p. 605.

89. Emilie Exner, *Emancipation*, p. 61.

made themselves into spectacles and grew accustomed to applause apparently acclimated poorly to domestic life. Such "exceptions" were frequently invoked in the discourse on sexual difference in Vienna at the time. Yet Emilie's statistical language suggested that these exceptions were not pathological monsters—"a form of mental hermaphrodite," in the words of one contemporary[90]—but outliers along a social continuum. By implication, such excesses could be prevented. Indeed, Emilie's goal was to allow women to pursue higher education while staying safely within the domestic sphere.

Upon becoming president of the Society for the Gainful Employment of Women in 1901, Emilie dedicated herself to expanding the society's course offerings to help young women make the transition to university. As she explained her goals at the end of her first year in office, "I would like just once to accomplish something that offers more than a pleasant pastime; rather to give those girls who take up an occupation a good preparation for a lifetime."[91] Emilie sharply distinguished her aims from those of female educators in the United States. At the new American women's colleges, female students were pursuing a broad humanistic education. Emilie warned that these young women became obsessed with the development of their own "personalities" and lost all interest in family life. Alarmed by this example, Emilie would make sure to steer women's education toward practical matters and to keep a firm focus on a woman's role as wife and mother.[92] For women to pursue education in the hopes of becoming "many-sided" or "universal" intellects would prove disastrous.

The model of success, in Emilie's eyes, was Japanese women. She read about Japan's schools for women in books for a German audience, such as the Leipzig publication *Our Fatherland Japan: A Sourcebook, Written by Japanese*. Emilie in turn reported to her bourgeois readers in Austria that the Japanese taught women classical languages, mathematics, and natural science alongside home economics (*Haushaltungskunde*) in both "practical" (cooking, washing, sewing) and "theoretical" (pedagogy, hygiene, ethics) forms.[93] In her view, this program promised to give women a sophisticated general education without "losing sight of the dangers that thereby arise for the future generation."[94] It only remained to find the "average woman" a suitable profession.

90. Tichy, "Geschlechtliche Un-Ordnung," p. 35.

91. Emilie Exner to Marie von Ebner-Eschenbach, 15 July 1902, folder 81082Ja, Stadtarchiv, Vienna.

92. Emilie Exner, *Frauen-Frage*, pp. 6–13.

93. Emilie Exner, *Frauen-Frage*, pp. 16–17.

94. Emilie Exner to Marie von Ebner-Eschenbach, 15 July 1902, folder 81082Ja, Stadtarchiv, Vienna.

Emilie decided on pharmacy. The job of pharmacist would allow women to work at home and demanded "neither Amazons nor geniuses."[95] It would keep women safely within Emilie's definition of average. Pharmaceutical studies at the University of Vienna opened to women along with medicine in 1900. Three years later, under Emilie's direction, the society introduced a course to prepare women for pharmaceutical studies. The course, which included only a bare minimum of Latin study, would enable women to pursue a productive career "without leaving the strictly feminine course of education.[96]

As things stood, Emilie observed, any intellectual achievements on the part of individual women tended to die with them, for learned women rarely married. Through her eyes, this situation took on a tragic cast: "The career woman typically remains unmarried, and all her hard-won gains are lost to the public; she is an abandoned advance post, who bravely stands her ground but who will fall victim to the enemy in the end, whose life's work will be destroyed with her, of whose store of thoughts and experiences no fertile seeds will fall into the receptive souls of children and there spring to life anew."[97] It need not be so, she argued. In ancient and medieval times, women were the authorities on the curative powers of herbs and extracts, and these remedies were "inherited from mother to daughter."[98] Here was a mode of intellectual inheritance in which women could participate without straining against the limits of the "average" or upsetting the social order. As a creative woman, it was understandable that she should seek a matrilineal component of intellectual inheritance. If men could achieve immortality through their ideas, she implied, why should women not transfer a morsel of knowledge to the next generation, provided they did so within the domestic sphere?

Overall, Emilie's reliance on the concept of the average woman was precarious. On one hand, she adopted the universalist language of statistical probabilities; on the other, she denied that women could be universal. Much as liberal educators at the university used the rhetorical figure of the "universal man," Emilie invoked the average woman to smooth out the diversity of the female population she sought to educate. Yet she insisted that women's allure lay in their inborn idiosyncrasies, in their stubborn resistance to the universal ideal.

95. Emilie Exner, *Weibliche Pharmaceuten*, p. 15.

96. Ibid., p. 12.

97. Emilie Exner, *Emancipation*, p. 54.

98. Emilie Exner, *Weibliche Pharmaceuten*, p. 5

THE reforms that the education ministry finally approved for the medical faculties in 1902 were a compromise that satisfied no one. The exams would be divided into two parts, spaced by up to three years, and they would be reoriented to test practical as well as theoretical knowledge. The various theoretical fields that formerly fell under natural history were grouped into a single course titled General Biology. General Biology was a diluted version of the old ideal of the "universal" power of philosophy to fertilize the sciences. As Emilie Exner put it, one could be grateful that the resolution had made things "only a little bit worse": "The examination regulation, with all that went with it, ruled our winter, and did not make it particularly enjoyable. Now after countless waverings the matter is halfway resolved, and only a little bit worse than before; people are a little richer in bitter experiences, and they are finally making an effort for once to bring it all to closure."[99] The compromise fell short of an endorsement of liberal ideals, but it also failed to institute truly progressive reforms, such as a stipend for interns.

Why did these reform efforts fail? The answer seems to lie in the incompatibility of the liberals' model of education with the real conditions of the Habsburg universities at the turn of the century. Liberals like Sigmund Exner continued to insist on the possibility of translating experiential knowledge into a "universal" form. They continued to claim that higher learning was a process of intellectual "inheritance," capable of producing "universal" thinkers. Meanwhile, universality had become in many eyes a questionable virtue. With ever growing aggressiveness, nationalists argued that education should constitute the transfer of a *particular* heritage. Women like Emilie Exner inverted the value of universality by casting idiosyncrasies as charms. Progressives like Hatschek suggested that the ideology of universality was the barrier to making higher education work in favor of social mobility. The very epistemology that liberals used to legitimate their authority had now fallen under attack. With the growing diversity of backgrounds among students at the universities, it became easy to doubt that *experience* could ever be a basis for "universal" knowledge.

WOMEN'S LEGACIES: A CODA

Particularity and universality, aristocracy and *Bürgertum*: perhaps the most famous Viennese treatment of these themes was the 1908 novel *Der Weg ins*

99. Emilie Exner to Marie von Ebner-Eschenbach, 27 March 1902, folder 81082Ja, Stadtarchiv, Vienna.

Freie (The Path into the Open) by Arthur Schnitzler, a Jewish graduate of Vienna's medical faculty. Schnitzler's story was an inversion of the *Bildung* narrative of the 1850s. The evocation of nature and freedom in the title was profoundly ironic. The main character's encounters with "chance" leave him unchanged, as weak-willed at the end as at the start. He tries repeatedly to understand the principal "accident" of the narrative, the stillbirth of his illegitimate child, within a probabilistic framework: so many babies per year *must* die at birth, he tells himself. In the postliberal world of the novel, probability stands not for freedom but for the hand of fate and the erosion of moral responsibility.[100]

In the same month that Schnitzler's novel began appearing in the *Neue Rundschau,* Emilie Exner was putting the finishing touches on her own meditation on Jewish assimilation and the relationship of bourgeois to aristocratic culture. This was an intimate biography of Vienna's most brilliant salonnières, Josephine von Wertheimstein and her recently deceased daughter Franzi.[101] The Wertheimsteins were a real-life version of the Ehrenbergs, the wealthy, cultured, Jewish family whose salon is the site of much of the intrigue in *Der Weg ins Freie*. These two stories, one by a Jew and one by a woman, each problematized the cultural ideal of universality. Yet Emilie suggested privately that her story was nothing less than a refutation of Schnitzler's.

Like so many of Schnitzler's non-Jewish readers in Vienna, Emilie was quick to disparage the novel:

> It is well observed but no work of art; besides, the society is, with one exception, so unlikeable, that I did not even once feel anything. The treatment of the Jewish question by a Jew is interesting. We've all discussed this theme ever so often with [Josef] Breuer [a Jewish colleague], until his sensitivity made touching on such questions impossible. And yet one still denies that racial peculiarities exist. But if Schnitzler's book counts as a novel of Vienna, then, in the name of the circle of *Bildung* and intelligence and standing on all fine *bürgerlichen* traditions, I would like to protest strenuously. A decadent, affected, coffee-house morality is still not Vienna.[102]

At stake in Emilie's memoir of the Wertheimsteins, then, was the authority to define who and what Vienna *was* at this transitional moment in the city's history.

100. Arthur Schnitzler, *Der Weg ins Freie* (Frankfurt: Insel Taschenbuch, 2002), esp. pp. 341–42, 349–53, 440–42.

101. On the Wertheimsteins, see Ernst Konau, *Rastlos zieht die Flucht der Jahre—: Josephine und Franziska von Wertheimstein, Ferdinand von Saar* (Vienna: Böhlau, 1997).

102. Emilie Exner to Marie von Ebner-Eschenbach, 25 July 1908, folder 81082Ja, Stadtarchiv, Vienna.

Josephine von Wertheimstein had been born into the Gomperz family, elite Viennese Jews who exemplified the nineteenth-century union of *Bildung* and *Besitz*.[103] Emilie traced the origins of Josephine's salon to the liberal intellectual atmosphere of Vienna in the 1840s. In those days, Josephine's home was the meeting place for men like Anton von Schmerling and Emilie's uncle Alexander von Bach, soon to be what Emilie called the "leaders of the masses." The chaos of the revolution of 1848 disrupted this social circle, but Josephine's salon resumed once order was restored. Indeed, as members of the educated middle class were given positions in the new government, Josephine's salon took on a more public role: "[S]o their circle too was drawn out of the narrow confines of the salon more into public life [die Öffentlichkeit] and, from then on, formed a characteristic feature in the physiognomy of the city."[104] In Emilie's hands, the Wertheimstein women became representatives of the fate of Vienna's *Bildungsbürgertum* over the course of the nineteenth century.

Emilie assured her readers early on that the Wertheimsteins were not common Jews. Remarking on the similarity between Franzi von Wertheimstein and her father, Emilie noted that the latter "looked almost just like an English diplomat of Disraeli's stamp, at least not like the top official in the Rothschild bank."[105] This was an observation astutely coded to the casual anti-Semitism of turn-of-the-century Vienna. Despite their Jewish and commercial background, Emilie drew the Wertheimstein women with all the grace, taste, wit, and sensitivity of the most refined countess. "In Josephine all qualifications were united in the highest degree. What the nobility acquired through age-old traditions of confidence and grace she possessed through the intuition and adaptability [*Anpassungsfähigkeit*] of a brilliant nature."[106] "Adaptability" was a virtue in a milieu that demanded assimilation. It also suggested a social Darwinist explanation of the ascension of the *Bürgertum*, a view Emilie frequently espoused. From this perspective, the bourgeoisie's adaptability was matched by the anachronistic and degenerative character of the aristocracy.[107]

103. Schorske, *Fin-de-Siècle Vienna*, pp. 298–89.

104. Emilie Exner (Felicie Ewart, pseud.), *Zwei Frauenbildnisse. Erinnerungen* (Vienna: self-published, 1908), p. 32.

105. Ibid., p. 14.

106. Ibid., pp. 33–34.

107. See, for example, Emilie Exner, *Emancipation*, pp. 32–33.

Still, Josephine was by any measure a rarity among bourgeois women. While Emilie's "average woman" devoted herself to her family, Josephine suffered a loveless marriage, focusing her full capacity for friendship and seduction instead on her guests. She and her daughter had little interest in the hopes and hardships of women at large, and Emilie, despite her activism, forgave them: "Through the goodwill of fate they had both been raised so far above their sisters that the troublesome path through rocks and thorny thicket, which may or may not have led to the goal, seemed to them not worth the sacrifice. Too much effort was necessary for the march, too much grace and elegance was lost in the struggle against wind and foul weather." Emilie excused the Wertheimsteins for straying from the duties of the average woman because they seemed not entirely of this world. They were "hothouse flowers,"[108] enduring everyday reality only in small doses. As Josephine flattered her good friend, the novelist Ferdinand von Saar, "You have the right to ennoble even the common [*dem Gewöhnlichen selbst den Ritterschlag zu verleihen*] and lift it into the ideal."[109]

As she linked the glory of the Wertheimstein women to the golden age of liberalism in Vienna, Emilie turned her memorial for mother and daughter into a tribute to a lost era. Symbolic of these former times was the Wertheimstein family home. The family lived in the hilltop suburb of Döbling, in a palatial classical villa adorned with frescoes by Schwind and paintings by Makart and Lenbach, who had themselves been frequent guests in this house. Like the buildings of the university or the physiology institute, the Wertheimstein house expressed the ethos of historicism. In Emilie's words, it was a "work of art," "built on the traditions of an even more dazzling time, which I know of only second-hand." Emilie's description of the house was an attempt to hold on to "the outlines" of times gone by: "The quick succession of events naturally produces the desire to capture what has been lost at least in pale outlines."[110] Josephine's salon had mirrored the history of *Bildungsburger* power in Vienna; now, in 1908, it was a relic of a bygone age.

With Franzi's death, the house would now be donated to the city. Emilie's view of its future was bittersweet. She was sure that "unknown mothers and children" would enjoy the grounds and that "those in need of education" would profit from the library. The house would thus be a resource for public education, but would the public understand its legacy? Emilie's memoir

108. Emilie Exner, *Zwei Frauen-Bildnisse*, p. 10

109. Josephine Wertheimstein to Ferdinand von Saar, 11 Feb. 1882, quoted in Rossbacher, *Literatur und Liberalismus*, p. 24.

110. Emilie Exner, *Zwei Frauen-Bildnisse*, pp. 8–9.

pointed to the shifting border between public and private life for the Viennese educated elite. After 1848, the Wertheimsteins' salon had been, as Emilie wrote, drawn into "public life"; it had become part of the public character of the city. For a generation, this "public" was entirely dominated by a bourgeois elite. From 1848 through the 1870s the public and private spheres of Vienna's *Bildungsbürgertum* merged indistinguishably. By 1908, however, Emilie characterized the "public" differently. The new "public" was made up of strangers, "those in need of education"—those to whom the liberals' own reforms had given a voice in the political domain and a right to education. Emilie could not suppress the "melancholy feeling" that the house's transformation amounted to the "collapse of a truly unique, singular work of art."[111]

Images of the Wertheimsteins haunted Emilie after Franzi's death. In the Istrian town of Lovrana, where she retreated to tend to her own ill health, their villa overlooking the Danube rose for her in the midst of the Mediterranean landscape. In writing their lives, Emilie hoped to preserve these ephemeral impressions: "Perhaps I will succeed in capturing the exotic scent of these hot-house plants in words." The images interspersed throughout the text—stiff portraits and photographs of empty rooms—only emphasized the gap between experience and representation. Emilie too acknowledged this distance. She and her friends carried their memories of the Wertheimstein women "in their hearts, as a precious possession, rescued from a poetic, noble past."[112] Precious as these memories were, they could not be passed on to others. Emilie forced her reader to wonder whether any of the media her memoir incorporated—art, architecture, or literature—was an effective memorial. As a literary exercise, her portrait of the Wertheimsteins (subtitled "Reminiscences") was an exploration of the power and workings of memory itself. The effort to capture half-conscious knowledge and visual memories in words was what linked Emilie's memoir to the psychological investigations of her husband. As she expressed in the opening pages, memories were the force that preserved the components of identity. This was a motif that integrated empiricist psychology and liberal values. Mill had expressed the same sentiment by calling memories the fabric of the self; in a political vein he wrote that it was "the privilege and proper condition of a human being . . . to use and interpret experience in his own way."[113] Written in the clutch of illness, this book was Emilie's own bid for immortality. And

111. Ibid., p. 8.

112. Ibid., pp. 8, 10, 91.

113. J. S. Mill, *On Liberty* (Indianapolis: Hackett, 1978), p. 55.

it did not go unheeded. Despite her conviction that women's legacies were best confined to the home, Emilie would be memorialized after her death for her literary oeuvre, and another woman would pen the tribute.[114] The power of language won this for her, but it was a language not of universality, but of the personal, the local, the particular.

114. Marie von Ebner-Eschenbach, "Exner, Emilie." Another obituary for Emilie was penned by the moderate feminist activist Marianne Hainisch in *Der Bund,* 4 May 1909, pp. 1–2.

CHAPTER SIX

The Education of the Normal Eye

Visual Learning circa 1900

IN THE summer of 1901 Emilie Exner sat at her desk at Brunnwinkl, pen in hand, peering up at Michelangelo's Sistine ceiling. The photograph, plastered to the ceiling of the cottage, had been a gift from Marie von Ebner-Eschenbach, recently back from Rome. Emilie was writing to express her thanks. She was also reflecting on her plan, as the newly elected president of Vienna's Society for the Gainful Employment of Women, to reform the city's Drawing School for Women and Girls. What the school needed, she had decided, was a "modern" makeover. Rote copying and stale conventions would make way for direct observation and a fresh, unmannered style. "It is remarkable," Emilie wrote of the Michelangelo, "how much one delights in something really good in daily interaction, without any aesthetic élan, almost unconsciously."[1] This, she made clear, was the best way to tutor young eyes.

Wherever one turned within the cottages at Brunnwinkl, one found models for viewing the natural world beyond the walls. Next door to Emilie, Serafin's wife Frederike had painted their cottage with scenes of her husband's voyages to the Orient and tropics. In the old miller's house, Marie Exner von Frisch displayed her watercolors of local landscapes, rendered in the naturalistic style in which better known painters had captured the beauty of this region in previous decades. A watercolor painted by Gottfried Keller on a visit to the Exners recalled a lesson from the Swiss writer's

1. Emilie Exner to Marie von Ebner-Eschenbach, 16 Aug. 1901, folder 81082Ja, Stadtarchiv, Vienna.

autobiographical novel *Der Grüne Heinrich*. Heinrich learned to forsake the facile conventions of landscape painting, to paint truly "from nature," and finally to move beyond "the look of common verisimilitude" toward an idealized realism. This last stage corresponded to what the Austrian novelist Ferdinand von Saar approvingly called in a letter to Ebner-Eschenbach "a healthy Real-Idealismus."[2] *Real-Idealismus*, as critic Karlheinz Rossbacher has recently observed, is a fitting description for the aesthetic preferences of the liberal Viennese of Sigmund's generation. Rossbacher associates the rise of this literary style with the golden age of Austrian liberalism in the 1870s. As Saar used the phrase in 1895, *Real-Idealismus* indicated a style equally opposed to "crass" naturalism and to modernist experimentation like that of the Jung-Wien circle.[3]

Keller and Ebner-Eschenbach won admiration at Brunnwinkl for their talent for verisimilitude. Sigmund Exner once wrote to Ebner-Eschenbach while traveling in Hungary that he felt as if he recognized the passersby. Characters from her stories, he explained, "float so vividly before my eyes . . . ; I delight in this beautiful possession of my imagination and feel the need once again to thank you for enriching my soul [*die Bereicherung meines Ich*]."[4] Sigmund seemed almost to prefer these fictions to the realities he encountered. His wife Emilie praised Keller's *Grüner Heinrich* on similar grounds. "How beautiful Grüner Heinrich is," she exclaimed in a letter to Ebner-Eschenbach, "and how gray and comfortless the life from which he grew. How unsympathetic most of the trades, the struggle to feed mother and sister, must have seemed. Next to that the majesty of [Keller's] poetry." Emilie read the novel as a tale of transcendence over the meanness of everyday life. In her eyes, Heinrich himself became "beautiful" through his search for beauty. An artist's highest goal, she implied, was to observe the world unflinchingly and draw out its hidden beauty. As Sigmund once explained to Ebner-Eschenbach, "I too have often declared myself an incurable idealist, not because I hope to see my ideals realized, but rather because

2. Saar to von Ebner-Eschenbach, 9 Feb. 1895, quoted in Rossbacher, *Literatur und Liberalismus*, p. 23.

3. Ibid. Rossbacher argues that Austrian realists like Ebner-Eschenbach shared with post-1848 realists in Germany the goal of verisimilitude. Unlike the Germans, however, they focused more often on characters outside of the bourgeoisie out of a resistance to current theories of environmental determinism.

4. Sigmund Exner to Marie von Ebner-Eschenbach, 10 Sept. 1910, folder 61028, Stadtarchiv, Vienna.

FIGURE 7a, 7b Watercolors of the Salzkammergut by Marie Exner von Frisch and Gottfried Keller. Sources: Hans Frisch, *50 Jahre Brunnwinkl*, frontispiece; Irmgard Smidt-Dörrenberg, *Gottfried Keller und Wien* (Vienna: Museumsverein Josefstadt, 1977), p. 25.

only on the basis of these will ideals be awakened in others."[5] "Ideal realism" is an apt description of the Exners' tastes.

Brunnwinkl's male scientists made these aesthetic models their own. As Karl Frisch would later write, it was the scientist's aim to describe nature "without any attempt to ornament the poetry of reality with fantasy."[6] With admiration for nature's humble beauty, Sigmund Exner wrote in an account of an optical experiment of a "small landscape framed by the microscope mirror . . . the white brick columns of the barn facing one of my windows, its red-tiled roof and brown planked walls . . . each of the delicate

5. Sigmund Exner to M. von Ebner-Eschenbach, 23 March 1896, folder 127.127, Stadtarchiv, Vienna.

6. Karl Frisch, *Aus dem Leben der Bienen* (Berlin: Springer, 1927), vi, signed "Brunnwinkl, Easter, 1927."

branches of a small plum tree against the blue sky."[7] This scene could have emerged whole from a page of Keller's or a Salzkammergut landscape. The same realist impulse was evident in various branches of the Exners' turn-of-the-century scientific pursuits. Together Serafin and Sigmund Exner founded the world's first archive of "phonograms," wax-plate recordings of human voices and other sounds. The brothers intended these artifacts to serve anthropological research, but they also prized them as faithful traces of individual voices, preserved for future generations.[8] Meanwhile, in 1895, Serafin Exner confirmed and publicized what was perhaps the most celebrated breakthrough yet in realist representation: x-rays.[9]

The Exners' was by no means a naïve version of realism. The photograph in figure 8, for example, is a view through a window, seen through the many-faceted eye of a firefly. Unlike the eyes of most insects, a firefly's eye remains in a lifelike state after being removed. To create this image, Sigmund Exner arranged a camera and a microscope with 120-power magnification just behind the retinal opening, where an image would ordinarily form. Decades after the photograph was published in 1891, Sigmund's nephew, biologist Karl Frisch, found the view at once disorienting and familiar: "You recognize the shape of the window, the window frame, you see the letter R pasted to a window pane and, admittedly somewhat blurred, a church tower farther in the distance."[10] In keeping with Sigmund's empiricist theory of vision, the scene hints that seeing is not a passive reception of information but instead an active interpretation of a visual language. The letter *R*, pasted to the window as a test figure (and reversed by the process of magnification), reminds us that we are "reading" the image. The window's frame is not a transparent opening onto the outside world but a murky boundary to our field of vision. It is a reminder of the mediation of our visual apparatus in our

7. Sigmund Exner, *Physiologie der facettierten Augen von Krebsen und Insekten* (Leipzig and Vienna: Franz Deuticke, 1891), p. 37

8. "Bericht über die Arbeiten der von der kaiserl. Akademie der Wissenschaften eingesetzen Commission zur Gründung eines Phonogrammarchives," *Anzeiger mathematisch-naturwissenschaftliche Klasse der österreichischen Akademie der Wissenschaften* 37 (1900), supplement pp. 1–6; Veronika Hofer, "Physiology Gains Space. On the Meaning of Sigmund Exner's Founding of the Phonogrammarchiv," lecture delivered at the History of Science Society Meeting, Denver, 2001.

9. Wilhelm Röntgen, Serafin's friend from his days at Kundt's laboratory in Zurich, notified him soon after producing his first image, and Serafin arranged to test the method at his institute. Among the observers was the son of the editor of the Vienna paper *Die Presse*, which announced the news on 5 January 1896 with a front-page article titled "A Sensational Discovery."

10. Karl Frisch, *Aus dem Leben der Bienen* (Berlin: Springer, 1927), 64.

experience of the world.[11] The shadowy church spires and the grainy texture might have reminded a contemporary of the Impressionists' experiments in recording their subjective impressions of color, light, and shadow. And yet this image *is* a photograph. Unlike the heightened effects of the Impressionists' landscapes, this is an unmediated record of light passing through organic and inorganic lenses. As Sigmund noted in a caption, the distance of the window from the eye was 225 cm, that of the church from the window 135 steps. The entire picture could be reconstructed using geometrical optics, as indicated by Sigmund's physical analysis of the compound eye.[12]

In the context of the 1890s, the photograph challenged the antirealist position of Sigmund Exner's opponents in the new field of experimental psychology. Ewald Hering and his followers sought an account of perception in terms of pure sensation, rejecting psychological research that began with physics, like that of Sigmund and his teacher Helmholtz. In this context, the photograph is a witty comment on the putatively solipsistic nature of vision. After all, it succeeds in showing us the world quite literally through someone (something) else's eyes. As Karl Frisch noted, the wonder of the picture is that it is so recognizable.

Within this realist framework and in the tradition of Fechner, Hermann Lotze, and Helmholtz, the Exners understood visual representations as symbols that "call forth visual memories in the viewer of that which they signify." Writing at the dawn of aesthetic modernism in the 1880s, Sigmund Exner had insisted that paintings should invite viewers to match their own memories to the common experiences represented on the canvas. It was the artist's task to "interact [*verkehren*] with the viewer through a language that speaks to the eye."[13] The verb "verkehren" (to associate or traffic with) underlined the social aspect of communication through and about visual representations. The universal language of art was of necessity "subjective." As Sigmund reasoned, "It lies in the nature of the matter that if the artist depicts what he sees, and the viewer demands to be reminded of what he has seen, then various subjective visual phenomena creep into the art work.

11. In *Techniques of the Observer* (Cambridge: MIT Press, 1990) Jonathan Crary recounts the nineteenth-century fascination with the physiology of vision, detailing a shift from an early modern conception of vision as a form of direct access to the world, to a quintessentially modern sense of vision as a complex process mediated by the body.

12. With the help of his brother Karl, a specialist in optics, Sigmund explained the geometry of insect vision in the first chapter of his *Physiologie der facettierten Augen von Krebsen und Insekten* (Leipzig and Vienna: Franz Deuticke, 1891), where this photograph was published.

13. Sigmund Exner, *Physiologisches und Pathologisches in den bildenden Künsten,* pp. 4–8.

FIGURE 8 View through a firefly's eye. Source: Sigmund Exner's *Physiologie der facettierten Augen von Krebsen und Insekten* (Leipzig and Vienna: Franz Deuticke, 1891), frontispiece.

This is not only natural but also entirely justified, as long as these subjective phenomena befit *all* people." Here "subjective" indicated not a personal idiosyncrasy but an experience shared by subjects who possessed a common sensory apparatus and faculty of reason. As in the empiricist philosophy taught at the Austrian *Gymnasium,* aesthetics was a matter of empirical investigation to be decided by consensus.[14] Consistent with this empiricist

14. Kurt Blaukopf, "Von der Ästhetik zur 'Zweigwissenschaft.' Robert Zimmermann als Vorläufer des Wiener Kreises," in *Kunst, Kunsttheorie und Kunstforschung im wissenschaftlichen Diskurs. In*

view, Sigmund applauded artists who mimicked the perceptual habits of normal vision; they were, in effect, skilled psychologists. It was another matter entirely to bring "subjective phenomena into art when these are not universally human, but rather limited to certain individuals." No one was guiltier of this transgression than the French. Impressionists and pointillists transmuted visual pathologies into painterly tricks. "When I walk through a modern art exhibition I am always tempted to make diagnoses of the so-called state of refraction of the eyes of each painter . . . It seems in fact that the manner in which the near-sighted painter sees the world . . . has grown into a fashion in modern painting . . ." Earlier masters had managed to overcome their visual idiosyncrasies. Raphael, for instance, had painted even the skinniest trees far in the distance in exquisite detail. "There must have been near-sighted people in those days as well, but they seem not to have had the boldness to let the errors of their eyes come to light in their paintings." By publicizing their pathologies, Sigmund charged, modern artists threatened to "injure the *normal eye*."[15]

Modern artists also seemed to Sigmund to rely too heavily on techniques for heightening reality. Pointillism, for instance, exploited the fact that the binding agent in mixed paint reflected a fair amount of white light. For that reason, colors appeared more saturated when "mixed on the retina instead of on the palette." Sigmund thought pointillism useful when applied with the utmost moderation to parts of landscapes that were hit by sunlight. Elsewhere, he found the method "wholly superfluous"; it made "only the impression of peculiarity."[16]

In short, Sigmund Exner insisted it was the duty of the artist to compensate for his own visual idiosyncrasies—through eyeglasses if necessary, but, more importantly, through self-discipline. Leave solipsistic fantasies to the French, Sigmund implied. What Austrians needed was a language of transparency.

This idealistic brand of realism ran like a thread through the Exners' aesthetic, scientific, and political commitments. In the arts it marked their taste for scenes of a humanized, *gemütlich* natural world. Scientifically, it characterized their affinity for appeals to the "probable" and "possible," for

memoriam Kurt Blaukopf, ed. Martin Seiler und Friedrich Stadler (Vienna: öbv & hpt, 2000), pp. 35–46.

15. Sigmund Exner, *Physiologisches und Pathologisches*, p. 10.

16. Sigmund Exner, *Studien auf dem Grenzgebiete des localisirten Sehens* (Bonn: Emil Strauss, 1898), pp. 168–71.

methods of approximation, and for heuristic models like that of the networked brain. And politically, it distinguished their utopian strain of liberalism. While there was nothing uniquely Austrian about the broad program to bridge realism and idealism, this set of resonances was distinctive. A generation earlier in Prussia, Sigmund Exner's teacher Helmholtz had forged a different link between realism in science and in art. In the aftermath of the failed revolution of 1848, liberal Prussians turned from utopian dreams of a new society to hard-headed plans for achieving unity through industrial and economic reform.[17] Among them, as Timothy Lenoir shows, were physicists and physiologists like Helmholtz and Du Bois-Reymond and physiologically minded artists like Adolph Menzel.[18] These allies wove together "political pragmatism" with realism in science, philosophy, and art. Helmholtz's deterministic realism in the sciences thus went hand in hand with a resolute antiutopianism in politics and a celebration of the man-made landscape in the arts.

Austrians accentuated their idealistic tendencies to distance themselves from this Prussian culture. When Hugo von Hofmannsthal schematized the differences between Prussians and Austrians in 1917, he stressed the Austrians' fondness for natural beauty against the Prussians' attachment to their built environment, the Austrians' emphasis on the private against the Prussians' business sense, and the Austrians' unique capacity for empathy.[19] Recognizing this emerging strain of national self-consciousness, we can see how the idealistic inflection of the Exners' realism marked distinctly Austrian-liberal values. The relations mediated by realist paintings—between artist and viewer or among viewers—were a means of avoiding solipsism without sacrificing autonomy. These "interactions" thus embodied a liberal ideal of sociability. At a time when nationalist movements threatened to fracture the multilingual Habsburg Empire, this was a vision of art as political unifier.[20] For the Exners, aesthetic realism epitomized a transparent language, an incarnation of the linden's communicative ideal.

17. Sheehan, *German Liberalism,* chapter 6.

18. Timothy Lenoir, "The Politics of Vision: Optics, Painting, and Ideology in Germany, 1845–95," in *Instituting Science* (Stanford: Stanford University Press, 1997).

19. Hugo von Hofmannsthal, "Preuße und Österreicher. Ein Schema," in *Gesammelte Werke. Reden und Aufsätze II (1914–1924),* ed. Bernd Schoeller (Frankfurt: Fischer Taschenbuch Verlag, 1979), pp. 459–61.

20. Schorske (*Fin-de-Siècle Vienna,* p. 237) argues that the Habsburg state embraced modern art circa 1900 as a unifying force.

THE EXNER WOMEN LEARN TO SEE

Twenty years after its founding, Brunnwinkl seemed to have achieved an aesthetic consensus. With the new century, however, arrived the first stirrings of dissent. The youngest Brunnwinklers were just then reaching adulthood. Karl and Felix were embarked on their scientific careers, Robert and Alfred were budding doctors, and Franz and Hans would soon be jurists. Emilie was pleased to see her daughter Ilse married. "Now she runs the house, cooks and shops independently," Emilie wrote her friend Ebner-Eschenbach, "with the same enthusiasm with which she used to sketch figures."[21] It seemed natural that Ilse should give up her artistic career and the studio she shared with her cousins Nora and Hilde to devote herself to her new household duties. The family could only hope that Nora and Hilde would soon follow Ilse's good example.

Yet Nora and Hilde would not let themselves be distracted from their artistic pursuits. In 1901 the cousins enrolled at the Vienna Arts and Crafts School, which had just opened its doors to women the previous year. This institute had been founded in 1868, under the ascendancy of Austrian liberalism, to provide a "more refined education of the taste for form and color."[22] Its first director, Rudolf von Eitelberger, had been hired to teach art history at the University of Vienna in the 1850s, making it the second university (after Berlin) to establish a chair in that discipline. Eitelberger intended to make art history an empirical field, founded on close observation of exemplary works of art, just as Franz Exner hoped to free philosophy and psychology from their former speculative tendencies and establish them on an empirical basis. On becoming director of the Museum for Art and Industry and the attached teaching school, Eitelberger had promised to guide society's tastes "on the right track."[23] Through exhibitions and the training of artisans, he aimed to establish an aesthetic unity among everyday objects that would transcend the empire's divisions of class and nationality. Eitelberger's advocacy of a universal aesthetic was similar to the rhetoric of French artisans during the Second Empire. The French likewise struggled to reconcile the demands of individual taste with a new understanding of

21. Emilie Exner to Marie von Ebner-Eschenbach, 15 July 1902, folder 81082Ja, Stadtarchiv, Vienna.

22. Quoted in Gottfried Fliedl, *Kunst und Lehre am Beginn der Moderne: Die Wiener Kunstgewerbeschule, 1867–1918* (Salzburg: Residenz Verlag, 1986), p. 80.

23. Ibid.

national style, one now defined not by the court but by the bourgeoisie.[24] In its quest for universality, the Arts and Crafts School soon commissioned a textbook on color theory from Ernst Brücke, Sigmund Exner's mentor in physiology.[25] The alliance between science and art continued with Sigmund himself, who lectured at the Museum for Art and Industry in 1882. His address paid tribute to aesthetic realism and defended stylistic continuity.[26] It was an argument calculated to appeal to Eitelberger's institution, where Renaissance art was virtually the definition of good taste, and the goal of a universal aesthetics was paramount.

This atmosphere changed abruptly in 1899, when the Arts and Crafts School came under the control of the Vienna Secession, a group of artists who had broken with the Academy of Fine Arts in 1898 in order to experiment with new styles. An article in the Secession's journal *Ver Sacrum* ("Sacred Spring") described the change of administration as the "storming" of the Arts and Crafts School.[27] One of the first reforms of the school's new director, Lucien Myrbach, was to accept female students. Although the school had been the first state-sponsored vocational institute to admit women at its opening in 1868, it had been effectively closed to women since the 1880s, when the administration decided to dam up the "flood of female pupils" in order to be able to focus on "the proper task of the school, namely, to train male forces for the varied needs of the arts and crafts."[28]

By the time Nora and Hilde Exner arrived in the first wave of female students, Myrbach had recruited to the faculty such rising stars as the Secessionist artists Alfred Roller, Kolomon Moser, and Josef Hoffmann. Roller had just returned from a tour of foreign art schools, which had left him deeply impressed with the teaching methods of the English arts and crafts movement. He praised the English educators for allowing the student's "individuality" to unfold fully without losing sight of the goal of "functional

24. Leora Auslander, *Taste and Power* (Berkeley: University of California Press, 1996), pp. 179–85.

25. Ernst Brücke, *Die Physiologie der Farben für die Zwecke der Kunstgewerbe auf Anregung der Direktion des kaiserlichen Oesterreichischen Museums für Kunst und Industrie* (Leipzig: Hirzel, 1866). On this textbook, see Lenoir, "Politics of Vision," pp. 161–65.

26. Sigmund Exner, *Physologie des Fliegens*. See chapter 3.

27. L. Hevesi in *Ver Sacrum*, quoted in Manfred Wagner, *Alfred Roller in seiner Zeit* (Salzburg: Residenz Verlag, 1996), p. 189.

28. Quoted from an official act in Barbara Doser, "Das Frauenkunststudium in Österreich, 1870–1935" (Ph.D. diss., Leopold-Franz-Universität, Innsbrück, 1988), quotation on p. 137.

beauty." Inspired by the English, Roller and his colleagues restructured the curriculum around "das Sehenlernen," learning to see.[29]

In Vienna, learning to see meant becoming a child again. At the Arts and Crafts School and at the Wiener Werkstätte, the production arm of the Secession, artists celebrated the carefree spontaneity, playfulness, and natural curiosity of children. Franz Cizek became known the world over for his children's art classes, where young people drew things "out of their heads, everything they feel, everything they imagine, everything they long for."[30] Later on, they would "lose their spontaneity and become ordinary . . . They see too much, they grow sophisticated."[31] Cizek and his colleagues prized the unselfconscious, even naïve character of a child's observations.

On these grounds, Myrbach and his colleagues rejected the curriculum of the art academies of the eighteenth and nineteenth centuries, with its reliance on copying from masterworks. While Rudolf von Eitelberger had set out to make art education "empirical" through direct observation of works of art, Myrbach's school shunned rote copying. In its place Myrbach instituted "Naturstudium," classes where students drew, painted, or sculpted directly from nature.[32] This method had already shaped the distinctive look of *Jugendstil* art in Austria. Austrians had seized on the illustrations of natural forms compiled by German zoologist Ernst Haeckel as models for their designs.[33] Likewise, the Secessionists celebrated the zoologically informed drawings of Hans Przibram, director of Vienna's Institute for Experimental Biology.[34] Echoing this naturalist stance, the Vienna-trained artist and educator Adolf Hölzl called in *Ver Sacrum* for the observation of nature and the search for the "formulae and laws" of sensation.[35]

29. Fliedl, *Kunst und Lehre*, pp. 164–65.

30. Francesca M. Wilson, *A Lecture by Professor Cizek* (Children's Art Exhibition Fund, 1921), p. 4.

31. Francesca M. Wilson, *The Child as Artist: Some Conversations with Professor Cizek* (Children's Art Exhibition Fund, 1921), p. 6.

32. See the review by Vienna art critic and teacher Camillo Sitte of Albert Kornhas, *Das Zeichnen nach der Nature*, in *Centralblatt für das gerwerbliche Unterrichtswesen in Österreich* 15 (1900), Heft 3–4, pp. 1–3.

33. Roland Franz, "Stilvermeidung and Naturnachahmung. Ernst Haeckels 'Kunstformen der Natur' und ihr Einfluß auf die Ornamentik des Jugendstils in Österreich," *Stapfia* 56 (1998): 475–80.

34. Deborah R. Coen, "Living Precisely in Fin-de-Siècle Vienna," *Journal of the History of Biology* 39 (2006): 493–523.

35. *Ver Sacrum* 2 (1901): 243–54.

Far from rejecting natural science, the Secession was part of a unique symbiosis of science and art in turn-of-the-century Vienna. This relationship was founded on common pedagogical principles. At the Arts and Crafts School, the student was to achieve a heightened consciousness of an object's materiality through independent reflection and direct contact rather than copying. He would learn to recognize purpose in an object's form, to answer the question, "why it is this way and not otherwise, and so must be. Only the student himself must find the answer to this question . . . In this way the student has found the foundational knowledge with guidance from the teacher, that is, [he] has truly made it his own."[36] As at Brunnwinkl, "learning to see" became an exercise in a style of observation that was at once self-reliant and attuned to a dialogue about the material world.

Heady celebrations of youth became a trademark of the Secession movement, and the pages of *Ver Sacrum* bloomed with images of childhood and vernal rejuvenation. These motifs can be traced back to Kolomon Moser's illustrations for children in the mid- to late 1890s.[37] In 1897 he lavishly illustrated an anthology of poems and stories for children by authors including Gottfried Keller and Marie von Ebner-Eschenbach.[38] The editor of this anthology was none other than Felicie Ewart, the pen name of Emilie Exner. As this collaboration suggests, there was a strong affinity between the visions of youth promoted by the Secessionists and by the Exner family.

The design of children's books and toys became another specialty of Vienna's *Jugendstil* movement, and Hilde and Nora Exner had great success with such exercises. In their first year at the Arts and Crafts School, they had taken Alfred Roller's introductory course in figure drawing.[39] Their aunt Emilie called Roller a "strict master," but he gave both young women good marks, and Hilde won one of three prizes in a class of sixty-five.[40] Years later, he would write to Hilde that of all his former students, she had a "special place" in his memory and his wife's. "But also when we speak in all seriousness of those colleagues of whom something good has come and even better

36. Max Eisler, *Österreichische Werkkultur* (Vienna: Kunstverlag Anton Schroll, 1916), p. 19.

37. Marian Bisanz-Prakken, *Heiliger Frühling: Gustav Klimt und die Anfänge der Wiener Secession, 1895–1905* (Vienna: Brandstätter, 1999), p. 63.

38. *Jugendschatz. Deutsche Dichtungen gesammelt von Felicie Ewart* (Vienna: R. v. Waldheim, 1897).

39. Information on their studies is found in the annual "Klassenkataloge" in the collection of the Oskar-Kokoschka-Sammlung, Universität für angewandte Kunst, Vienna.

40. Emilie Exner to Marie von Ebner-Eschenbach, 15 July 1902, folder 81082Ja, Stadtarchiv, Vienna.

awaits, you stand at the front."[41] In 1902–3 the Exner cousins advanced to Kolomon Moser's class in "decorative drawing and painting," where, with classmate Franz Fiebiger, they illustrated a children's alphabet. Their boldly colored animal-themed prints were soon reprinted in *Ver Sacrum*, quite an honor for beginning students. Hilde continued to design children's books, producing a foldout story of Noah's ark with geometrical, silhouetted figures, as well as a fairy tale illustrated with stark black and gold landscapes. She also built a toy "Trojan horse" with moveable soldiers that fit inside a trap door.[42] Nora's most celebrated sculptures depicted adolescents at play. These objects had a pedagogical function as an extension of the aesthetic transformation of everyday life into the world of the child.[43] For the artist, meanwhile, the design of these objects was an exercise in seeing through the eyes of a child.

The Secession was leading what might be called an empiricist revolution in artistic education. Austria's elite educators had recognized early the value of training the eye to observe closely, as suggested by Stifter's link between drawing and *Bildung* in *Der Nachsommer*. As Austrian art teachers "naturalized" their methods, science teachers simultaneously embraced a newly aestheticized model of learning to see. One Viennese *Gymnasium* teacher argued that the value of studying natural history lay largely in the "sensibility for the beautiful and fine," which was stimulated "through observations of [*Hinweise auf*] proportionality, symmetry, through observations of the gradations of colors (for instance in insects, etc.), further through observations of details from the lives of animals."[44] Educators in imperial Germany denigrated drawing classes as too workmanlike for the *Gymnasium* and thus relegated them to the *Realschule*.[45] In Vienna, by contrast, the art student was learning to think like a biologist, just as the student of natural history was learning to view nature as a work of art.

Among those fostering this convergence was Friedrich Jodl, the Exners' colleague in the philosophy faculty, who sat on the board of the Vienna Art

41. Alfred Roller to Hilde Exner, 28 Oct 1906. Collection of Dr. P.C. Dijkgraaf, Den Helder, The Netherlands.

42. All in the collection of P. Dijkgraaf.

43. *Traum der Kinder, Kinder der Träume: Wien um 1900*, exhibition catalog (Vienna: Galerie bei der Albertina, 1984).

44. Alois Heilsberg, "Bemerkungen über modernen Betrieb des naturgeschichtlichen Unterrichtes am Gymnasium," *Jahresbericht des k.k. Staatsgymnasiums im XIX. Bezirk von Wien für das Schuljahr 1902/3* (Vienna: k.k. Staatsgymnasiums im XIX. Bezirk, 1902–3), pp. 29–40, quotation on p. 35.

45. Albisetti, *Secondary School Reform*, p. 24.

School for Women and Girls. Jodl was a committed liberal and empiricist. A future leader of the Vienna Monist League and of the progressive adult education movement, he was somewhat more radical than the Exners. Like them, however, he disputed the existence of absolute laws of nature and rejected appeals to natural laws in jurisprudence. And like Emilie Exner, Jodl was a strong proponent of women's education, which led to his support for the founding of the Art School for Women and Girls in 1897. At the festivities marking the end of the school's first year, Jodl addressed the students: "Take care to look at nature as if no painter before you had ever painted, as if you were the first to glimpse it; take care to see not with the eyes of other painters but with your own eyes."[46] This empiricist principle linked Jodl's views of art and science. Direct observation and independent analysis were the keys to success in either field.

Like Jodl, Emilie Exner was drawn into the world of modern art through her commitment to the cause of women's education. Jodl's school was only one of a number of educational institutions for women founded in late nineteenth-century Vienna, usually by private charities or associations. In her capacity as president of the Society for the Gainful Employment of Women, Emilie oversaw the schools managed by the society, including its Drawing School for Women. Consistent with the society's broader mission, the Drawing School's purpose was the "training of [the students'] taste for the tasks already open to them in all fields."[47] Not by coincidence, the school's first director was the wife of Rudolf von Eitelberger, the director of the Arts and Crafts School. Its classes centered on the drawing of ornaments, flowers and plants, the themes that had been considered appropriate for women since well before the beginning of the nineteenth century. Portraiture and historical painting were reserved for the academies of fine arts, whose students were almost exclusively men.[48] During the 1880s and 1890s, while the Arts and Crafts School severely restricted the number of female students it admitted, the Drawing School grew proportionately in stature. In step with the progression of art nouveau, the Drawing School shifted its focus to crafts and ornamentation. In keeping with the trend in the Austrian arts and crafts movement, from 1892 drawing was taught almost exclusively "from nature."[49]

46. *Jahresbericht über das I. Vereinsjahr 1897/98*, p. 7, quoted in Doser, "Frauenkunststudium," p. 170.

47. Doser, "Frauenkunststudium," p. 134.

48. M. Friedrich, "Versorgungsfall Frau," *Jahrbuch des Vereins für die Geschichte der Stadt Wien* 47–48 (1991–92): 263–308, quotation on pp. 286–87.

49. Doser, "Frauenkunststudium," pp. 134–35.

Emilie was determined to make the Drawing School into a gateway to the Arts and Crafts School for ambitious young women. She therefore looked to the Secessionists for guidance. As she wrote to Ebner-Eschenbach in 1901, "I too find the modern movement [*die moderne Bestrebungen*] for artistic education for children to be thoroughly justified and promising."[50] In 1904 she enlisted Myrbach to inspect the school and produce recommendations for transforming it into a "Vorbereitungsschule für Kunstunterricht," in other words a preparatory school for Myrbach's own institute.[51] The reforms Myrbach subsequently recommended made clear the gap between traditional and modern art study. For the first year course, he advised "naturalistic study from plants, inanimate nature (*not* still life) and small living animals. *No* so-called stylizing [*Stilisieren*]. In the second year, the teaching of life drawing should work only from live models: "*no* anatomy." Following the inspection, Emilie Exner wrote to the education ministry to express her belief in the "necessity" of "reorganizing the society's Drawing School according to the demands of modern drawing instruction, indeed according to the perspective that your young inspector Herr L. von Myrbach recommended for particular consideration. The Drawing School shall serve in the future as an elementary school for those girls who aspire to admission to the Arts and Crafts School and so respond to what, according to gathered opinions, is a real need."[52] Having secured financial support from the ministry, that year the school was able, on Myrbach's recommendation, to hire a student of Kolomon Moser's to teach its afternoon course. The following year Emilie informed the ministry that the Drawing School required "expert superintendence" and that she hoped Franz Cizek would take on that role. With these moves, Emilie had made the first steps toward bringing the Drawing School under the wing of the leading figures of Viennese modernism and ensuring that all doors would be open to its graduates.[53] As she wrote to Ebner-Eschenbach toward the end of her term as president, "The society must either be made into a model of a modern institution, or *collapse*. Naturally that is a gigantic task."[54]

MEANWHILE, Brunnwinkl buzzed with talk of modern art. "I hear much here about the most modern art," Emilie Exner wrote in the summer of

50. Emilie Exner to Marie von Ebner-Eschenbach, 16 Aug. 1901, folder 81082Ja, Stadtarchiv, Vienna.

51. AVA Unterricht 3703: 6790/1904, Österreichisches Staatsarchiv, Vienna.

52. Ibid.

53. AVA Unterricht 3703: 5036/1905, Österreichisches Staatsarchiv, Vienna.

54. Emilie Exner to Marie von Ebner-Eschenbach, 29 July 1905, folder 81082Ja, Stadtarchiv, Vienna.

1902. “Both my nieces and my son-in-law are stuffed full of interests and work so intensely that it is a great joy to me to chat with them.”[55] Others recalled a rather different reception. Adolf Exner’s son Franz wrote of the “large debates” that broke out as Nora and Hilde introduced modern art to the Exner circle, and the “lively battles” that the beginnings of the Vienna Secession unleashed among them.[56]

When Nora Exner arrived at Brunnwinkl fresh from her first year at the Arts and Crafts School, her aunt Marie entrusted her with a small task. Nora was to paint an altarpiece of angels playing music, a familiar Baroque theme paying tribute to one of the family’s favorite joint pastimes and to their good taste in music and art. At first Marie reported happily that Nora was “making it very sweet. She has learned a great deal indeed.” But the following day, she complained:

> Unfortunately Nora ruined the very striking sketch yesterday with her crazy colors. The children were wearing bathing suits, which fell in pretty folds, which she captured in her sketch; but then she covered the entire surface with a red-white checker pattern, the other angel blue-white, so that I gave a cry when I came upon it. What’s more, she raged like a savage in the paint box and uses as much color for a single square-foot surface as I use for ten. To be sure, it has all been done with talent, only her taste seems to me spoiled to the core, despite Italy.[57]

Nora and Hilde were apparently given a chance to redeem themselves in another of the cottages. Once again, however, they left the bare white walls “patterned with garish colors, which were utterly unsuited to the rustic character of the house.”[58]

The family’s scorn for Nora and Hilde’s choice of colors expressed a dense web of anxieties. The late nineteenth-century fashion for vivid colors, made possible by advances in the dye industry, seemed linked to a broader social transformation. The Wiener Werkstätte’s use of color provoked rapture and outrage perhaps in equal measure. The walls of its ambitious Cabaret Fledermaus, unveiled in 1906, were covered in over seven hundred thousand multicolored majolica tiles, accented by crisp black-and-white-tiled floors. Depending on which reviews one read, the room was either “chromatic

55. Emilie Exner to Marie von Ebner-Eschenbach, 15 July 1902, folder 81082Ja, Stadtarchiv, Vienna.

56. Franz Exner, *Exnerei*, p. 22.

57. Karl Frisch, *Fünf Häuser am See*, p. 65.

58. Karl Frisch, *Fünf Häuser am See*, p. 65.

chatter," "a colorful chamber of horrors," or a "formal concord crystallized out of the purpose in hand."[59] The Werkstätte's wallpapers sent at least one critic into chromatic ecstasy, as he "wreathed" himself "in these color chords as though in silk . . . whether it be blue with silver and white, or ochre with gold and broken green . . . Now it is as though the notes of a harp quivered within my ear, soft and tender as though melting into each other, now like the sound of trumpets and drums, strong and assertive, but never harsh in their contrast."[60]

These excesses seemed to some observers to throw into question the very concept of "taste." "The question 'What is Beautiful?' becomes harder to answer," wrote Alois Riegl, a curator and critic at the Museum of Art and Industry, circa 1895. "Everything that exists is beautiful, or at least everything that is colored."[61] Social critics of the day expected that such an aesthetically revolutionized environment would transform its inhabitants, not necessarily for the better.[62] "Woe to the lady who would enter such a room in a dress that was not artistically suitable!" wrote one critic of the Cabaret Fledermaus.[63] "How can they live like parrots in such bright parrot-colors?" quipped Karl Lueger, the leader of Vienna's Catholic conservatives. Lueger blamed this social ill, like most others, on "Jewish bankers," suggesting that behind his disapproval lay fear of the racial and sexual exoticism he perceived in the new fashions.[64]

Emilie Exner, despite her sympathy for the modern art movement, also saw cause for alarm in this new rage for color. She warned of the psychological and moral effects of overexposure to garish dyes and fabrics:

> On fine, warm summer evenings there gathered a few years ago in one of the most majestic parks of our metropolis a sizeable quantity of shimmering silk, flowers, ribbons, and feathers, whose wearers would have been deeply ashamed had they been able to observe their reflection unguardedly [*unbefangen*]. It had become absolutely impossible to distinguish between good society and

59. Quoted in Werner J. Schweiger, *Wiener Werkstätte: Design in Vienna, 1903–1932* (New York: Abbeville Press, 1984), p. 186.

60. Rudolf Haybach circa 1929, quoted in Schweiger, *Werkstätte*, p. 187.

61. Unpublished manuscript, quoted and translated by Wolfgang Kemp in his introduction to Riegl's *The Group Portraiture of Holland* (Los Angeles: Getty, 1999), p. 3. See too Margaret Olin, *Forms of Representation in Alois Riegl's Theory of Art* (University Park: Pennsylvania State University Press, 1992), p. 101.

62. See Silverman, *Art Nouveau.*

63. Karl Ernst Osthaus, 1906, quoted in Schweiger, *Wiener Werkstätte,* p. 222.

64. Quoted in Rossbacher, *Literatur und Bürgertum*, p. 298.

> demi-monde, and a sense of horror gripped the sober [*kühl*] observer at the sight of this regalia, which had gathered there in the street.
>
> Our men have thus become unresponsive to the appeal of girlish, untouched youth, they have had too much make-up, too much ornamentation before their eyes, and colors must grow ever bolder to be able to please their fatigued eyes.[65]

As Emilie saw it, modern fashion's brash colors made it virtually impossible to distinguish between well-born ladies and kept women. She meant this literally. Her essay targeted the problem of prostitution, which, she warned, "grips ever wider circles, it reaches into the highest classes." The scene she painted was carnivalesque not just in its colors and sensuality but in its inversion of social hierarchies. These costumed women had lost the ability to see themselves "unguardedly," without pretense, as the "sober observer" of this scene did. The sight of them in turn spoiled the eyes of young men, making them insensitive to the more muted charms of chaste young women.

The essay in which Emilie Exner sketched this scene took the form of a series of letters from a mother to her family's physician—likely modeled on the Exners' friend and former doctor Josef Breuer, Freud's early collaborator on the treatment of hysteria. In a medical context, Emilie's reference to the "fatigued eyes" of the young men was a sign of what contemporaries called cultural degeneration. Emilie concerned herself in this letter with the cause of "the degeneration of the male youth" of the Italian aristocracy, "which must of course leave its tragic print on the life of the mind." In the sketch of a Vienna street scene quoted above, Emilie implied that her compatriots were just steps behind on the degenerative trajectory of their southern neighbors. By contrast, in the cities of Holland, women were "sincere" and "modest," and prostitution, though admittedly no less rampant, was successfully cordoned off from good society, such that it "did not need to be *seen* by young men and women."[66]

For Emilie, moral corruption was a contagion that passed through the eyes. In the context of the empiricist theory of vision, "visual fatigue" was a pathological state in which "normal" color vision was suspended. As we will see shortly, this was an important explanatory concept, since it accounted for experiences that deviated from the empiricists' model of "normal" vision.

65. Emilie Exner, *Emancipation*, pp. 32–33.

66. Ibid. On the theme of ornament and deviance in Vienna, see Jimena Canales and Andrew Herscher, "Criminal Skins: Tattoos and Modern Architecture in the Work of Adolf Loos," *Architectural History* 48 (2995): 235–56.

In Emilie's usage, visual fatigue was not just a pathological state; it was a psycho-physiological symptom of cultural decay.[67] And it was a specifically male ailment. The eyes of women and young men were equally at risk of corruption, but men were expected to be the connoisseurs of virtue. Male vision set the standard against which to measure aesthetic and moral degeneration.

Emilie's husband Sigmund was equally critical of a trend in the fine arts toward heightened colors, gaudy effects, and the representation of pathological states, of which pointillism was but one manifestation. Painter Gabriel Max, for instance, was popular with members of the Secession for his otherworldly portraits of female saints in the throes of religious ecstasy. With the authority of a physician, Sigmund identified Max's subjects as "hysterics." He insisted that anyone would agree who briefly compared Charcot's photographs of his patients. Charcot's diagnosis of hysteria in women purportedly possessed by demons can be seen as a stridently anticlerical move.[68] Exner's attack on Max was equally hostile to religion, but it was also a defense of sober realism. "A normal person plunged deep in thought does not look like that," he insisted. Max's images captivated viewers as faces that "might appear to one in a dream."[69] Sigmund warned that the public, with its "sensitive nerves," was developing a taste for the "breathtaking," a disturbingly feminine weakness.[70]

In their vague sense that modern art, women's tastes, and social change were somehow linked, Emilie and Sigmund Exner were not far from the mark. The Secessionists were responsible for reopening the Arts and Crafts School to women, and women would outnumber men among the school's students during the First World War. Emilie Exner celebrated this reform—as long as it did not distract young women from the goal of marrying and having children. But Nora and Hilde Exner had other plans. Hilde never married, while Nora's marriage remained childless. Both women pursued their careers right up to their untimely deaths. Even their choice of sculpture over painting was rebellious. There were few women sculptors in Austria before the First World War,[71] and it was considered a particularly difficult

67. For this theme, see Anson Rabinbach, *The Human Motor* (Berkeley: University of California Press, 1992).

68. Goldstein, "The Hysteria Diagnosis," p. 236.

69. Sigmund Exner, *Physiologisches und Pathologisches*, pp. 25–26.

70. On the theme of neurasthenia, gender, and aesthetics in this period, see Silverman, *Art Nouveau*, pp. 79–83.

71. On this topic, see Sabine Plakolm-Forsthuber, *Künstlerinnen in Österreich, 1897–1938: Malerei, Plastik, Architektur* (Vienna: Picus, 1994).

medium for women. As a reviewer of Nora's sculptures would later put it, the plastic arts did not allow women to express "their mood" in the way that landscape painting could. Sculpture required a more sober gaze and the aggressive capacity "to capture movement."[72]

Undeterred by such popular wisdom, in November 1905, Hilde gave notice that she was withdrawing from the Arts and Crafts School in order to pursue her sculpture in Rome. As teenagers she and her sister Priska had traveled to Rome with their father Serafin, an experience that "contributed immeasurably to our education, the love of art and beauty became second nature to us . . . "[73] Since childhood their father had communicated to them his own love of travel, telling them bedtime stories of the voyages of "little Hilderl." Their cousin Nora harbored similar dreams. At sixteen, she had prized Otto Benndorf's gift of an ancient stone that he had excavated himself at Ephesus. Her letter of thanks expressed but one regret: "Had I been a *boy*, I would long ago have accompanied you as an uninvited guest on these interesting journeys."[74]

In 1905, in their mid-twenties, Nora and Hilde finally embarked on a foreign adventure of their own. Hilde found herself a studio in the Via Nomentana, on the edge of Rome, where Nora soon joined her (perhaps without her husband, a physician and artist's son). "When it comes to art, we're eating the Romans out of the kitchen," Nora reported happily. "Hilde and I are also fortunately quite full of ideas and the urge to create."[75] When their cousin Karl von Frisch visited the artists in 1906, he found them to be "vibrant girls, filled with ambition and with their art."[76] Hilde, however, refused to be known as a "woman artist." She declined an invitation to join the Association of Austrian Women Artists (Vereinigung bildender Künstlerinnen Österreichs), founded in 1910, explaining that she would be interested only in joining an organization of "artists, no matter whether men or women." Her reason was, as she put it, "artistic." Male and female artists "all pursue the same goal, even if perhaps their ways and means appear quite different."[77]

72. From a review by critic Arthur Roessler, "Plastiken von Nora von Zumbusch," 1914, quoted in Plakolm-Forsthuber, *Künstlerinnen*, p. 220.

73. Priska Exner Dijkgraaf, "Aus meinem Leben."

74. Nora Exner to Otto Benndorf, 7 May 1895, 666/6, Manuscript Collection, Österreichische Nationalbibliothek, Vienna.

75. Hilde and Nora Exner to Otto Benndorf, 20 Jan. 1906, 640/37-2, Manuscript Collection, Österreichische Nationalbibliothek, Vienna.

76. Karl Frisch, *Erinnerungen eines Biologen* (Berlin: Springer, 1973), p. 22.

77. Quoted in "Nicht nur mit Lippenstift," *Wiener Zeitung*, 9 April 2004.

One reason, then, why modern art became so controversial among the Exners was because of the transformation of gender roles it presaged. The older generation fostered the independence of the young Exner women, but they still feared its consequences. At the root of this conflict, however, lay a broader anxiety, one that emerged when the Exners became engulfed in the great Viennese scandal of the year 1900.

FROM KLIMT'S "PHILOSOPHY" TO THE SCIENCE OF COLOR

Gustav Klimt's mural for the philosophy faculty (frontispiece to the introduction) was to represent the illuminating light of knowledge, a theme that would complement the symbolism of the University's Renaissance façade on the Ringstrasse. To the majority of professors, Klimt's dreamscape was an outrage. Even Friedrich Jodl, who sympathized with the Secession's efforts in the field of art education, condemned the mural's "dark, obscure symbolism."[78] Sigmund Exner signed a petition protesting the mural's installation from an aesthetic standpoint. Similarly, Serafin Exner publicly argued that the mural was objectionable on "purely aesthetic grounds":

> Our proceedings are not directed against the Secession, against the modern art movement; we too can appreciate Klimt as an artist, but we cannot regard this painting as a work of art. We do not find this manner of symbolic representation artistic, aside from the fact that the picture does not suit the architectural character of the university building.[79]

Like Jodl, Serafin objected to the "symbolic" character of the mural, which he contrasted with the university's Renaissance architecture. Where the Exner brothers and their colleagues had expected to find an allegory of enlightenment, they saw no more than the nightmare of a troubled mind.

These scientists were not aesthetic reactionaries demanding a return to Renaissance style. Nor was the rational ideal they felt Klimt scorned a dogmatic standard of absolute certainty. It was the fantastical, dreamlike quality of the work they resisted. The Exners deemed the representation of pure fantasy an unworthy goal for an artist. Sigmund Exner condemned Hugo von Hofmannsthal, the literary giant of modernist Vienna, for precisely this offense. "I too recognize it as valuable when an author, whose work aims to depict a [illeg.] dream, develops this depiction with the virtuosity

78. Quoted in Schorske, *Fin-de-Siècle Vienna*, p. 233.

79. Quoted in Hermann Bahr, *Gegen Klimt* (Vienna: J. Eisenstein, 1903), pp. 23–4.

of Hofmannsthal. I do not, however, find such depictions justified as ends in themselves... But I must admit that H. has some ability, since it was his achievement that last night I slept worse than I had in a long time."[80] Sigmund could admire Hofmannsthal's skill in representing the dark depths of a sick mind, but he could not approve of this use of the writer's talent. The philosophy faculty's objections to the Klimt mural were similarly motivated.

The root of this opposition was best expressed two years later in the pages of Vienna's liberal daily. The author, Alois Riegl, had joined the university's philosophy faculty in 1897. In 1902 he published a radically original study of the early modern Dutch group portrait. As Riegl saw it, this genre was remarkable for its success in establishing a relationship between the viewer and the work of art. Against this standard, Riegl condemned modern artists and viewers alike. His 1902 article in the *Neue Freie Presse* was, as historian Margaret Olin writes, a "critique of solipsism":

> As Riegl saw it, humanity had lost confidence in the shared, communicable "reality" of the visible world. Riegl valued baroque art for its reconciliation of a new "subjective" element with the respectful separation which unites subject and object while preserving their individual identities. Modern viewers, by subsuming everything into themselves immediately and "subjectively," isolate themselves.[81]

For Riegl, baroque portraiture expressed a social as well as aesthetic ideal. Seventeenth-century Dutch society had successfully balanced personal freedom against corporate spirit, and this success was visible in its most characteristic genre of painting. The tendency of modern society toward solipsism, on the other hand, was reflected in modern art. Like Sigmund Exner, Riegl pointed to Hofmannsthal's *Jung Wien* movement as an illustration of this problem. Moreover, Riegl's praise for the Dutch group portrait mirrored Sigmund Exner's account of the workings of art. Exner had described art as an "interaction" between artist and viewer, based on a visual "language" grounded in the experience of a common reality. Like Riegl, Exner made clear that art, like science, should be a conversation about a shared world. As such it had the power to foster an observational attitude that was at once independent and related, resisting both dogmatic conformity and solipsistic skepticism.

Such was the aesthetic ideal of Vienna's liberal intellectual elite, and it was this ideal that Klimt's mural insulted. The Secessionists (from whom Klimt

80. Sigmund Exner to Marie von Ebner-Eschenbach, 16 March 1910, folder 57619, Stadtarchiv, Vienna. It is not clear to which of Hofmannsthal's works Exner referred.

81. Olin, *Forms of Representation*, p. 168.

now distanced himself) were in this respect the professors' allies, not foes. Far from embracing radical subjectivity, Vienna's *Jugendstil* artists prided themselves on their proximity to the natural world. They went so far as to reject the "stylizing" of nature. "In close alignment with nature, striving ahead towards freedom and new independence"—so did the movement describe itself.[82] Crucially, then, the professors' position in the Klimt scandal was not "antimodernist." In fact, much of what Vienna's artists themselves considered "modern," such as direct contact with nature and the imitation of childhood spontaneity, reflected the ideals of the liberal elite. "Modern" artists sought a "freedom and independence" anchored by their attention to the material world.

Still, Klimt's supporters had raised the specter of relativism with their insistence that the new age demanded a new aesthetics. "To the age its art" ran the motto of the Secession. "What Is Ugly?" was the title of a lecture in 1900 by art historian Franz Wickhoff. His answer, ultimately, was nothing—aesthetic norms were the fleeting products of their age.[83] To liberals like Jodl, Riegl, and the Exners, such insistence on the relativity of beauty threatened to cast doubt on the possibility of communication about a shared material world. They faced the challenge of setting aesthetic standards without descending into dogmatism.

SO IT was that in 1900 Serafin Exner visited Vienna's Museum for Art and Industry armed with a photometer. His target was the museum's trove of Persian carpets. Under the lens of his instrument, Serafin would extract the secrets of these richly colored yarns, yielding quantitative values of hue, brightness, and saturation. As a young man, Serafin had journeyed to Persia, intent on studying an exotic climate. Now, in the vibrant patterns woven in that distant land, he saw an answer to the question that had recently riven Viennese society in two: the question of what is beautiful. Perhaps echoing Wickhoff's "What Is Ugly?" Serafin would title his 1902 report on this research "On the Characteristics of Beautiful and Ugly Colors." As his old friend Gottfried Semper had shown, the patterns of these carpets had adapted over the course of hundreds of years to the ultimate arbiter: human taste. How else to explain the appeal of these rugs here, in the bourgeois

82. Quoted in Franz, "Stilvermeidung," p. 478.

83. Schorske, *Fin-de-Siècle Vienna*, p. 232. Wickhoff was a popular lecturer, and Emilie Exner admired his addresses to the Society for the Gainful Employment of Women. (E. Exner to Marie von Ebner-Eschenbach, 24 Jan. 1901, folder 81082Ja, Stadtarchiv, Vienna.)

households of turn-of-the-century Vienna, where it was "absolutely a point of honor to own at least one authentic Oriental carpet"?[84] Ironically, Serafin's quest for beautiful colors would lead to the Orient only by way of a middle-class parlor. This was a hunt for the universal that began and ended at home.

Color theory at the time was a field hotly contested by scientists and artists, physiologists and psychologists, conservatives and modernists, where several competing theories waged battle. The two principal accounts of color perception were attributed to Hermann von Helmholtz and Ewald Hering.[85] Helmholtz, building on the work of James Young, had postulated in the 1860s that three spectral curves describe the sensitivities of a "normal" human eye in the red, green, and blue ranges. He called the peaks of these curves the "fundamental sensations." Although the fundamental sensations themselves lay outside the visible spectrum, any visible color could be expressed mathematically as a linear combination of the three fundamentals. Supposing there to be three types of receptor in the retina, one for each fundamental color, visible colors corresponded to simultaneous excitations of all three receptors in varying degrees.

Against Helmholtz's three-color theory, Hering argued that there must be six fundamental sensations. The sensation of yellow was fundamental, not a composite of red and green. White was likewise "a sensation of its own nature," not an equal mixture of all other colors; and black too was a sensation in its own right, not a mere absence. As in Goethe's theory, Hering's account centered on polar oppositions. What one sensed were the perceived weights between oppositional colors: red/green, blue/yellow, white/black. While Helmholtz measured colors as wavelengths, in Hering's theory colors were specifiable only at the level of sensation.

Hering and other critics argued that Helmholtz's theory hinged on his classification of his experimental subjects. In Helmholtz's terms, his color-blind subjects were dichromats (either "red-blind" or "green-blind"), implying

84. Alois Riegl, *Altorientalische Teppiche* (Leipzig, 1891), p. iv. Riegl was the curator of the museum's tapestry collection at the time of writing. Semper had analyzed early carpet patterns as clues to the material history of weaving, introducing an evolutionary and materialist approach to art history. Serafin borrowed Semper's insight years later, in the never-published cosmic history to which he devoted the last years of his life, where he presented carpets as examples of historical survivals containing evidence of biological, material, and cultural evolution.

85. On this research, see R. Steven Turner, *In the Eye's Mind: Vision and the Helmholtz-Hering Controversy* (Princeton: Princeton University Press, 1994), Richard L. Kremer, "Innovation through Synthesis: Helmholtz and Color Research," in *Hermann von Helmholtz and the Foundations of Nineteenth-Century Science*, ed. David Cahan (Berkeley: University of California Press, 1993), 205–58.

that they lacked one kind of receptor. Hering attacked these categories as artificial, claiming instead that color sensitivity varied continuously from one individual to another (as well as between different parts of the retina). This continuous variation seemed to Hering to doom the three-color theory, because it implied that the three fundamental sensitivity curves were not fixed. Citing these "individual differences in color perception," Hering accused Helmholtz's followers of overlooking the variability of "normal" color vision.[86]

By the turn of the century, popular writers had taken up this critique. The broad attention to color theory at this time can be attributed in part to a revival of interest in Goethe's *Farbenlehre*. Goethe had presented his experiments on his own vision at the beginning of the nineteenth century as a revolutionary challenge to Newtonian optics, a liberation from the "tyranny" of physical dogma.[87] Many late nineteenth-century readers were introduced to Goethe's scientific writings by the Romantic conservative Julius Langbehn in his 1890 *Rembrandt as Teacher*. The poet's experiments in turn became a model of intellectual inquiry to Central European modernists and antimodernists alike, to thinkers as diverse as Rudolf Steiner, Oswald Spengler, and Italo Svevo.[88] The renowned physical chemist Wilhelm Ostwald, who began to work feverishly on color theory during the First World War, hailed Goethe as a "prophet."[89] Perhaps the most successful popularizer of Goethe's color theory was Houston Stewart Chamberlain—a botanist by training, a self-described "philosopher of life," and by any account a racist.[90] Chamberlain never failed to capture an audience in Vienna. He was even a member,

86. Ewald Hering, *Ueber individuelle Verschiedenheiten des Farbensinnes* (Prague: Tempsky, 1880), reprinted in *Wissenschaftliche Abhandlungen* (Leipzig: Georg Thieme, 1931), section 45.

87. Myles Jackson, "A Spectrum of Belief: Goethe's 'Republic' versus Newtonian 'Despotism,'" *Social Studies of Science* 24 (1994): 673–701.

88. Steiner, *Nature's Open Secret: Introductions to Goethe's Scientific Writings*, trans. by J. Barnes and M. Spiegler (Great Barrington, Mass.: Anthroposophic Press, 2000); Spengler refers to Goethe throughout *The Decline of the West* (Oxford: Oxford University Press, 1991), but particularly in chapter 3; Svevo discusses the *Farbenlehre* in *Confessions of Zeno*, trans. by Beryl de Zoete (New York: Vintage, 1989), p. 391.

89. Ostwald, *Goethe, Schopenhauer und die Farbenlehre* (Leipzig: Unesma, 1918), *Goethe der Prophete* (Leipzig: Hegner, 1932). On Ostwald's color theory, see chapter 10.

90. Chamberlain's *Foundations of the Nineteenth Century* was selected for the library of Emilie Exner's Society for the Gainful Employment of Women (*Jahresbericht Mädchenlyzeum des Wiener Frauen-Erwerb-Vereines*, 1901–2). In Emilie's estimation, the book was "a hymn to the Germans as a race, an enthusiastic call to culture of the future based on ethics and art. All constructed on the basis of natural scientific and historical fact; the work of a refined, keen, honorable dilettante,

along with Sigmund Exner, of the university's Philosophical Society, where he lectured in 1897 and 1898 on Indian thought and on Wagner's philosophy. In his 1912 *Goethe and the Sciences* Chamberlain presented Goethe's color theory as a new model for the study of nature and as an antidote to mechanistic science.

Chamberlain was not the only turn-of-the-century critic to try, as Goethe had, to wrest color theory from the hands of the physicists. Others approached this conflict from the perspective of educators intent on reinvigorating the aesthetic culture of the nineteenth century. Alfred Lichtwark, the director of the Hamburger Kunsthalle and an advocate of art education reform, published "The Education of the Color Sense" almost simultaneously with the appearance of Serafin Exner's study of beautiful colors. Lichtwark's work in Hamburg had become a reference point for educators in Austria, and Emilie Exner, among others, pointed to Lichtwark as a model for reforms in Vienna.[91] Yet Lichtwark insisted that physicists were not equipped to teach the aesthetic side of color. The principal failing of the physicist's color theory, he charged, was its inattention to individual differences in color perception: "In practice the starting point lies indeed in the color sense of the individual and not in the physical foundations of light and of the color problem . . . When one speaks of color one should never lose sight of the fact that it is nothing but a process in the activity of our eyes, and that this process, which depends on our corporal and psychological constitution, unfolds differently in each individual."[92] Contra Helmholtz, Lichtwark emphasized the variability of color perception across populations and historical eras in support of innovation in the arts.

Lichtwark's contempt for a physics of color was matched by that of Camillo Sitte, a Vienna-based architect and design teacher. Sitte was no enemy of the natural sciences. To the contrary, he was an enthusiastic Darwinian, an avid reader of Haeckel and Spencer, and even an early supporter of Helmholtz's empiricist psychology.[93] Yet in the year of the Klimt scandal

as he calls himself." Emilie Exner to Marie von Ebner-Eschenbach, 7 Dec. 1899, folder 81082Ja, Stadtarchiv, Vienna.

91. Vera Vogelsberger, "Sequenzen aus Kunsterziehung und Geisteswissenschaften," in *Kunst—Anspruch und Gegenstand: Von der Kunstgewerbeschule zur Hochscule für angewandte Kunst in Wien, 1918–1991*, ed. Erika Patka (Vienna: Residenz Verlag, 1991), pp. 274–314, quotation on p. 276; Emilie Exner to Marie von Ebner-Eschenbach, 16 Aug. 1901: "Lichtwark in Hamburg thut viel in dieser Richtung." Folder 81081Ja, Stadtarchiv, Vienna.

92. Alfred Lichtwark, *Die Erziehung des Farbensinnes* (Berlin: B. Cassirer, 1902), p. 9.

93. Michael Mönninger, *Vom Ornament zum Nationalkunstwerk: Zur Kunst und Architekturtheorie Camillo Sittes* (Braunschweig and Wiesbaden: Seweg 1998), pp. 117–18. Mönninger describes

Sitte blamed the shortcomings of nineteenth-century arts and crafts on the "horrid recommendations" of the physicist's color wheel. Green and yellow, for instance, could not be defended as harmonious on any physiological grounds; yet lately they had become "one of the most fashionably popular Secessionist combinations." This, he insisted, was a development to be explained exclusively "in the realm of contemporary taste, of pure artistic sensibility, and not within the walls of physical or physiological laboratories."[94]

The Exners were quick to argue that the revival of "Goethe the scientist" and the accompanying attacks on the Newtonian tradition rested on misconceptions. Shortly after the publication of Chamberlain's *Goethe and the Sciences*, for instance, Sigmund Exner penned a note on the *Farbenlehre* for a Vienna medical journal. He reminded readers that, despite many useful observations, Goethe had by no means reached an " 'explanation' in the modern sense of the word." Here explanation meant an account based on optics and physiology. After diagramming one of Goethe's optical experiments, Sigmund concluded that the poet must have been quite near-sighted. He had failed to correct for his own pathology, and thus fell short of the Exners' intersubjective ideal.[95]

Serafin Exner could not resist addressing Chamberlain directly. In a letter from Brunnwinkl in the summer of 1913, Serafin introduced himself as an "admirer of Goethe," then turned to correcting Chamberlain's "false and meaningless" statements about physics. Careful not to sound pedantic, he explained, "I know of course that *in the place in question* the content of Newton's law is entirely irrelevant, but it is not irrelevant if a reader sees that he cannot place his trust in the author to fulfill his tasks." Serafin closed with an apology for the liberty he had taken in addressing Chamberlain in this manner and assured him that he had read the book with "great interest."[96]

Chamberlain replied without delay, presenting himself as a man of taste answering a narrow-minded pedant. He painted Exner as a Beckmesser,

---

Sitte as a follower of Helmholtz's physiological theories, but, in his discussion of Sitte's "Über Farbenharmonie" (1900) he fails to see that this article was a direct attack on the perceived anti-individualism of Helmholtz's color theory.

94. C. Sitte, "Über Farbenharmonie," *Centralblatt für das gewerbliche Unterrichtswesen in Österreich* 18 (1900): 196–227, quotation on p. 207.

95. Sigmund Exner, "Ein Versuch aus Goethes Farbenlehre und seine Erklärung," *Wiener Klinischer Wochenschrift* 25 (1912): 22–23.

96. Franz Serafin Exner to Houston Stewart Chamberlain, 2 Aug. 1913, Chamberlain-Nachlass, Richard-Wagner-Gedenkstätte, Bayreuth.

Wagner's absurdly doctrinaire character in *Die Meistersinger von Nürnberg* who claimed to have reduced music to systematic rules:

> Respected sir,
> Since I may presume a friendly intent despite the coarse form, I thank you for your lines of yesterday and hope that you will not take it badly if I say to you, as Hans Sachs said to Beckmesser:
> Herr Merker, was doch solch' ein Eifer!
> Was doch so wenig Ruh'?
> Euer Urtheil, dünkt mich, wäre reifer,
> hörtet ihr besser zu.[97]
> [Mister Marker, why such zealousness?
> Why so little calm?
> Your judgment, it seems to me, would be more mature
> if you listened more carefully.]

Even more cutting was Chamberlain's next comparison:

> The similarity—even identity—between many professors and the priests who decree the truth strikes me again in reading your note: always attending only to words, to definitions, never considering the thoughts of another with respect, mindfulness, compassion, at least until one has penetrated them and obtained a vivid sense of a differently constituted *Weltanschauung*.
> But nothing can be done about that.

Nothing could have been more of an affront to Serafin's self-conception as a scientist and liberal Austrian than to be accused of dogmatism and likened to a priest. Chamberlain drove this attack home by quoting scientists against science's authority in the realm of visual phenomena. He noted that French mathematician Henri Poincaré had admitted that the modern theory of light was merely a useful "custom." In the words of Russian physicist O. D. Khvol'son, Chamberlain declared, "The study of light does not belong to physics." (Neither citation was entirely apt. Khvol'son's challenge had been directed less against physicists than against the extreme materialism of Ernst Haeckel. And Poincaré had argued that there were objective grounds, such as the criterion of simplicity, for accepting one theory over another.)[98]

97. Wagner, *Die Meistersinger von Nürnberg,* act 1, scene 3.

98. Houston Stewart Chamberlain to Franz Serafin Exner, 2 Aug. 1913, 293/30-1, Manuscript Collection, Österreichische Nationalbibliothek, Vienna. On Poincaré, see Galison, *Einstein's Clocks*.

Serafin Exner could not let stand Chamberlain's caricature of himself as a Beckmesser, a priest of scientific dogma. Promptly seizing his pen, he presented himself as the broad-minded one:

> If I once again lay claim to your time it is because I do not want to worsen your already poor opinion of professors—which, unfortunately, I share in many respects. I know these Beckmesser natures only too well and for just that reason I did not feel insulted by H. Sachs's verse. You are mistaken if you believe I had begun my reading without trust in the author or even with the intention of picking holes in the author. To the contrary: I have seldom read a book with so much pleasure and, I believe, to such profit as yours.

Serafin refused to be labeled a dogmatist or pedant. He also assured Chamberlain that he knew Poincaré and Khvol'son "both personally and through their works and know myself to be in full agreement with them." Finally, though he admitted that Goethe's attitude to Newton remained to him "a psychological puzzle," he showed that he too had taken to heart Goethe's critique of the sciences:

> If I may be permitted to say what in your book . . . pleased me the most, it was the emphasis on Goethe's position on the principle of causality, that is, his replacement of "causes" with "conditions." Here he truly presents a *Weltanschauung*, indeed one founded on experience, and one cannot bow deeply enough before the genius who grasped this in those days.[99]

Intent on dispelling Chamberlain's image of himself as dogmatic, Exner made his position on causality the case in point. He agreed with Goethe that the language of causality lacked a grounding in "experience." Exner may have been referring to Goethe's 1798 essay "Empirical Observation and Science," where the poet used the terms "circumstances" and "conditions" in the place of "causes," and where he suggested that experimental observations and natural laws should only be expected to agree approximately. Goethe's goal in this essay, like Exner's in his letter to Chamberlain, was to portray science as neither dogmatic nor merely speculative.[100]

In his questioning of causality, still radical in 1913, Exner found common ground with the critics of mechanistic science. But this was no "capitulation to hostile forces," as Paul Forman has described support for an acausal

99. Franz Serafin Exner to Houston Stewart Chamberlain, 8 Aug. 1913, Chamberlain-Nachlass, Richard-Wagner-Gedenkstätte, Bayreuth.

100. In *Goethe: The Collected Works,* vol. 12, *Scientific Studies,* ed. and trans. Douglas Miller (Princeton: Princeton University Press, 1995), pp. 24–25.

physics in Weimar Germany.[101] On the contrary, this probabilistic stance was constitutive of the liberal culture in which Exner's science was rooted. Like Chamberlain, historians have misrepresented the position of the Exners and their colleagues in the aftermath of the Klimt scandal. They have been painted as dogmatists and pedants, when in fact the version of rationalism they were defending was as self-consciously antidogmatic as it was hostile to the undisciplined subjectivity they saw in Klimt's mural.

A CONSENSUS OF NORMAL EYES

In another account of the origins of the Exners brothers' color theory, the one they would later tell their scientific colleagues, their preoccupation with color begins instead on a summer afternoon. We can picture Sigmund Exner hiking alone up the Schafberg, the peak that rose behind the Exners' cluster of cottages. With a few strokes of his pen, Sigmund removed himself from the battlefield of cultural politics and resurfaced in the calm of the summer retreat. Assuming the habitus of the *Sommerfrischler*, he cast himself as the consummate connoisseur of nature, sharp-eyed and untainted by dogmatism. Sigmund recalled "experiments"—"which I set up without taking notes, almost as a game on hot summer days. It was on a mountainside covered with countless flowers and low bushes, where all sorts of insects buzzed around, flying from flower to flower . . . "[102] In this fertile clearing, Sigmund "amused himself" by playing a little trick on the insects. He lay down slips of colored paper in shades to match the surrounding flowers and watched as the insects flew to them, "apparently deceived." He began to wonder about the appeal of flower colors to insects and about the "aesthetic pleasure" that flowers offered humans. How was it that bees, with their multifaceted eyes and microscopic brains, were attracted to the very same flowers as poets?

Perhaps Sigmund shared his observation with Serafin at supper that night. Serafin would have countered with his own query: why was it that in the history of art and industry throughout the ages and across cultures, not only certain colors but certain shades appeared to dominate? The physicist and physiologist would now have seen the outlines of a project. For Sigmund,

101. Forman, "Weimar Culture."

102. Franz Serafin Exner and Sigmund Exner, "Die physikalischen Grundlagen der Blütenfarbungen," *Sitzungsberichte der Österreichischen Akademie der Wissenschaften I* 119 (1910): 191–245.

it would be a return to an investigation he had begun in the 1860s, under the influence of Helmholtz's physiological optics. For Serafin, it would mean embarking on a new field, one lying at the contested nexus where physics and physiology intersected with the new experimental psychology. For both men, it would be an effort to resolve the aesthetic disunity that had cut a rift through both their professional and domestic worlds.

As the brothers emphasized, the perception of color varied widely, not only from one species to another but even among members of a single species—even among "apparently normal people." Who exactly qualified as normal? Most immediately, the Exners were following contemporaries in assuming that there existed distinct classes of observers: normal, dichromat color-blinds, and monochromat color-blinds. "The first difficulty" in constructing a theory of colors, Serafin later explained, is that the correlation of stimulus to sensation varies even from one "normal" (in this sense) person to the next "according to the individual nature of the receiving organ."[103] Consistent with the empiricist approach to aesthetics in the Austrian *Gymnasium*, Serafin proposed to begin empirically.

Employing two hundred volunteers and papers of several different shades within each color family, Serafin asked his subjects to choose those colors which, "without regard to any practical application, stimulate the most or least pleasant sensation in the eye."[104] The subjects confirmed his suspicions. Despite the vagaries of "personal taste, idiosyncrasies of the eyes, and other contingencies," they showed a significant preference for certain shades of red over others, and likewise for particular blues and greens. This trifold classification of colors hinted at the theoretical explanation Serafin would pursue. It suggested that Serafin's first task was to measure what Helmholtz called the fundamental sensations.

This was more difficult than it might seem. First, one needed observers. Performing photometric experiments on two hundred observers was impractical. The scientists would have to rely on their own eyes, with the aid of an assistant or two. Second, they needed an experimental method that did not assume more than was known. Back in 1868, Sigmund had measured his own fundamental sensations by relying on the principle of sensory "fatigue": when the eye is exposed to a given color of light for an extended period,

103. Franz Serafin Exner, *Vorlesungen über die physikalischen Grundlagen der Naturwissenschaften* (Vienna: Franz Deuticke, 1919), p. 616.

104. Franz Serafin Exner, "Zur Characteristik der schönen und hässlichen Farben," *Sitzungsberichte der Österreichischen Akademie der Wissenschaften IIa*, 3 (1902): 901–21.

its sensitivity to that color decreases temporarily. Sigmund reasoned that the eye would show the least sign of fatigue when exposed sequentially to wavelengths close to the fundamental sensations, since primarily one type of receptor would respond at a time, allowing the others to recover. By testing the sensitivity of his eyes to blue, red, and green light after long exposure to light from another region of the spectrum, he was able to estimate the primary sensations. This approach was powerless, however, to determine the *shapes* of the eye's response curves in the three color ranges. This is where Serafin believed he could improve on his brother's method (see fig. 9). His approach yielded three fundamental wavelengths: a blue close to indigo, a green verging on blue-green, and a red "similar to carmine, only further toward purple, somewhat like very dark roses."[105]

Serafin now believed he had glimpsed the foundations of a universal aesthetics. Within each group of reds, greens, or blues, the ones his subjects had deemed most beautiful were those closest to the corresponding fundamental sensation. Conversely, the further a color lay from the primary, the "more unpleasant" the observers found it. The "beauty" of a color was therefore determined by its saturation and by the ratio in which it excited the eye's red, green, and blue receptors. Serafin concluded that beautiful colors did indeed exist, not as romantic ideals or rational principles, but as the consequences of spectral physics and the physiology of the normal eye.

The Exner brothers now had the tools to explain why the colors of flowers were equally appealing to bees and botanists. While Serafin applied his photometer to a rainbow of roses, sunflowers, and clematis, Sigmund analyzed the physical structures through which plants transformed white sunlight into the myriad refracted colors that dotted an alpine hillside in high summer. Sigmund argued that the "liveliness" of flower colors was due to a light-reflecting layer of the petal, which acted like the metal backing fixed to gems. It gave rise to complex refractive and reflective effects, and it reduced the amount of white light reflected at the surface, making the petal color highly saturated. According to Serafin's measurements, flowers, alongside fine gems, possessed the most highly saturated colors in nature. The brothers were satisfied that they now knew why the reddest rose or the most violet violet was so *universally* dazzling.[106]

105. Franz Serafin Exner, "Young-Helmholtz'schen Farbensystem," *Sitzungsberichte der Kaiserlichen Akademie der Wissenschaften zu Wien IIa* 111 (1902): 868.

106. Franz Serafin Exner and Sigmund Exner, "Die physikalischen Grundlagen der Blütenfarbungen," *Sitzungsberichte der Kaiserlichen Akademie der Wissenschaften zu Wien I* 119 (1910): 191–245.

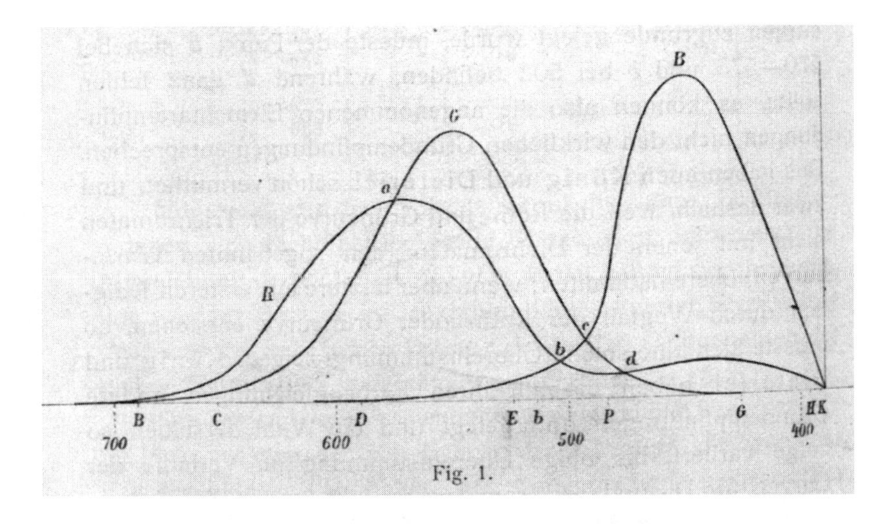

FIGURE 9 Visual sensitivity to light of different wavelengths, based on empirical measurements and on the theory that the retina contains three distinct types of receptors. To construct this, Serafin assumed merely that the curves overlapped in roughly the manner pictured above. His task was to fix the points *a, b, c, d*. Consider point *a*. Here, the green and red fundamental sensations are equally stimulated, while the blue response is insignificant. The eye's response here will be independent of the absolute intensity of the light within a wide range. For light shifted slightly redder than *a*, lowering the intensity will make the color appear redder, since the green response will fall, while for light slightly greener than *a*, lowering the intensity will make the light appear greener. (Point *c* would be the hardest to observe, since all three primary responses were active there.) For each point, Serafin arranged his prism, spectral lamp, and double lens so that he could see a narrow band of the spectrum in both fields, and could vary the intensity of each by narrowing the aperture. Source: Franz Serafin Exner, "Young-Helmholtz'schen Farbensystem," *Sitzungsberichte der Kaiserlichen Akademie der Wissenschaften zu Wien IIa* 111 (1902), p. 868.

Yet the full weight of this claim rested on the behavior of just a few pairs of eyes. Years later, Serafin's own students would fault him for using his aging eyes as a standard.[107] Even more problematic for the Exners' concept of the normal eye was the objection raised by Franz Hillebrand. Like Alexius Meinong and Alois Höfler, Hillebrand had studied philosophy and natural

107. Eduard Haschek, "Quantitative Beziehungen in der Farbenlehre," *Sitzungsberichte der Kaiserlichen Akademie der Wissenschaften zu Wien IIa* 136 (1927): 461–68, on pp. 466–67; Erwin Schrödinger, "Die Gesichtsempfindungen," in *Gesammelte Abhandlungen* (Vienna: Verlag der Österreichischen Akademie der Wissenschaften, 1984), vol. 4, pp. 183–288.

science with Franz Brentano. In the late 1880s, Hillebrand made a fateful move when he sided against Brentano and in favor of Hering's opponent-process theory of color. He then worked for several years as an assistant in Hering's Prague laboratory, where he also studied physics with Mach. After teaching briefly in Vienna, he founded and directed the Institute for Experimental Psychology in Innsbruck until his death in 1926. His initial venture into color theory began as a challenge to Helmholtz's three-color theory. Helmholtz had postulated that every visible color could be specified by three variables—hue, brightness, and saturation—and these variables were independent in the sense that varying any one of them would not alter the others. Helmholtz gave each of these variables a physical definition. Hue was proportional to the color's position along the spectrum, corresponding to a wavelength of light; brightness was proportional to the intensity, a purely physical measure of the quantity of light; and saturation was inversely proportional to the amount of diffuse daylight, or "white" light, the color contained. On the basis of this system, Helmholtz could in principle uniquely specify any color within an imaginary space defined by axes corresponding to hue, brightness, and saturation, and by a suitable unit of measurement or, in the language of geometry, a line element.

Yet one well-known phenomenon defied Helmholtz's system. As Jan Purkinje had first noted in 1825, at dusk the colors blue and green seem to increase in brightness. For instance, red and blue pigments that appear equally bright in full illumination will begin to differ in brightness as the light dims, with the blue gradually appearing brighter than the red. According to Helmholtz's theory, brightness should correlate directly with intensity, yet here was evidence that their relationship was more complicated.

This problem threatened the Exners' color theory in its most basic aspiration. It suggested that, in the case of brightness, there was no unambiguous, one-to-one correlation between the physical stimulus and the subjective sensation. Without such a simple relationship, an account of perception would be possible only at the level of sensation, not of physical phenomena. There would be no way to translate precisely between mental experience and material reality, no way to overcome solipsism.

Circa 1900 the most promising solution to this problem was that of the Freiburg physiologist Johannes von Kries. Kries preserved Helmholtz's theory and its physical definition of brightness by employing a hypothesis about the physiology of the eye ascribed to Arthur König. According to König, the retina contains two kinds of receptors, rods and cones; the first dominated in dim light, the second in bright light. Kries suggested that the rods were more sensitive to long wavelengths than the cones. Therefore, long-wavelength blues and greens appeared brighter in dim light.

In his 1889 paper, Hillebrand challenged Kries's theory first by demonstrating that the Purkinje phenomenon occurred even for light striking parts of the retina that had been shown to be void of rods. He went on to dispute the building blocks of Helmholtz's theory. Hillebrand argued that color sensations were of a different sort than sensations of tone or pressure, which simply disappeared as the stimulus decreased in intensity. With color sensations, decreasing intensity produced a qualitatively different sensation, that of black. On this basis, Hillebrand rejected Helmholtz's equation of the sensation of brightness with the physical intensity of light.

Thirty years later, at the opposite end of his career, Hillebrand returned to this line of argument in his dispute with Serafin Exner.[108] His evidence came from experiments performed with a rotating disk. The disk was divided into inner and outer regions, with each region split into three unequal segments, covered with papers either of a color (C), white (W), or black (B). When the experimenter rotated the disk at an appropriate speed, the observer would perceive a single color. In this way, spinning the disk effectively "mixed" the colors for the observer. The experimenter could adjust the proportions until the color of the inner region appeared to have the same "white value" A as the color of the outer region:

$$C + W + B = A = C' + W' + B'$$

where the unprimed variables refer to proportional areas of the inner region and the primed ones to the outer. Hillebrand began an experiment by determining the "white value" of his colored papers. His results hinged on the definition of this term. Hillebrand defined it as the proportion of gray that a color contained. When a dark-adapted observer looks at a colored object in very dim light, it appears to be a shade of gray, which can be compared to a true gray on a gray scale. Once Hillebrand had determined the white values in this way, the experiment was straightforward. He set the proportions of the color, white, and black on the inner and outer ring such that they differed in brightness and in saturation. He then directed a polarimeter with two fields for comparison toward the disk, one field aimed at the outer ring and the other at the inner ring. Finally the angle α of the disk's plane could be

108. Franz Hillebrand, "Purkinjesches Phänomen und Eigenhelligkeit," *Zeitschrift für Sinnesphysiologie* 51 (1920): 46–95. The rest of the dispute between Hillebrand and Exner unfolded in the pages of this journal and that of the Vienna Academy of Sciences.

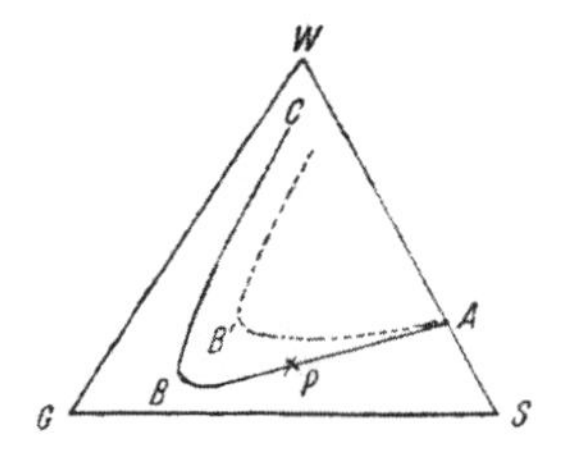

FIGURE 10 Hillebrand's diagram of the relationship of saturation to brightness. The curve *A-C* represents the change in saturation as the intensity of the light is changed from zero (*A*) to a very high (but still tolerable) intensity. (The point *A* depends on the surrounding light and adaptation of the eye.) As the intensity increases the sensation becomes brighter and more saturated, until it reaches the maximum of saturation at *B*. (At this point, other means, such as juxtaposing a contrasting color, could make the color appear even more saturated.) As the intensity is increased further, the sensation becomes ever brighter but less saturated, until it becomes close to white (*C*). (The dotted curve represents the same process with diffuse daylight mixed into the original light.) Source: Franz Hillebrand, "Purkinjesches Phänomen," *Zeitschrift f. Sinnesphysiologie* 51 (1920): 61.

adjusted, changing the proportion of light it reflected, until the colors in the two fields were equal. Then the white value of that particular color could be found in terms of W, W', C, C' and α. According to Helmholtz's theory, mixtures of any two colors with equal proportions of black and white should appear of equal brightness, but Hillebrand found this not to be the case. For instance, the more blue a mixture contained, the brighter it appeared. In order to explain this anomaly Hillebrand proposed that each color has its own "specific brightness," with blue inherently brightest and red darkest (see fig. 10). In one strike, Hillebrand stripped brightness of its status as an independent variable, pulling the rug out from under the empiricist theory of color.

Serafin Exner responded vigorously to this attack, focusing on the problem of defining a color's "white value." Rejecting Hillebrand's procedure for defining it as an equivalent gray, Exner measured the white value of a given colored paper as the ratio of its brightness to the brightness of a standard white paper. (Like Hillebrand, Exner worked with colored papers produced by Hering's lab.) Exner's white values differed markedly from Hillebrand's for colored papers. When he inserted his own white values into Hillebrand's expressions, the need to introduce the notion of specific brightness vanished.

Exner turned this refutation into an opportunity to attack the Hering school more broadly. He labeled their approach "phenomenological," in opposition to his own "empiricist" approach, the same contrast that "divides Goethe's color theory from Newton's." His introduction made clear the stakes of the disagreement as he saw them:

> Justifiably, pure phenomenology—in the field of color theory, the description of our sensations—can claim to be of great interest; but the goal of research still remains above all the origin of these sensations, the knowledge of the physical and physiological phenomena which condition [*bedingen*] the psychic ones. A relation between objective and subjective phenomena must be produced which best satisfies experience.[109]

Serafin saw no other goal for a theory of color than the correlation of sensations with physical stimuli. In contrast to monists like Mach or Ostwald, he was convinced that it was possible to demarcate physical from psychic phenomena. From his perspective, a "clearer conception of the concepts used" necessarily meant supplying *physical* definitions. The phenomenologists, on the other hand, remained wholly on the side of the "subjective." They were content to furnish an explanation of appearances rather than reality, just as, in his analogy, pre-Copernican astronomers had been content with the Ptolemaic theory. Serafin implied that the notion of specific brightness was, like medieval epicycles, an ad hoc attempt to rescue a doomed theory.

The problem of defining brightness brought into relief the difference between the "phenomenological" and "empirical" approaches. It forced researchers to choose whether to begin their analysis with the evidence of sensations or of physical measurements. Hering and his followers rejected light intensity as a measure of brightness because the zero-point of intensity did not correspond to "no sensation"; rather, it corresponded to the sensation of black. This argument struck Exner as absurd. One might just as well demand that physicists abandon the concept of temperature because zero degrees corresponded to the sensation of cold rather than "not warm." Serafin recognized that the "phenomenologists" had made great efforts to quantify colors. They had, for instance, defined a shade of gray as the proportion W/B *or* $W/W + B$. But Serafin argued that quantification was impossible without units of measurement, and these could only be applied

109. Franz Serafin Exner, "Einige Versuche," pp. 1829–30.

to objective physical phenomena, not to sensations.[110] Exner drew a distinction here between two ways of interpreting quantitative descriptions of sensations or "psycho-physical laws." Fechner had been the first to formulate such laws, and the most famous of them bore his name; it stated that if a stimulus grew arithmetically, the corresponding sensation grew logarithmically. The accuracy of this result had since been challenged, but Exner accepted this result as a rough empirical correlation between sensations and physical phenomena. He denied, however that Fechner had succeeded in measuring the strength of sensations directly. Sigmund Exner had made the same point years earlier, when he disputed Fechner's interpretation of this law as a statement "about the relationship of the nervous system to the soul."[111] As Serafin put it, that reading was no more than an "illusion."[112] Unlike Fechner, the Exners preferred to "leave the question of the soul out of consideration."[113]

Serafin offered the following illustration: consider two surfaces, one illuminated by ten candles and one by a single candle. An observer could say only that the first was brighter than the second. Physical methods, on the other hand, could arrive at a precise quantitative measurement of the relative intensities. For instance, by placing an opaque disk with a sector of thirty-six degrees removed over the first surface, the brightnesses could be made to appear equal. When we say that one surface is ten times brighter than another, "This statement thus expresses the objective phenomenon, not however the subjective sensation to which it gives rise." Serafin argued further that the same interpretation had to be applied to the statement that one colored surface was twice as bright as another. Such a conclusion could only be reached "by obtaining such objective measurements through some kind of physical methods."[114]

Exner was arguing that the clarity of scientific language depended on the possibility of defining subjective concepts in terms of measurable physical effects. He made the same demand of the concept of probability, as we will see in the next two chapters. Probability could not be a measure of degree of belief, since Serafin was sure that mental states could not be measured.

110. Franz Serafin Exner, "Einige Versuche," pp. 1835–36.

111. Sigmund Exner, "Grosshirnrinde, in *Handbuch der Physiologie*, ed. L. Hermann, vol. IIa: *Allgemeine Physiologie* (Leipzig: F. C. W. Vogel, 1879–83), p. 238.

112. Franz Serafin Exner, "Einige Versuche," p. 1836.

113. Sigmund Exner, "Grosshirnrinde," p. 238.

114. Franz Serafin Exner, "Einige Versuche," p. 1836. Compare F. S. Exner, *Vorlesungen*, p. 644.

Probabilistic statements had clear and universal meaning, just as quantifications of brightness did, because they were measurements of objective states of the physical world. Kries had worked with this same analogy between colors and probabilities in the 1880s. He began by criticizing Fechner's claim that the psychophysical law was a measure of sensations, and he ended by arguing for an objective interpretation of probability, though not a frequentist one.[115] From Exner's perspective, to define either color or probability as a mental rather than a physical state would mean surrendering the goal of transparent communication.

Hillebrand understood these stakes well. In the case of specific brightness, he claimed that the relation between stimulus and sensation was not simple, and therefore that a physical description of color was impossible.[116] Hillebrand was disputing the dualist position from which the Exners argued. Carrying the torch of Ernst Mach, he implied that there could be no strict separation between mind and matter, sensation and stimulus. There was thus no inherent difference between measurements of psychic and of physical phenomena. Anyone who, "like Exner," denied the possibility of psychic measurements, "deludes himself if he simultaneously believes that a measurement by indirect (in this case, physical) methods is possible."[117]

Hillebrand charged that Exner and other Helmholtzians distorted their account of perception by presupposing a one-to-one correlation between stimulus and sensation. In cases where a one-to-one correlation obtained, they spoke of "normal" perception, but they judged all other cases "abnormal." Then they searched for higher-order "psychological" explanations for these anomalies. Thus, in cases where perceptions conflicted with physical measurements, as in so-called optical illusions, the Helmholtzians resorted to the notion of "errors of judgment" or "unconscious inferences."

This charge was not unfounded. Helmholtz had portrayed color perception as a process of gauging an observer-independent reality on the basis of past experience. Individuals learn through experience to judge the inherent colors of objects independently of the surrounding illumination. Eventually, they become unconscious of the sensations that determined their judgment.

115. See Michael Heidelberger, *Die Innere Seite der Natur: Gustav Theodor Fechners wissenschaftlich-philosophische Weltauffassung* (Frankfurt am Main: Klostermann, 1993), p. 264; Stöltzner, "Exner's Indeterminist Theory," p. 17.

116. Hillebrand, "Purkinjesches Phänomen," p. 51.

117. Hillebrand, "Grunsätzliches zur Theorie der Farbenempfindungen," *Zeitschrift für Sinnesphysiologie* 53 (1922): 129–33, quotation on p. 133.

Helmholtz had explained incongruities between perceptions and physical reality as the result of this learning process. A gray surface, for instance, appears whiter when viewed next to a black surface, an effect known as "simultaneous contrast." In this case, the observer draws on past experience to make an "unconscious inference" that the black surface is an indication of dim illumination. The eye judges the gray surface to be inherently whiter than it was. Helmholtz never doubted that such "illusions" could be corrected by physical measurement. He even claimed that an observer could train himself to overcome optical illusions, to see not the subjective appearance but the physical reality.[118]

By the time of Exner and Hillebrand's dispute, a growing number of critics were forcefully rejecting the concept of unconscious inferences. Already in 1886 Hering had attacked Sigmund Exner for his reliance on this notion.[119] In 1913, in an article that was foundational for the Gestalt psychology movement, German psychologist Wolfgang Köhler likewise criticized the concept as an invalid ad hoc hypothesis.[120] These critics insisted that labeling certain phenomena "illusions" merely masked the inability of Helmholtz's theory to account for a variety of common visual experiences. In the place of explanations on the basis of higher-order mental functions, Hering, Hillebrand, and their followers sought to explain perception in terms of the functioning of the peripheral sensory apparatus. For instance, instead of classifying the phenomenon of simultaneous contrast as an error of judgment, Hering explained it as a "reciprocal action" [*Wechselwirkung*] between regions of the visual field.

These critiques of the "empiricist" approach to perception highlight the work that the concept of "normal" vision did for the Exners. Hillebrand's accusation cannot be refuted: when perception conflicted with physical measurement, the empiricists inevitably cried "error." They stacked the cards in favor of a physicalist theory of vision. But from their perspective, this bias made perfect sense. They believed that we learn to perceive the world through experience, correcting our mistakes and gathering the basis for future judgments as we go. Vision in this sense is purposeful and adaptive, and it should therefore be expected to agree with the evidence of other modes of investigating our surroundings, including physical instruments.

118. Helmholtz, *Handbuch der physiologischen Optik* (Leipzig: L. Voss, 1867), pp. 388–417.

119. Hering, *Wissenschaftliche Abhandlungen*, section 47.

120. Wolfgang Köhler, "Über unbemerkte Empfindungen und Urteilstäuschungen," *Zeitschrift für Psychologie* 66 (1913): 51–80; Ash, *Gestalt Psychology*, pp. 135–37.

While the "nativists" claimed that the visual faculties were rooted in innate physiology, the "empiricists" believed that we fine-tune our vision to our environment. It was from this perspective that the "empiricists" equated "normal" vision—in the sense of health, regularity, and typicality—with a simple correlation between stimulus and sensation.

By connecting mind and world in this way, the Exners believed they could overcome the strictly personal conditions of perception. The first problem with empiricism, as Serafin Exner acknowledged, was that, "as we know from experience, by no means do the same external stimuli correspond to the same sensations in all individuals; that this correlation is therefore in no way universal, but rather varies according to the individual nature of the receiving organ."[121] Unlike many of their contemporaries, the Exners did not simply replace "subjective" observers with "objective" instruments.[122] They never allowed themselves or their audience to forget that, on the other end of a laboratory instrument, there was inevitably a human being, whose senses were implicated in any calibration or reading of the instrument. "All that we can attempt for now," Serafin wrote, "is a purely empirical coordination of the processes in our consciousness with the physical variations in the external world to which our sensory organs react."[123]

The Exners' solution to the problem of variability was not mechanical but statistical and normative. Leaving aside the "exceptions" of blind and color blind individuals, "one can ascribe to the normal eye . . . on the whole an entirely definite [*bestimmt*] reaction to a given [*bestimmt*] external stimulus. The following discussion can only apply to this normal average [*normalen Durchschnitt*]."[124] Like the "average man" introduced by the Belgian astronomer Adolphe Quetelet in the 1840s, the Exners' "normal eye" was a mathematical fiction, an ideal as well as a distillation of facts. As a statistical mean, it was descriptive of a population, but it was also a normative definition of "health," like the French pathologist Claude Bernard's celebrated idea of an organism's "normal" state.[125] As an avowed approximation, this concept

121. Franz Serafin Exner, *Vorlesungen*, p. 616.

122. Lorraine Daston and Peter Galison identify the nineteenth-century ideal of "mechanical objectivity" in "The Image of Objectivity," *Representations* 40 (1992): 81–128.

123. Franz Serafin Exner, *Vorlesungen über die physikalischen Grundlagen der Naturwissenschaften* (Vienna: Franz Deuticke, 1919), p. 615.

124. Franz Serafin Exner, *Vorlesungen*, p. 616.

125. On Quetelet, see Porter, *Statistical Thinking*, and Hacking, *Taming of Chance*; on Bernard, see Georges Canguilhem, *The Normal and the Pathological* (New York: Zone, 1991).

avoided any tinge of dogmatism or the implication of externally imposed conformity.

The Exners were not alone in relying on the notion of a normal eye. Many of their contemporaries likewise based their empirical conclusions on the assumption that there existed distinct classes of observers: normal, dichromat color-blinds, and monochromat color-blinds. The Exners, however, applied this concept far beyond the laboratory. They specified not only the wavelengths of light to which the normal eye was sensitive but also its favorite colors, its tastes in the fine arts, and even in women. The normal eye encapsulated their standards of friendship, family, and sexual relations. As a definition of the universally human, the normal eye did not spring fully formed from mathematical calculations or laboratory investigations. Like nineteenth-century ideas of the nation, the normal eye was a labor of the imagination.[126] I suggest that we see this concept as its authors' themselves saw it: as an account of the possibility of communication between autonomous beings about a shared world. What these scientists deemed universal originated as a vision of what they held in common on a far smaller scale.

The concept of the normal eye was rooted in the experience of social intimacy, of private "Verkehr." Of course, laboratory measurements constrained the contours of the Exners' vision of universal taste. But what breathed life into this abstraction was their self-consciousness as members of a private community. Consider this remark from Sigmund and Serafin's joint publication on color theory in 1910: "We know of course that we can associate [*verkehren*] with a person for an entire lifetime, have innumerable conversations with him about paintings and other colored objects, and only late in life discover that this person is color-blind, of which he too was ignorant, until specially designed experiments demonstrated it."[127] The point of this vignette is that intimate relationships rest on the perpetuation of a conversation about a shared world and thus on the willingness of each individual to adjust his perceptions to a common standard. The essence of intimacy is the attempt to see through the eyes of another.

IN Serafin and Sigmund Exner's decades-long defense of this definition of normal vision, the years 1898–1902 marked a turning point. Perhaps never before had Vienna, city of aesthetes, had such high expectations for the

126. Benedict Anderson, *Imagined Communities: Reflections on the Origin and Spread of Nationalism* (New York: Verso, 1991).

127. Franz Serafin and Sigmund Exner, "Physikalische Grundlagen," p. 192.

visual arts. It was a time when works of art were believed to "speak a common language," and painters were thought capable of uniting a fractured land.[128] With their ardent attention to nature and their celebrations of the untutored eye, the Secession seemed committed, like the Exners, to the faithful reproduction of "the poetry of reality." Klimt's mural therefore came as a shock: a dark parody of the professors' ideal of transparent intellectual exchange. In the ensuing dispute, the scientists' efforts to recover a standard of intersubjectivity were sorely misinterpreted. The dust was still settling as Serafin's daughter Hilde and her cousin Nora arrived at Brunnwinkl, fresh from their first year at the Arts and Crafts School, only to deface the pristine walls of the cottages with their brash colors. As these affronts to the self-disciplined gaze of the normal eye followed in quick succession, it seemed to the elder Exners that a solipsistic vein of modern art was eating away at the very fabric of the liberal project. Liberalism, in their eyes, depended on the possibility of speaking unambiguously about a shared world. This meant rendering language transparent, making experience universalizable, on the model of the mathematization of the natural sciences. In this way the Exners tied their political aspirations to the laboratory project of correlating subjective sensations with measurable physical phenomena. Both projects depended on the cultivation of a certain observational attitude, one that was independent yet empathetic, sober but not unmoved by beauty—a realism, in short, modified by idealism.

Even as the Exners used science and art to communicate their utopian vision to the next generation, they sought to impart a disciplined imagination, one that would not stray beyond the limits of the possible. They would teach the young to discriminate between "probabilities" and "mere possibilities," using the probability calculus as their model.

128. Schorske, *Fin-de-Siècle Vienna*, p. 237.

CHAPTER SEVEN

Citizens of the Most Probable State

The Politics of Learning, 1908

IN 1906 the Exners celebrated the twenty-fifth anniversary of their summer colony. The children who had spent their first summers at Brunnwinkl were now starting families of their own. In fact, Brunnwinkl was also the site of a wedding that summer, the fruit of a local romance. Marie's son Otto married Jenny Richter, whose parents owned a summer villa nearby. On the traditional Brunnwinkl holiday of the twenty-eighth of August (the birthday of Goethe and of the first Franz Exner), family and friends surprised Brunnwinkl's founders Marie and Anton von Frisch with a cornucopia of gifts. Their son Karl had assembled a photo album, his brother Ernst had researched a historical "chronicle" of Brunnwinkl, and Emilie Exner had composed Brunnwinkl's "family history."

It was a time for giving thanks and taking stock, and Emilie Exner used it in particular to reflect on the effects of a Brunnwinkl childhood. In her opinion, the colony had always been "paradise" for the young Exners. "Nowhere was there a barrier or obstacle; behind the house were the meadow and forest, before the house the lake."[1] A busier if no less contented picture of youthful summers on the Wolfgangsee appeared at about this time in letters from Emilie's nephew Karl von Frisch to a classmate. Accounting for his time, Karl explained that at Brunnwinkl, "If you aren't actually part of a game, you're usually watching someone somewhere doing something

1. Emilie Exner, *Der Brunnwinkl,* p. 12.

or other, and the rest of the time I hide out in the museum."[2] The next summer, between working on his natural history "museum," an upcoming conference in Salzburg, and the imminent visit of one of his teachers to Brunnwinkl, Karl swore to his friend that nothing less than "a Zeppelin ride to the North Pole" would lure him away from St. Gilgen.[3] Brunnwinkl offered all a young biologist could want, from companions and games, to a lake full of plankton, to a stream of illustrious visitors. Karl's descriptions of the summer colony, like Emilie's, evoked a happy convergence of learning and play.

Emilie and Karl were not, of course, impartial observers. But local notable Marie von Ebner-Eschenbach was duly impressed with the young Exners. After a talent show staged by the children in 1893, Ebner-Eschenbach had remarked in her diary that while other parents fretted about their offspring's progress in school, the Exner and Frisch youngsters "wander effortlessly [*spielend*, literally "playfully"] through *Gymnasium*," all the while honing their talents in the arts and athletics.[4] So Ebner-Eschenbach was in a position to understand Emilie's concerns in the summer of 1907, as she sat at her desk in the lake house, thinking bitterly of the psychological damage wrought by the middle schools of the day. Outside, her six-year-old grandson Heinz scampered about, collecting specimens for Karl's museum. As Emilie explained in her letter to Ebner-Eschenbach, the little boy was the inspiration behind the essay she was at work on. "My little grandson Heinz is supposed to be sent to school already this fall and the thought that this little mind, a fountain of interests and thirst for knowledge, will now be stopped by a classroom grind often makes me quite gloomy."[5]

Emilie's concern for young Heinz was aggravated by a recent spate of coming-of-age novels whose young protagonists were crippled with anxiety. Worse still were newspaper reports of suicides by *Gymnasium* students.[6] At Brunnwinkl, on the other hand, Heinz seemed to be thriving. He spent half the day in Karl's museum and had begun his own collection with "a

2. K. v. Frisch to Friedrich Paneth, 8 Aug. 1908, Paneth Nachlaß—K. Frisch, Abt. III, Rep. 45, Mappe 37, Archiv zur Geschichte der Max-Planck-Gesellschaft, Berlin-Dahlem.

3. K. v. Frisch to Friedrich Paneth, 1 Sept. 1909, Paneth Nachlaß—K. Frisch, Abt. III, Rep. 45, Mappe 37, Archiv zur Geschichte der Max-Planck-Gesellschaft, Berlin-Dahlem.

4. Marie von Ebner-Eschenbach, diary entry for 25 Aug. 1893, *Tagebuch*, vol. 4, p. 240.

5. Emilie Exner to Marie von Ebner-Eschenbach, 13 Aug. 1907, folder 81082Ja, Vienna Stadtarchiv, Vienna.

6. Discussed in Emilie Exner, "Der Mittelschüler in Literatur und Wirklichkeit," *Österreichische Rundschau* 12 (1907): 123–30.

few snail shells and tree fungi." "The little guy goes his own way," Emilie reported to Ebner-Eschenbach, "always has a question, and enjoys himself very much. The children have it really good here. This freedom and alongside so many adults who care for them with reason and love."[7] Emilie's reflection belongs to a late moment in the history of Austrian liberalism, a time when the educated middle classes could still speak, in a single breath, of freedom, reason, and love—political, intellectual, and domestic ideals—as the essential elements of a good home. In late July Emilie sent a draft of an essay on the state of Austria's middle schools to Ebner-Eschenbach, commenting, "An old account that has long weighed on my heart has now been settled. Of course one writes to free oneself; that's why sitting at one's writing desk is lovely."[8]

In this article, which soon appeared in a popular weekly, Emilie contrasted the typical student at an urban *Gymnasium*—thin, pale, and nervous—with the natural state of children in "mountain villages." In the country, the benefit of "contact with nature" overcame the debilitating effects of classroom drills. What children needed most of all was the chance to "play"—meaning crafts, sports, collections, the care of plants or animals, "all that brings children into contact with nature."[9] On the virtues of play, Emilie approvingly cited the Swiss psychologist Karl Groos, who pictured young children as born empiricists and described play as a form of "experimentation."[10] Emilie also insisted that children be given time and space for "dreaming," "that is, for independent reflection on all manner of problems, which have nothing to do with school . . . " Like other Austrian liberals, Emilie argued that teachers could not hope to foster ethical characters by formulating absolute rules: "Rigid rules know no notion of the thousand contingencies of life . . . "[11] Freedom was thus the essential condition for healthy development. Adults could encourage this process not by imposing strict rules but by allowing children to develop their own skills for managing contingency. As Groos argued, childhood play cultivated just these skills.

Similar proposals to integrate learning and play had been made in the 1890s and had met resistance from critics of the liberals' broader educational

7. Emilie Exner to Marie von Ebner-Eschenbach, 9 July 1907, folder 81082Ja, Vienna Stadtarchiv, Vienna.

8. Emilie Exner to Marie von Ebner-Eschenbach, 22 July 1907, 81082Ja, Vienna Stadtsarchiv.

9. Emilie Exner (Felicie Ewart, pseud.), "Der Mittelschüler in Literatur und Wirklichkeit," *Österreichische Rundschau* 12 (1907): 123–30, quotation on p. 123.

10. Groos, *Die Spiele des Menschen* (Jena: Gustav Fischer, 1899), pp. 91, 122.

11. Emilie Exner, "Mittelschüler," p. 124.

goals. In 1891 a front-page article on the Austrian middle schools in the Catholic conservative paper *Das Vaterland* complained, "We now stand . . . under the banner of children's games, namely, of official, so to speak obligatory, children's games, as a pedagogical aid of the middle school . . . If this continues the *Gymnasien* will indeed be reduced to play-schools for older children." As this protest suggests, children had become pawns in the political battles of the splintering empire.[12]

By 1907 three political groups had launched separate assaults on the education system that had flourished under liberal control since the days of the first Franz Exner. These challenges came to a head in that year. In the ensuing clashes, the significance of probabilistic reasoning to the liberals emerged with particular clarity. Against competing conceptions of childhood, the Exners waged a concerted defense of the pedagogical value of freedom.

THE POLITICS OF CHILDHOOD

Nothing proves the necessity of the Gymnasium so much as all the silly attacks launched by its critics.[13]

Critics of the liberals' program of learning on the far left, ranging from progressives to socialists, charged that the *Gymnasium*'s classical curriculum was an obsolete and irrelevant prerequisite to a university education. Their leader was the journalist Robert Scheu, whose journal on school reform popularized the catchphrases of this movement. Scheu's supporters called for the "individualization" of the curriculum in order to make each student optimally "productive." This position was a combination of progressive psychology and socialist ethics. In 1906 Scheu founded the Cultural-Political Society, which conducted an independent "Middle School Inquiry" from late 1906 to early 1907. Among the proposals this group considered was one to create middle schools that would entirely exclude the teaching of foreign languages, ancient and modern. Another proposal would have established a "minimal curriculum," allowing students to choose their own courses according to taste and ability. The themes of individualization and efficien-

12. Tara Zahra, *Your Child Belongs to the Nation: Nationalization, Germanization, and Democracy in the Bohemian Lands, 1900–1945* (Ph.D. diss., University of Michigan, 2005); "Das österreichische Mittelschulwesen und der Mittelschultag," *Das Vaterland,* 16 April 1891, Nr. 104, p. 1, Weisinger Collection.

13. Franz Serafin Exner, "Lebenslauf" (1917), p. 3, Personalakt, Archiv der Österreichischen Akademie der Wissenschaften, Vienna.

cy drew together the progressives' diverse concerns. They explicitly rejected the liberal goal of producing "universal" thinkers.[14]

The liberals' second group of opponents in the field of education were anti-Semitic pan-German nationalists, a growing force in Austrian politics during the 1880s and 1890s. Not surprisingly, these critics demanded a greater emphasis on German language and history at the expense of foreign and classical languages. Nationalists dismissed the liberals' claims that the value of science study was to be measured in terms of their contributions to students' characters, not by the substance of the knowledge acquired alone. The nationalists had no appreciation for the liberals' project of fostering "many-sidedness," aiming as they did to produce "German" rather than "universal" thinkers.[15]

As much as the progressives, nationalists, and liberals differed in their goals, they all had a political interest in keeping the Catholic church out of the Habsburg schools. Their common opponents were the Catholic conservatives of the Taaffe administration and then, in the late 1890s, the new Christian Social party. The Christian Socials were a conservative, clerical movement with their power base in the bourgeoisie, though their rhetoric was radical and populist.[16] The education system was a prime target of their proposed reforms. The Christian Socials' charismatic, provocative, and virulently anti-Semitic leader Karl Lueger, mayor of Vienna from 1897 to 1910, is still memorialized in Vienna today for his support for the construction of schools. He called schools "shrines for children," emphasizing their religious function. Lueger's anti-Semitism was far more than a pragmatic stance.[17] He typically expressed his antipathy to Jews in attacks on science, medicine, and higher education—fields in which Jews, and liberals, dominated in Austria. While historians disagree about Lueger's actual intentions for Austrian education, there is no question that liberals perceived

14. On Scheu's middle school reform movement, see Engelbrecht, *Geschichte des Österreichischen Bildungswesens*, vol. 4.

15. See too Gustav Uhlig, *Die Entwicklung des Kampfes gegen das Gymnasium*, Vortrag in der Wiener Festversammlung der deutschen Gymnasial-Vereine am 2. Oktober 1909 (Vienna and Leipzig: Carl Fromme, 1910).

16. John W. Boyer, *Political Radicalism in Late Imperial Vienna: Origins of the Christian Social Movement, 1848–1897* (Chicago: University of Chicago Press, 1981), Boyer, *Culture and Political Crisis in Vienna: Christian Socialism in Power, 1897–1918* (Chicago: University of Chicago Press, 1995).

17. Richard S. Geehr, *Karl Lueger: Mayor of Fin-de-Siècle Vienna* (Detroit: Wayne State University Press, 1990).

the Christian Socials to be threatening the "clericalization" of the schools, implying a reversal of the liberal reforms that had followed 1848 and 1867.[18] The Christian Socials won greater control of the public schools through new legislation in 1904, and in 1905 they made attendance at church mandatory for pupils. A modest liberal demonstration against these new laws was, to one sympathetic observer, the "death-throes of a drowning man—namely, of liberalism in Austria."[19]

The Christian Social party envisioned education, on the model of military training, as a means of teaching respect for authority. The party sponsored military-style youth groups for the children of its elite members.[20] Lueger's public addresses often invoked the rights of inheritance and stressed the importance of a son's loyalty to his father. This conservative conception of youth is essential to understanding the Christian Socials' academic politics. There was no room in their project for a moral education that would involve independent investigation and the critical questioning of received opinions.

The Christian Socials and the radical left shared the populist ambition of making education as public a responsibility as possible. Their youth groups were designed to place education in the broader sense of *Erziehung* in the hands of the community rather than the parents. In this sense, they drew the line between public and private differently than the liberals, who had long stressed that education was as much a domestic responsibility as a duty of the state. To the Exners, it still seemed that education must begin at home. "One can not say loudly enough that a *Gymnasium* education is not suited to everyone, that it unconditionally presupposes a certain education level of the families from which the students come, and that, if this foundation is missing, the results cannot live up to expectations." Indeed, the Exners claimed authority in these matters based on their own family heritage—based, as Serafin Exner put it, on "the traditions of the author of the organizational proposal of 1849, which are still alive in me."[21] The Exners implied that the *Gymnasium*'s lessons were somehow effective only in combination with the family life of the *Bildungsbürgertum*.

By 1907, then, liberal educators faced a tripartite challenge. Each group

18. Boyer, *Culture and Political Crisis*, p. 174.

19. Philosopher Friedrich Jodl to jurist Karl von Amira, 27 Oct. 1905, in Jodl, *Vom Lebenswege. Gesammelte Vorträge und Aufsätze* (Stuttgart and Berlin: J.G. Cotta, 1916), vol. 1, p. 515.

20. Geehr, *Karl Lueger*, p. 291. On Lueger's conception of youth, see Boyer, *Culture and Political Crisis*, p. 9.

21. Franz Serafin Exner, "Ein Wort zur Schulreform," *Neue Freie Presse*, 18 Feb. 1908.

of challengers—progressives, nationalists, and Christian Socials—had their own plan for the moral education of the empire's youth, none of which was compatible with the liberals' primary goal of promoting critical reasoning and the ability to speak from a universal viewpoint. For the liberals, the most important condition of learning was freedom, although implicit in their image of freedom was the counter-balancing force of self-discipline. Socialists deemed these goals elitist and inefficient; German nationalists judged them insufficiently German; and Christian Socials found them dangerously subversive. Sensing that liberal power was weakening, these critics intensified their attacks during the last few months of 1907.

THE MIDDLE SCHOOL INQUIRY

Faced with these various calls for reform, the liberal minister of education Gustav Marchet consented to hold a public hearing on the matter of the secondary schools. Following a model used by the German government in 1890, Marchet invited various authorities to contribute prepared comments. This structure was expected to give the government control over the proceedings. Sympathetic to liberal education as it stood in Austria, Marchet staged the inquiry to appease critics, not to launch reforms. He forbade references to politics, made no effort to represent the empire's minorities, and invited only three women. Thanks to her recently published article, Emilie Exner was one of them.

The hearings gave Emilie a chance to bring to the attention of state officials and pedagogical experts insights drawn from her domestic experience, especially at Brunnwinkl. She repeated her conviction that children thrived in freedom and learned from playing. Emilie's remarks were seconded by Marianne Hainisch, the advocate of women's education whose agitation had first prompted this public hearing. Hainisch approved particularly of Emilie's emphasis on the vital role of the family in a young person's education.[22] A critical note came from a parliamentary representative named Petelenz, who accepted Emilie's expertise on the state of the *Gymnasien* but insisted that she knew nothing of the *Realschulen*, the more practically oriented middle schools that led to vocational or technical careers. Petelenz allowed that the children of the upper classes might learn best when left

22. *Die Mittelschule-Enquete im k.k. Ministerium für Kultus und Unterricht, Vienna 21–25 Januar 1908, Stenographisches Protokoll* (Vienna: Alfred Hölder, 1908), p. 57; see too the words of approval from Freiher von Pidoll, ibid., p. 18 and 36.

to their own devices, but they were the exception. "One should not imagine though that the young boy entering school is wild about learning and being taught, that he has learning alone at heart. (Laughter.) That would be counter to the nature of boys." Emilie had described young pupils as "plastic," drawing on the Herbartian psychology that was deeply rooted in the Austrian *Gymnasien*.[23] Petelenz insisted to the contrary that the material with which a *Realschule* teacher had to work was often "indifferent" and "vexing." "One student can be compared to wet plaster, another to hard marble, a third to impenetrable granite."[24] The implication was that Emilie's image of learning was elitist and inapplicable to the population at large.

Emilie, who had once proclaimed her allegiance to the "aristocracy of intellect,"[25] hardly recognized this elitism as a shortcoming. Like Serafin Exner, she believed that a *Gymnasium* education was rightly reserved for a small segment of the population. Indeed, she blamed the *Gymasium*'s failings partly on the more modest backgrounds of the instructors. As she wrote to another academic wife, "Our *Gymnasium* professors are mostly peasants' sons and their . . . field of view quite small."[26] Her ideal of a learning environment was the vacation home of a privileged family, and its relevance to the education system of the empire was questionable.

SHORTLY before the opening of the Middle School Inquiry, Wilhelm Ostwald arrived in Vienna from Leipzig to add his voice to those of Austrian critics of the classical *Gymnasium* curriculum. Europe's leading physical chemist, Ostwald was not easy to define politically. Like Friedrich Jodl, he was a member of the anticlerical Monist League. But Ostwald belonged to the Berlin-based branch of this movement, which was known for its German nationalism, while Vienna's chapter embraced pacifism.[27] Ostwald was neither a liberal nor a socialist. Better put, he was a technocrat, a strong believer in social engineering. This set him far apart from Austrian liberals, who placed their hopes in individual freedom and the free market. With the coming of the First World War, Ostwald also became a fervent German nationalist.

23. Ibid., p. 11.

24. Ibid., p. 340.

25. Emilie Exner to Marie von Ebner-Eschenbach, 29 July 1905, 9 July 1907, 22 July 1907, folder 81082 Ja, Stadtarchiv, Vienna.

26. Emilie Exner to Helene Bettelheim, undated, 909/35, Manuscript Collection, Österreichische Nationalbibliothek, Vienna.

27. Hacohen, *Karl Popper*, p. 43.

Tellingly, Ostwald had disliked his colleague Serafin Exner from the moment they met in Vienna in 1889. The two men soon became embroiled in a dispute over techniques for measuring contact potentials, in which substantive disagreements were overshadowed by issues of experimental skill and scientific integrity.[28] Serafin accused Ostwald of "superficiality," of changing his position from one moment to the next, and of falsely attributing opinions to him.[29] Meanwhile, Ostwald described Exner in his autobiography as "lacking scientific objectivity."[30] Between the lines there emerged a clash of values.

Unlike the Exners, Ostwald came from a lower middle-class background and was raised in the Baltic provinces of the Russian Empire, far from the centers of German culture. There he had been forced to study Russian and Latvian, followed by French, Italian, and Dutch, on top of Latin and Greek. Speaking in December 1907 in the capital of the multilingual Habsburg Empire, Ostwald declared that all these classes had been a waste of time.[31] In Vienna, Ostwald denounced the teaching of foreign languages in the name of efficiency and progress. The goal of schooling was to turn children into adults who were "rational, logically thinking, and work effectively with their brains." "The arbitrariness in the construction and form of language, which finds its expression in countless exceptions from the rule, kills in the young

28. Ostwald and Exner disagreed over the implications of measurements made with a "mercury drip electrode," an instrument developed by William Thomson and used to determine the potential of liquids and of air (hence its importance for Serafin Exner's research on atmospheric electricity). Since both Ostwald and Exner were opponents of the "contact" theory of electricity, there was no theoretical disagreement at stake. Their dispute was nominally over the value of the potential difference between the mercury electrode and the acid whose potential was to be measured. Serafin took the difference to be zero, while Ostwald took it to be 0.9 V. Each one accused the other of deriving his value from the discredited contact theory. Franz Serafin Exner and J. Tuma, "Studien zur chemischen Theorie des galvanischen Elementes," *Sitzungsberichte der kaiserlichen Akademie der Wissenschaften zu Wien IIb* 97 (1889): 917–57; Wilhelm Ostwald, "Über Tropfelektroden," *Zeitschrift für physikalische Chemie* 3 (1889): 354–58; Franz Serafin Exner and J. Tuma, "Ueber Quecksilber-Tropfelektroden," *Repertorium der Physik* 25 (1889): 597–614; W. Ostwald, *Zeitschrift für physikalische Chemie* 4 (1890): 570; 597–614; Franz Serafin Exner and J. Tuma, "Ueber Ostwald'sche Tropfenelektroden: Zweite Erwiderung," *Repertorium der Physik* 26 (1890): 91–101.

29. Franz Serafin Exner and J. Tuma, "Ueber Quecksilber-Tropfelektroden," pp. 597–99.

30. Wilhelm Ostwald, *Lebenslinien,* vol. 1, p. 256.

31. Wilhelm Ostwald, *Naturwissenschaftliche Forderungen zur Mittelschulreform* (Vienna: Verein für Schulreform, 1908); reprinted in *Die Forderung des Tages* (Leipzig: Akademische Verlagsgesellschaft, 1910), pp. 517–37. On the role of the multilinguistic state in shaping Austrian philosophy, see Allan Janik and Stephen Toulmin, *Wittgenstein's Vienna* (New York: Touchstone, 1973).

mind the feeling for the regularity and for the magnificent order that we find in nature again and again by studying it. It kills the causal sense [*Kausalitätsempfinden*]."[32] Ostwald did not explain whether this causal sense was of physiological or psychological origin. But he made clear that it was a feature of a rational and efficient mind, which could either be sharpened by a training in natural science or dulled by exposure to grammar.

Ostwald's speech outraged Vienna's Association of Friends of the Humanistic Gymnasium (Verein der Freunde des Humanistischen Gymnasiums). Among the members of this group, founded in 1906 to protect the classical curriculum, were Serafin, his brother Sigmund, and Sigmund's wife Emilie. The VFHG's president, Salomon Frankfurter, was the author of a celebratory history of the reform program of the first Franz Exner. These liberal humanists met just weeks after Ostwald's appearance. Their responses focused on the German chemist's characterization of language. Turning his argument on its head, they insisted that the very irregularity of language made it an essential tool for developing a rational mind. Wilhelm Jerusalem, a cofounder with Friedrich Jodl of the anticlerical Ethical Movement, insisted that language must be viewed "psychologically" rather than "logically." "Then this study offers insight into the nature of actual thought and feeling."[33] A less prominent member seconded Jerusalem's point, adding that "the strict regularity of natural laws cannot teach us to understand the illogical in man, his emotional life."[34] From this perspective, the "causal sense" was a poor guide to understanding human nature.

Other Austrians responded to Ostwald by emphasizing the proximity rather than the distance between language and natural science. A pamphlet by philosophy professor Georg Albert argued that both the natural and human sciences rested on empirical observation, "quick intuition," and "thoughtful reflection." Both fields of study trained "the student's sense for faithful observation, precise measurement, and objective thought."[35] Albert opened his essay with a long quotation from Adolf Exner's 1891 rectorial address (see chapter 4), reminding his readers of Adolf's call for a new, antideterministic

32. Ostwald, *Mittelschulreform*, p. 10. Ostwald made a similar point in *Der Energetische Imperativ* (Leipzig: Akademische Verlagsgesellschaft, 1912), p. 394.

33. Wilhelm Jerusalem, *Mitteilungen des Vereins der Freunde des humanistischen Gymnasiums* 5 (1908): 21. On Jerusalem, see Friedrich Stadler, *Studien zum Wiener Kreis* (Frankfurt am Main: Suhrkamp, 1997), esp. pp. 94–95.

34. *Mitteilungen des Vereins der Freunde des humanistischen Gymnasiums* 5 (1908): 25.

35. Georg Albert, *Ein Wort für das humanistische Gymnasium zur Erwiderung an Geheimrat Ostwald* (Leipzig and Vienna: Deuticke, 1908), pp. 29–30.

approach to political education that would teach students to analyze complex social phenomena. A few weeks later, in the education ministry's official hearings on the middle school question, parliamentary representative Robert Pattai likewise stressed that language study sharpened the skills of empirical observation. Pattai argued that languages acquired their irregularities through a natural growth process, just as crystals grew surfaces of unequal areas. In language as in crystals, closer observation revealed the underlying structure.[36]

These proponents of the classical curriculum defended the study of language as a training in empirical reasoning on par with the study of nature. This was an argument characteristic of Austrian liberalism. In Baden in the 1860s, by contrast, even a liberal defender of the classical curriculum like Hermann von Helmholtz had stressed that grammar fell short as a pedagogical tool because it lacked "general laws of unexceptionable validity and of an extraordinarily comprehensive character." Such laws, unique to the experimental sciences, demonstrated "the conscious logical activity of the mind in its purest and most perfect form."[37] Helmholtz, like Ostwald, recommended the study of nature's absolute laws in preparation for what he famously referred to in this address as "the intellectual mastery of the world." Vaunting the value of science for an industrializing society, Helmholtz and Ostwald characterized science in terms of prediction and control. Austria's liberal scientists at the turn of the twentieth century were far more alert to the threat of religious dogmatism. In Austria, teaching students to analyze systems that were stubbornly *unlawlike* was a defense against intellectual rigidity.

Vienna's liberal humanists also rejected Ostwald's call to "individualize" the middle school curriculum.[38] As the VFHG's president Frankfurter tried to make sense of it, Ostwald's argument led from the laws of thermodynamics, to the conditions of happiness, to the structure of education. It was something along the lines of "all the conditions of our lives, if they are ordered normally, must stimulate normal sensations; for that reason work must be designed such that it is accomplished with pleasure." Out of this hodgepodge of utilitarian rhetoric, Frankfurter drew the conclusion that Ostwald's principle of individualization was a destructive form of relativism. He countered: "The school must accustom the student to adapting himself to

36. Reprinted in Robert Pattai, *Reden und Gedanken* (Vienna: self-published, 1909), p. 14.

37. Hermann von Helmholtz, "On the Relation of Natural Science to Science in General," in Cahan, *Science and Culture*, pp. 76–95, on 86–88.

38. See too Uhlig, *Entwicklung des Kampfes*, p. 17.

something larger and to fulfilling duties toward the whole without consideration of his personal comfort."[39] This aspiration to a universal perspective was typical of Austrian liberals, with their emphasis on self-discipline as a counterweight to skepticism and their defense of the supranational principle of the Habsburg Empire.

This dispute with Ostwald highlighted the convictions that set liberals in Austria apart from left-leaning intellectuals in Germany. Ostwald's technocratic inclinations made causal reasoning central to his education program. Austrian liberals, hounded by clerical conservatives, insisted instead that students must learn to come to grips with uncertainty. Against Ostwald's charge that languages killed the causal sense, the Austrians argued that causal reasoning was only one narrow function of a well-trained mind. More important was learning to analyze systems that did not conform to idealized laws, like those actually found in nature and society. Ostwald, like Helmholtz, believed that language's contingent and irregular character made it an insufficient model for the exercise of reason. His Austrian opponents valued language precisely as a system fraught with contingencies, which could nonetheless be analyzed. But the very prominence of uncertainty in Austrian education made it imperative to steel students against the pitfall of solipsism. To their ears, Ostwald's principle of individualization verged on relativism. Echoing the supranational politics of imperial Vienna, they therefore demanded that students reason from a perspective beyond the merely personal.

As we will see, discussion of the ethical status of uncertainty also demarcated Austrian liberals from their domestic political opponents to the left and right. Socialists intended to train students to recognize that natural and social laws were absolute. Catholics meant to teach students to recognize the evidence of design. Only liberals like the Exners treated the confrontation with contingency as a vital element of moral education.

LUDWIG WAHRMUND AND THE "CONQUEST OF THE UNIVERSITIES"

While Ostwald warned against exposing young minds to phenomena that could not be explained on the basis of absolute laws, Austria's Catholic conservatives were also working to safeguard young minds against uncertainty. They insisted that students must not be allowed to contemplate complex phenomena such as cosmic structure or consciousness without the security of a teleological explanation. This position hardened in July 1907, when

39. *Mitteilungen des Vereins der Freunde des humanistischen Gymnasiums* 5 (1908): 10.

Pope Pius X issued his syllabus against the errors of "modern" thought. The list of "errors" concluded as follows:

64. Scientific progress demands that the concepts of Christian doctrine concerning God, creation, revelation, the Person of the Incarnate Word, and Redemption be readjusted.
65. Modern Catholicism can be reconciled with true science only if it is transformed into a nondogmatic Christianity; that is to say, into a broad and liberal Protestantism.[40]

With the pope condemning modern science and rejecting a more flexible Catholicism, liberal university instructors in Austria saw an urgent need to protect their academic freedoms. In September, professors gathered in Salzburg for an annual meeting and resolved to preserve "the independence of the universities." Then, at the Sixth General Catholic Congress in Vienna in November, Mayor Lueger countered with words that resonated in the halls of the universities as a declaration of war. He called for "the conquest of the universities. The universities must not continue to be a soil for subversive ideas, a soil for revolution, a soil for the rejection of the fatherland and religion."[41] Liberal professors answered with a new slogan: "Against the clericalization of the universities!"[42] Shortly thereafter, liberals in parliament sponsored a "motion of urgency" that called for the government to defend the universities against the Christian Socials. This passed after being rephrased more generally as a guarantee of the protection of "academic freedom." Nonetheless, the climate at the universities was now volatile.

The final spark came in January 1908, when an obscure professor of canon law at the law faculty of the University of Innsbruck gave a lecture

40. Pope Piux X, *Syllabus Condemning the Errors of the Modernists*, 3 July 1907.

41. *Neue Freie Presse*, 18 Nov. 1907, quoted in Matthias Höttinger, "Der Fall Wahrmund" (Ph.D. diss., University of Vienna, 1949), p. 18. This dissertation remains the most detailed analysis of the Wahrmund affair. The author, Matthias Höttinger, described the episode as an intrusion of politics into the academy. He accused conservatives and liberals alike of feeding the flames of a debate that would otherwise have quickly burned itself out. Höttinger was deeply critical of Wahrmund and his liberal apologists and sympathetic to their Catholic critics. Höttinger matriculated at the University of Vienna in October 1945, so perhaps his sympathy for Wahrmund's Catholic critics reflected an association between anti-Catholicism and National Socialism.

One of the most interesting contemporary comments on the affair came from Arthur Schnitzler, whose drama *Professor Bernhardi* considered what might have happened had Wahrmund been Jewish. See Schnitzler's letter to R. Charmatz, reprinted in the Schauspielhaus edition, p. 149. On this episode see also Engelbrecht, *Bildungswesen*, pp. 248–50, and Boyer, *Culture and Political Crisis*, pp. 200–1.

42. *Neue Freie Presse*, 21 Nov. 1907, quoted in Höttinger, "Fall Wahrmund," p. 19.

entitled "Catholic Worldview and Free Science," soon published as a pamphlet. The author, Ludwig Wahrmund, had begun to anger his Catholic colleagues as early as 1902 with arguments against religious instruction in public schools. Now he responded to the papal antimodernity decree. Where German physiologist Emil Du Bois-Reymond had attempted to reconcile science and dogma by leaving accounts of consciousness and final causes to theologians, Wahrmund insisted that the two worldviews were, as the pope claimed, simply incompatible.[43] Wahrmund refused to admit any limits to the explanatory power of natural science. That scientists could not yet explain consciousness or final causes was no reason to believe in a cosmos governed by angels and devils:

> Certainly, we are still in obscurity about the essence of force and matter; but that we must therefore believe in Gods who personally descend from heaven, or in men, who personally ascend, no reasonable man will maintain.
>
> Certainly, we are not able to explain the facts of consciousness; but that in one and the same man divine omnipotence and mortal finitude could not be simultaneously joined, is nonetheless entirely certain to us.[44]

Wahrmund continued in this vein and drew such ire from the Innsbruck community that Tyrolean peasants descended on the university to demand his removal. Throughout the spring, German-national students, hostile to the influence of Rome, clashed with Catholics on campuses throughout the empire. In March the papal nuncio in Vienna intervened to have Wahrmund fired.

The next move fell to the academic senate of the University of Vienna, of which Serafin Exner was an elected member for the academic year 1907–8. The Senate concluded unanimously that the nuncio's attempt to control the composition of the law faculty contradicted the principles of the university as they had been established by the Constitution of 1867 and the school law of 1869. These had made the secular faculties independent of clerical influence and protected the freedom of research and teaching.[45] But the dispute was far from over. In May, peasants stormed the university in Graz, to which Wahrmund was scheduled to be transferred. The following month German

43. On Du Bois-Reymond's argument, see Keith Anderton, "The Limits of Science: A Social, Political, and Moral Agenda for Epistemology in Nineteenth-Century Germany" (Ph.D. diss., Harvard University, 1993).

44. L. Wahrmund, *Katholische Weltanschauung und freie Wissenschaft* (Munich: J. F. Lehmann, 1908), p. 5.

45. Höttinger, "Fall Wahrmund," p. 55. See Akademischer Senat, G. Z. 1389 ex 1907/8, Archiv der Universität Wien, Vienna.

national students in Vienna went on strike to express their opposition to the clerical forces. These two events unleashed a flood of debate in parliament. Liberal senators, like the Vienna academic senate, treated the affair as an issue of a freedom guaranteed by the constitution. Christian Socials, on the other hand, argued that Wahrmund had violated the section of the criminal code that prohibited the slander of recognized religions.[46] Minister of Education Marchet took a conciliatory stance, defending the "freedom of scientific research," yet calling on scholars not to contradict "without need" others' religious convictions. As a temporary solution, the government asked Wahrmund to stop teaching for two months. He wisely left the country for a holiday in Spain.

The Exners watched these events unfold with a mixture of satisfaction and frustration. "As much as I lament the conflicts," wrote Emilie Exner of the Wahrmund affair in early June of 1908, "which will likely end badly for an excellent minister of education, nonetheless I must express my pleasure that the academic youth defends itself so energetically against the violations of the clergy."[47]

At stake in Wahrmund's heresies, as in the reforms of the secondary schools, were competing visions of adolescence. Should youth be a period of indoctrination, as the Christian Socials would make it? Of focused specialization, as progressives and technocrats insisted? Or of "freedom," as liberals continued to insist, even as they struggled to define the meaning of this concept? Unlike other historians who have treated the Wahrmund affair exclusively as a political controversy, I mean to take seriously the moral dimension of this conflict. Certainly, as John Boyer argues, the scandal marked the moment of a break between the Austrian bourgeoisie and the intelligentsia and therefore a lost chance for liberalism in Austria. Never again would professors serve as spokesmen for the middle classes, as they had in the 1860s and 1870s.[48] But for Austria's liberal educators academic freedom was not an abstract civil right. It was a necessary condition for the moral development of scholars and citizens.

Wahrmund's inflammatory pamphlet was, as philosopher Friedrich Jodl noted, "a diatribe, not a scientific work."[49] As the dean of Vienna's philosophy

46. Höttinger, "Fall Wahrmund," p. 6.

47. Emilie Exner to Marie von Ebner-Eschenbach, 4 June 1908, folder 81082Ja, Vienna Stadtarchiv, Vienna.

48. Boyer, *Culture and Political Crisis,* pp. 200–1.

49. "Der Fall Wahrmund," *Das Freie Wort* 2 (April 1908).

faculty for 1907–8, Jodl judged this case firsthand. Like many liberal academics, Jodl overlooked Wahrmund's crude arguments and rabble-rousing tactics and defended him in the name of academic freedom. In March 1908, as Wahrmund's fate was being determined, Jodl commented on the proceedings in Vienna's liberal newspaper. Jodl insisted that a professor's freedom of speech in the lecture hall was a necessary condition for the university's task of instructing students "in the most many-faceted manner . . . A right of students to be protected from any views whatsoever . . . cannot exist, because it contradicts the essence of the university."[50] For these words Jodl was attacked as a "diabolical atheist" by a member of the theology faculty.[51]

Jodl's anticlerical views were well known. He was a prominent advocate of full secularization of Austrian schools, committed in particular to establishing ethical education on a secular basis. In the wake of the Wahrmund affair and the Middle School Inquiry, he significantly expanded his theory of ethical education. As in the literary narratives of *Bildung* of the nineteenth century, Jodl attributed formative power to a young person's confrontations with contingency. In his view, "chance" was a pervasive element in human lives. Like most moral philosophers of the period, Jodl meant by chance not events without causes, but our own subjective experience of ignorance about those causes. According to Jodl, we have nothing to fear from chance and everything to learn. Events that appear random tend to reveal regularities under repetition—if they did not, Jodl noted, "Today's experience would be able to teach us nothing about tomorrow. We would always remain children." Moral development, on Jodl's account, was a process of mastering chance. Ethical education was "nothing else but a struggle against chance, the chance of the moment of anger, of mood, of passion."[52] Ethics could indeed be taught, but not as a set of "absolute rules," "because the multiplicity of life's situations leads each absolutely rigid rule or habit of action inevitably ad absurdum."[53] Like the laws of physics, ethical rules were abstractions from experience, necessarily incomplete attempts to grapple with the unexpected.

50. *Neue Freie Presse,* 3 Nov. 1908, p. 8.

51. Margarete Jodl, *Friedrich Jodl,* p. 231.

52. F. Jodl, "Der Begriff des Zufalles: Seine theoretische und praktische Bedeutung" (1904), reprinted in *Vom Lebenswege,* pp. 515–33, quotation on p. 533. See too Jodl, "Zufall, Gesetzmäßigkeit, Zweckmäßigkeit," speech delivered in the Academy of Sciences, Vienna, May, 1911, in *Vom Lebenswege,* pp. 533–48.

53. F. Jodl, "Die Lehrbarkeit der Moral," in *Das Problem des Moralunterrichts in der Schule: Zwei Vorträge* (Frankfurt am Main: Neuer Frankfurter Verlag, 1912), pp. 30–45, quotation on p. 33.

Another interpretation of the lessons of the Wahrmund affair came from Carl Menger, the liberal economist and member of the upper house of parliament. In early 1908 Menger delivered an address in parliament on Lueger's threatened "Conquest of the Universities." Menger's speech was a sketch of the conditions of intellectual progress. Human history had advanced from a blind trust in authority, a "rigid dogmatism," to "free progressive science;" from the age of "patriarchy," the "childhood of humanity," to the age of "objective research."[54] In the Enlightenment tradition of Auguste Comte, Menger compared human history to an individual life, designating empirical science as the full flower of intellectual maturity. The path to adulthood, whether individual or collective, required the rejection of received ideas in favor of the evidence of reason and experience. Like Jodl, Menger treated Wahrmund's case as an opportunity to defend a model of moral development central to Austrian liberalism.

In this liberal vision of adolescence, the experience of uncertainty played a pivotal role. As liberals like Jodl and Menger intervened in defense of academic freedom, they invoked a familiar trope of *Bildung* as a confrontation with chance. Through an ever widening sphere of experience, the individual came to grips with life's contingencies and renounced the false comforts of religion and authority. The Wahrmund scandal was, fundamentally, a conflict over how to train future citizens. The Christian Socials hoped to use education to pass on a sense of responsibility to the state and an acceptance of the social order. The liberals, by contrast, wanted to use education to guide students from a blind acceptance of received knowledge to a critical engagement with the world. They understood the university's "essence" to be an open environment in which students could mature into independent thinkers.

THE PROBABLE AND THE MERELY POSSIBLE

A few months before Wahrmund burst onto the scene, Serafin Exner had released a petite, popular book on astronomy, which, at first glance, was a world removed from the concurrent turmoil of academic politics. Its few copies were printed in an old-fashioned font with archaic spellings, strewn with Latinisms, and embellished with rhyming verses. Published independently, the book was dedicated to an anonymous young woman, identified

54. Menger, "Die Eroberung der Universitäten," reprinted in A. J. Peters, *"Klerikale Weltauffassung" und "Freie Forschung": Ein offenes Wort an Professor Dr. Karl Menger* (Vienna: Georg Eichinger), pp. 7–9.

by other Brunnwinklers as the family's friend Grete Conrad. The title page identified the author only with Greek initials: Ω. Σ., that is, "Onkel Serafin." This playful style reflected the Exners' love for riddles and costumes, and it suited the typical tone of after-dinner conversation when the family gathered on summer evenings. This was, in fact, the setting in which Serafin first read the book aloud.[55] Earlier discussions at Brunnwinkl of Boltzmann's natural philosophy and Einstein's theory of special relativity had primed Serafin's younger relatives to appreciate the work.[56]

Part 1 of the *Simple Astronomy* was an account of the present state of the solar system. It began, however, with a coming-of-age story:

> When man reaches the age when an inclination awakens in him to regard his surroundings critically he has usually not seen much of the world; his village or his town with the neighboring mountains . . . are as a rule all that he knows and from his elders he hears from childhood on that things were no different in the past . . . Since the fathers and grandfathers have known the same mountains and rivers as the grandsons, it is understandable that one gets used to considering the earth as an unshakeable fundament. No wonder then that nothing gives a man such a panicked fright as the first earthquake that he experiences, since with the first quiver of the ground crumbles too the inherited feeling of security and peace that the earth affords.[57]

As in a *Bildungsroman,* Exner introduced the reader to a young man impatient to leave his birthplace in search of new experiences. It is to be expected that the hero will face a loss of certainty, a descent into doubt, and a break with the thinking of his parents. This story continued at the opening of part 2, an account of the solar system's evolution:

> As in an earthquake, what begins to shake is just what we have since childhood grown used to regarding as the one solid thing in the world, and so in an instant our previous faith proves to be false and one-sided; just so, when setting foot in a distant land, it appears that the new world does not agree with the old and familiar one . . . [G]radually we will find our way, then see with some surprise that what appeared to us a new world is really just the old one, if in another form; but the core has remained the same. For whom would such a recognition not yield a certain satisfaction?[58]

55. According to Hans Frisch, *50 Jahre Brunnwinkl,* p. 10.

56. Franz Exner, *Exnerei,* p. 22; Karl Frisch, *Fünf Häuser am See,* p. 66.

57. Franz Serafin Exner, *Der Schlichten Astronomia* (Vienna: self-published, 1908), p. 42.

58. Franz Serafin Exner, *Astronomia,* p. 138.

This account of an intellectual coming of age employed two vivid metaphors. The first, travel, was a familiar trope of the *Bildungsroman* and a central theme in Serafin's account of his own life. The second was the earthquake, an image of the uncertainty one experiences when leaving behind the familiar world of accepted beliefs. Serafin's father had likewise written of the moment in a *Gymnasium* education when the "ground of experience . . . becomes shaken by doubts." The earthquake was an aptly chosen metaphor. Serafin's contemporaries judged that the emergence of a view of earthquakes as natural rather than supernatural was a key stage in the evolution of scientific thought.[59] At the start of the twentieth century, earthquakes represented a final, uncolonized frontier of the campaign to bring nature under human calculation and control. The earthquake stood for an element of uncertainty that could be neither eliminated nor ignored.

Part 2 of *The Simple Astronomy* outlined a strategy for coming to grips with such uncertainty. Here Serafin introduced the reader to the stochastic perspective of Maxwell and Boltzmann. The universe consists of bits of matter in random motion; the rise and fall of mountains, the flow of rivers, all these arrangements of matter are merely "chance states."[60] The reader must now be taught to reason in the wide world where certainty is unattainable. In the final chapter, after proposing to lead the reader through the solar system's history, Serafin issued the following warning:

> In turning to such questions I ask you, dear reader, to consider that we are leaving the ground of facts and of certain knowledge . . . Yet in doing so by no means do we want simply to give our imagination free rein; on the contrary we will take twice as much care with it here, on uncertain ground; and so we want to produce, if not always what is certain, then indeed only what is thoroughly probable.[61]

Now that the reader had accepted a mature view of the universe as contingent, she would have to take care to distinguish between degrees of certainty, between the probable and the merely possible, in order not to fall into blind speculation. A few pages later Serafin remarked, "There we are admittedly reliant on conjectures, nonetheless a few things may be said with a certain probability."[62] And shortly thereafter: "As we promised, we do not want to overreach the very probable and lose ourselves in the merely possible."[63]

59. Mach, *Analyse der Empfindungen*, 2nd edition (Jena: G. Fischer, 1900), p. 210.

60. In "Exner's Indeterminist Theory" Stöltzner perceptively notes that Exner shifted here from a Newtonian to a Boltzmannian framework.

61. Franz Serafin Exner, *Astronomia*, p. 235.

62. Ibid., p. 240.

63. Ibid., p. 246.

The cosmic theory that Serafin treated so gingerly in this chapter was the nebular hypothesis, first proposed by Laplace at the end of the eighteenth century. Laplace theorized that the solar system originated in a primordial cloud of matter, in which gravitational attraction caused bits to lump together into chunks and finally into planets. Laplace had proposed this theory to counter Daniel Bernoulli's claim that the organization of the planets revealed the hand of God. Appropriately for Serafin's purposes, Laplace had used the mathematics of probability to calculate that the planetary orbits could in fact be a product of pure chance. Serafin's text thus recapitulated a historical as well as individual progression from certainty to doubt to probability. Serafin used the concept of probability to guide the young reader safely through her formative encounter with contingency.[64]

The Simple Astronomy appeared in February 1908, shortly after Wahrmund's controversial pamphlet and the close of the Middle School Inquiry. It was a subtle and skillful intervention in the education politics of the day. Like Jodl and Menger, Serafin worked with an implicit analogy between individual development and human history. He associated childhood with blind trust in authority and adulthood with empirical science. At another level, his textbook was a defense of the status of scientific hypotheses, both as a general principle and in the specific cases of evolution, atomism, and the nebular theory of the solar system. Each of these theories contradicted Catholic theology and risked being banished from classrooms by the Christian Socials. As one contributor to the Middle School Inquiry remarked, it was necessary to introduce students to hypotheses as a mode of scientific reasoning—without, of course, losing sight of their merely probable status.[65] Similarly, the *Simple Astronomy* showed young readers how to handle hypotheses conscientiously. Following a characteristically Austrian-liberal model of moral development, the book led the reader through the formative experience of uncertainty, then provided her with the philosophical tools to emerge from the pit of skepticism.

Four months later, Serafin made up his mind to intervene more directly in the politics of education. In June he was unanimously elected Rector Magnificus of the University of Vienna. As he wrote in response to a note of

64. On the politically progressive connotations of the nebular hypothesis in nineteenth-century Britain, see Simon Schaffer, "On Astronomical Drawing," in *Picturing Science, Producing Art*, ed. Peter Galison and Caroline Jones (New York: Routledge, 1998), pp. 441–74.

65. Quoted in Salomon Frankfurter, *Verlauf und Ergebnisse der Mittelschulenquete des Unterrichtsministeriums 21.–25. Jänner 1908 und andere Beiträge zur Geschichte der österreichischen Mittelschulreform* (Vienna: Carl Fromme, 1910), p. 116.

congratulations from Friedrich Jodl, "it was really the difficult times, from which no one can escape, which moved me to take over an office that I would perhaps rather have avoided, and the fact that I see myself met with confidence from all sides; may it be merited!"[66] Thanks to his sense of duty and the trust of his colleagues, Serafin now held responsibility for defending academic freedom against Lueger and his followers. The faculty had lived through months of armed clashes and student strikes, and they looked to the new rector to guarantee a peaceful academic year.

KANTIAN MOLECULES AND THE LIBERAL STATE

Serafin's inauguration took place on 15 October 1908, "in the usual festive manner and with the highly unusual lively participation of the students." The university's large ceremonial hall was packed with "the full academic Senate, numerous University professors, and many ladies."[67] As the press made clear, Serafin's speech was hotly anticipated. Those who awaited a defense of academic freedom would not be disappointed. But most would be surprised by the form it would take.

Sixteen years earlier, Serafin's brother Adolf had delivered his own rectorial address from that same podium. Adolf had charged that the study of natural science had accustomed students to analyzing deterministic systems, leaving them helpless before the intricate causal web of the social realm. He had called on the universities to provide a new basis for "political *Bildung,*" founded on the ability to analyze complex, historical phenomena—"the probability calculus on whose correct use all practical statecraft rests."[68] Adolf had died suddenly three years later, at the height of his career, but liberals continued to cite his rectorial address.[69] Now, using language strikingly similar to Adolf's, Serafin used this opportunity to defend the value of his own field for training the citizens of a liberal state.

Serafin's argument rested on an insight due to the latest Viennese physics: that an orderly system could arise out of an aggregate of strictly independent

66. Franz Serafin Exner to F. Jodl, 28 June 1908, folder IN 13340, Stadtarchiv, Vienna.

67. "Rektorsinauguration an der Wiener Universität," *Neue Freie Presse* (evening edition) 15 Oct. 1908, p. 9.

68. Adolf Exner, *Über politische Bildung,* p. 12.

69. For instance, just two days after Serafin's inauguration a writer in the *Neue Freie Presse* cited Adolf's speech in an article on civics education in progressive schools (*Neue Freie Presse,* 17 Oct. 1908, p. 26). The humanist Georg Albert likewise cited Adolf in his defense of the classical *Gymnasium* curriculum (see above).

and unconstrained individual components. This was the lesson of Boltzmann's statistical reformulation of the second law of thermodynamics. Using this new understanding of the laws of physics, Serafin turned Adolf's critique of natural science on its head. Adolf had insisted that political factors are often inaccessible to measurement, requiring analytical methods distinct from those of the sciences. Serafin agreed that in the social sphere "visible events are not the only factors, rather all those imponderables that cannot be captured; but the physicist finds himself in the same situation. By imponderables we mean events whose physical correlate remains unknown to us, and the physicist is all too often in a position to come to terms with these."[70] The physicist, like the historian or statesman, had to learn to work with imperfect knowledge. Serafin was now in a position to argue that physics, far from inculcating rigid modes of thought, was actually an ideal training ground for grappling with uncertainty.

Where Adolf had attacked science for obstructing political education, Serafin demonstrated that the study of physics actually taught crucial political lessons. He began, like his brother, by stressing the similarity between physical and juridical "laws." The laws of physics "do not exist in nature," he claimed; "they are formulated by man, who uses them as a linguistic and arithmetic aid, and he implies merely that events in nature proceed as if matter, like a rational being, would obey these laws."[71] Exner had begun to develop this analogy between laws in science and statecraft in an earlier popular address.[72] In 1890 he had given the comparison a Bismarckian cast, stressing that a gas, like a *Rechtstaat* (a constitutional state), was subject to inexorable laws. Now, after a decade of municipal government by the Christian Socials, he drew from the same analogy distinctly libertarian implications. According to Boltzmann's version of the second law, "every ordered state" is "counter-natural," possible only on account of "an intervention against the law of chance." But social order was equally counter-natural: "All juridical laws, written and unwritten, which keep society on an orderly course, are in fact nothing other than such interventions."[73] A noninterventionist government was therefore the most natural form, allowing society to evolve toward its most probable state.

70. Franz Serafin Exner, *Über Gesetze in Naturwissenschaft und Humanistik, Inaugurationsrede* (Vienna: Hölder, 1909), p. 71. He dated the speech 28 Aug. 1908, St. Gilgen.

71. Franz Serafin Exner, *Über Gesetze*, p. 49.

72. Franz Serafin Exner, "Über unsere Atmosphäre," speech delivered 19 Nov. 1890, published in *Populäre Vorträge* (Vienna: Verein zur Verbreitung naturwissenschafticher Kenntnisse, 1891), pp. 129–63.

73. Franz Serafin Exner, *Über Gesetze*, p. 63.

Which state would that be? Serafin insisted that this equilibrium state would not be homogeneous. He explained that the circulation of money depended on the self-interest of innumerable independent individuals and thus had the character of a chance phenomenon. The "most probable state" therefore was not an equal distribution of wealth—"that would instead be highly improbable." There should instead emerge a large middle class and smaller upper and lower classes, with all deviations from the average income distributed along a normal curve. Exner dismissed legislative attempts to counter this natural effect as wholly irrational. The most highly evolved society implied the perfect independence of each individual. The liberal (nation-) state was the most probable (physical) state.

Compared to representatives of statistical thinking elsewhere at the time, Exner was unusual in taking the collective as the basis for his analysis, yet attributing freedom and individuality to its members. As he would later put it, "Where herdlike uniformity ends, where the individual emerges from the mass, there begins culture."[74] This move reflected Vienna's liberal tradition of economic thought. By 1908, Carl Menger's students Eugen von Böhm-Bawerk and Friedrich von Wieser were leaders of a distinct "Austrian School" of economics. Serafin's link to the Austrian School was personal as well as intellectual, since Wieser's daughter Anna married Adolf Exner's son Franz in 1910. In a strategy characteristic of Austrian liberalism, Serafin's rectorial address thus took aim at two sets of opponents. While defending academic freedom against clericals on the right, he also defended economic freedom against socialists on the left.

In his closing remarks, Serafin brought the weight of his statistical worldview to bear on the inescapable issue of the day. Culturally as well as economically, he explained, a free society evolves toward a state of maximal differentiation. Serafin praised his alma mater as an illustration of such an evolutionary path. Broaching at last the matter of academic freedom, he portrayed the modern university with its multifarious specializations as a system ideally adapted to a state of freedom:

> Is the university as an institution itself not the result of a thousand-year evolution? Certainly this too is only the accumulation of a large number of chance events, under the influence of which the university as it is today emerged as the most functional [*zweckmässigste*] state. But we know that such an evolution continues only so long as the external conditions do not change. Now, for the

74. Franz Serafin Exner, *Vom Chaos zur Gegenwart. Eine kulturhistorische Studie*, Sig. I888.089 MS, Library of the University of Vienna, p. 433.

> university there has always been essentially only one such condition to which its entire constitution and its deepest essence is adapted, that is the absolute freedom of research and teaching. So long as this is preserved for us, and it will be preserved, our alma mater will be an ever faithful reflection of the entire world.[75]

Throughout his address, Exner had built up a series of nested models of a liberal society, packed together like a set of Russian dolls. He had compared the university to a "large" and "many-branched" family and to a society in miniature, and had modeled these all on an ideal gas. A gas in equilibrium, a family at leisure, a modern university—the structure of each system depended on the full independence of its members. But could the lessons of smaller scales apply to larger ones? This was a question that would plague Exner during his term as rector.

Liberals, at least, were pleased with Exner's performance. He had shown liberalism to be the only "natural" state for the university and society at large. According to a report in the liberal *Neue Freie Presse,* "The ingenious discourse of the renowned physicist was received with tumultuous, prolonged applause."[76]

ACADEMIC FREEDOM AT WHAT PRICE?

During his first term as rector Serafin Exner mediated some of the most violent student clashes since 1848. He also faced ongoing conflicts with the Christian Socials. In one case, an impoverished student at the law faculty, publicly identified only as "R. R.," had been offered a scholarship sponsored by the municipal government. This was an exceptional opportunity for R. R., who until this point had to work in a law office to support his studies. Before receiving the scholarship, however, R. R. was made to sign the following oath, which one commentator in the *Neue Freie Presse* called "Dr. Lueger's formula":

> R. R. makes it known herewith that he, first, is not presently a Social Democrat, nor will be one in the future; secondly, is not presently a German nationalist [*Alldeutscher*], nor will be one in the future; thirdly, did not participate in the strike and in the future will not participate in any strike.[77]

R. R. had apparently never taken part in the June student strike nor in any other political movement. Yet after signing the declaration, R. R. regretted

75. Ibid., p. 87.

76. "Rektorsinauguration an der Wiener Universität," *Neue Freie Presse* (evening edition) 15 Oct. 1908, p. 10.

77. *Neue Freie Presse,* 2 Nov. 1908, p. 6.

having sworn never to participate in the future. He asked to have the form back, but the city government refused to return it. To the *Neue Freie Presse*'s commentator, this had all the signs of a pact with the devil. According to the writer ("an expert"), the "principal thing" was that the student was being forbidden ever to change his opinions. The oath was at odds with a student's ability to "develop freely."[78] To the liberals, the principle of self-directed development was inviolate. The Christian Socials, on the other hand, viewed freedom at such a young age as an invitation to rebellion.

Student protests, even violent ones, had long been excused as a rite of youth, and were even celebrated in the memorials to 1848. Yet events early in Exner's term as rector shook this complacent view. A group of students in Vienna from the Habsburg lands in northern Italy had petitioned in parliament for a university in Trieste, but they had received no response. They approached the new rector and received his promise to raise the matter at the next meeting of the academic senate in November, which he did.[79] Meanwhile, on the tenth of November, violence erupted between Zionist and German-national students, and Exner was hard pressed to mediate a standoff. This was his first experience as peacekeeper, and he discovered that the university's autonomy from the state entailed a burden as well as a privilege. The fighting had taken place on the palatial circular ramp leading from the Ringstrasse up to the university's main entrance. The ramp was officially public land, not part of the university's grounds, and therefore subject to police jurisdiction. As Exner commented in an interview with the *Neue Freie Presse,* the students had at least respected the university grounds. But the police had been reluctant to intervene because the ramp was only nominally within their jurisdiction. The university's independence from the state powers could thus leave it vulnerable. Its unique autonomy rested on the tenuous assumption that the academic community would keep politics outside its walls.[80]

Exner had barely recovered from this episode when two weeks later Italian students lost patience with the stalemate over the issue of a university in Trieste.[81] They received the rector's permission to demonstrate in the university courtyard on the condition that they refrain from singing

78. *Neue Freie Presse,* 2 Nov. 1908, p. 7.

79. See his letter to the Ministry of Culture and Instruction, 7 Nov. 1908, Akademischer Senat G.Z. 311, ex 1908/9, Archiv der Universität Wien, Vienna.

80. *Neue Freie Presse,* 11 Nov. 1908, p. 12.

81. In between, he had held off a fight by negotiating separately with the German and Italian students. See his letter to the Ministry, 18 Nov. 1908, S.Z. 311 ex 1908/9, Archiv der Universität Wien, Vienna.

patriotic songs. According to the German nationalist students, who had likewise filled the courtyard, the demonstrators eventually launched into a hymn to Garibaldi. Deciding that the Italians had "broken their promise," the German-nationalists tried to push them out of the courtyard. In the scuffle, over a dozen students were wounded, and three Italian students were arrested on suspicion of having shot revolvers. This was one of the most violent incidents at the university in decades. Liberal commentators were particularly shocked at the use of revolvers. The entire university and all other *Hochschulen* in Vienna were closed for the rest of the week. In an interview with the *Neue Freie Presse* Serafin Exner called the fighting "acts of the grossest barbarism."[82] In an official announcement released the same afternoon, he denounced the incident as "a deep wound to the sanctity of academic ground, a mockery of academic order, a disgrace for the perpetrators. What certain Italian students carrying deadly weapons have undertaken here, in the cause of culture, was not a demonstration within the bounds of the concepts of honor of citizens and students. These were crimes."[83] Exner's message was that students had to earn the privilege of academic freedom by demonstrating a bourgeois sense of honor, founded on self-discipline.[84]

In his final address as rector, Exner reminded his audience of the defense of academic freedom he had offered at his inauguration. Reflecting on the lessons of the previous months, he drew the year to a close on a more circumspect note. He addressed himself to the students in the audience:

> What counsel shall I give you as I take my leave? I can only recall to you the words which I once before had occasion to address to you: In your hands the most precious possession of the university has been laid: academic freedom. Protect it for your own well-being and for the well-being of those who come after you; but do not believe that the professors or the academic authorities or any power of the world whatsoever can preserve it for you, if you do not do so yourselves. But you can do it; you can preserve this freedom uniquely if you do not abuse it. Take this advice to heart, follow it and you will be certain of success, notwithstanding all those to whom this freedom is a thorn in the eye.[85]

Over the course of this challenging year, Exner's support for academic freedom had not wavered, but his tone had changed. In admonishing the students not to "misuse" their freedom, Serafin recalled his message of

82. *Neue Freie Presse*, 24 Nov. 1908, p. 4.

83. Ibid.

84. On *bürgerliche* notions of honor, see Frevert, *Men of Honour*.

85. *Bericht über das Studienjahr 1908/1909* (Vienna: Adolf Holzhausen, 1909), p. 21.

November, when he spoke of the "sanctity" of the university and the "concepts of honor" appropriate for its members. Freedom was not simply an "external condition" for the development of young intellects; it was a privilege they would have to earn for themselves. In doing so, they would have the satisfaction of confounding those opponents of liberalism to whom academic freedom was as irritating as "a thorn in the eye."

THE events of 1907–9 forced the Exners to be explicit about what they meant by "freedom." Freedom was the fertile soil of the liberal society the Exners cultivated at Brunnwinkl. It was the precondition for an adolescent's formative experiences of doubt and uncertainty. Only in conditions of freedom could young people learn to contend with the foreign, the unexpected. Only through freedom would they learn to act responsibly in an uncertain world. The alternative was a life spent clinging to the child's trust in received ideas, a life of dogmatism.

For this reason, Serafin Exner attributed moral value to a stochastic worldview. In a universe of randomly colliding particles, no description at a macroscopic scale could be more than an approximation. To borrow a metaphor from the Exners, the recognition that determinism was an illusion would pull the firm ground of certainty out from beneath a student's feet. But a foothold would be found in the theory of probabilities. With probability as her guide, the student could see that order in the system as a whole depended on the freedom of each individual. This was the statistical physicist's ultimate lesson to future citizens of a liberal state.

In 1908 the Exners seized opportunities to apply the lessons of their utopian experiment at Brunnwinkl to the empire at large. Bourgeois confidence drove them to present the family as a model for the university and the university as a model for the empire, just as the physicist saw society mirrored in a flask of gas. Yet the results of the Middle School Inquiry and of Serafin Exner's year as rector cast doubt on this analogy of scale. Lessons drawn from a bourgeois household did not necessarily apply to an empire in the throes of democratization.

CHAPTER EIGHT

Into the Open

Measuring Uncertainty, 1900–1918

DORMICE and pine martens lodged in the attics at Brunnwinkl; a woodpecker made a cageless home for himself in one cottage, along with a fox, brought by a relative from Germany "to give him his freedom."[1] A pet donkey was housed in a stall attached to one of the cottages, which Karl von Frisch later converted into quarters for his "museum." Nature entered through every crevice.

The Austrian summer home's distinguishing physical feature was its lack of insulation from its immediate surroundings. The open, rustic architecture was designed to complete the romantic spell cast by the Alpine landscape. From the many windows and long balconies of an alpine cottage, it was said, one could "take in the image of the beloved landscape."[2] The permeability of the cottages was suited to Karl von Frisch's passion for collecting. "Karl, a creature!" relatives called to alert him to a beetle or wasp. Within the community, people and animals alike roamed freely. "Pests" as such did not exist.[3] This absence of internal boundaries depended on the myth of external ones. Like the Rose House in Stifter's *Nachsommer,* the summer home seemed to its inhabitants to be naturally insulated from the transient modern world.

In the history of science, remote domestic settings are not unknown. Indeed, like the fiefdom of an early modern aristocrat or the country house of a

1. Karl Frisch, *Fünf Häuser am See,* pp. 71–72.

2. Karl Kraus, quoted in Hans Haas, "Die Sommerfrische," p. 368.

3. On the social construction of "pests," see Sarah Jansen, "An American Insect in Imperial Germany: Visibility and Control in Making the Phylloxera in Germany, 1870–1914," *Science in Context* 13 (2000): 31–70.

FIGURE 11 Karl von Frisch and Sigmund Exner in the "museum" at Brunnwinkl, probably before the First World War. Source: Karl Frisch, *Fünf Häuser am See. Der Brunnwinkl: Werden und Wesen eines Sommersitzes*, fig. 20, p. 62. © Springer Verlag, Berlin, 1980. With kind permission of Springer Science and Business Media.

nineteenth-century English gentleman, Brunnwinkl's pastoral situation appeared to protect it from the vagaries of politics or fashion.[4] But the Austrian summer home was unusual in its permeability to the elements, as well as its intimate style of social interaction. As a site of science, the Austrian summer home was unique.

By contrast, since the seventeenth century the typical laboratory had been "a disciplined space," "a better space in which to generate authentic

4. On the former, see Owen Hannaway, "Laboratory Design and the Aim of Science: Andreas Libavius versus Tycho Brahe," *Isis* 77 (1986): 585–610; and J. R. Christianson, *On Tycho's Island* (Cambridge: Cambridge University Press, 2000). On the latter, see Simon Schaffer, "Physics Laboratories and the Victorian Country House," in *Making Space for Science: Territorial Themes in the Shaping of Knowledge*, ed. Crosbie Smith and Jon Agar (New York: St. Martin's Press, 1998), pp. 149–80.

knowledge than the space outside it in which simple observations of nature could be made."[5] The new nineteenth-century urban laboratories were designed to be maximally insulated in order to give experimenters optimal control over the conditions of measurement. When Louis Pasteur sought to prove the efficacy of a vaccine at Pouilly-le-Fort, an undisciplined space far from his laboratory, he did all he could to control the boundaries of this "natural" site.[6] Science at the *Sommerfrische* inverted this strategy. At Brunnwinkl, insulating measuring instruments from external "disturbances" was as inappropriate as eliminating "pests" from the summer cottages or excluding them from Karl Frisch's museum.[7] This chapter follows the movement of experimental strategies from the summer home to the urban laboratory, tracing along the way the history of a probabilistic turn in Austrian physics.

TAKING NATURE'S PULSE

In the 1880s, as the Exners were setting up their lives at Brunnwinkl, Serafin Exner devised a research program in atmospheric physics that would bring him and his students from their urban laboratory into the open, "*ins Freie*." Their topic was the electrical potential of the earth's atmosphere, part of the broader European program of electrodynamical research at the end of the nineteenth century. Exner argued that atmospheric research could not be conducted in Vienna, since in cities such measurements furnished "only a picture of the local pollution."[8] He was not alone in this judgment, as Europe's major mountaintop observatories were built in the 1880s.[9] Exner insisted that not only the location but also the design of the observing station was crucial. Best of all was a shelter maximally exposed to the elements.

5. Steven Shapin and Simon Schaffer, *Leviathan and the Air Pump* (Princeton: Princeton University Press), p. 39.

6. Bruno Latour, *Science in Action* (Cambridge, Mass: Harvard University Press, 1987). Pasteur's tactics went beyond the management of the site's borders, as Gerald Geison showed in *The Private Science of Louis Pasteur* (Princeton: Princeton University Press, 1995).

7. On field science, see Henrika Kuklick and Robert Kohler, eds., "Science in the Field," *Osiris* 11 (1996). On the "borders" between field and laboratory, see Robert Kohler, *Landscapes and Labscapes* (Chicago: University of Chicago Press, 2002).

8. Franz Serafin Exner, "Über transportable Apparate zur Beobachtung der atmosphärischen Elektrizität," *Repertorium der Physik* 23 (1886): 657–69, quotation on p. 658.

9. On British "cloud physics" and the Ben Nevis observatory, see Peter Galison, *Image and Logic: A Material Culture of Microphysics* (Chicago: University of Chicago Press, 1997), chapter 2. On other aspects of field physics in the late nineteenth and early twentieth centuries, such as observations of cosmic rays and earthquakes, see ibid., chapter 3, and Galison, *How Experiments End* (Chicago: University of Chicago Press, 1987), chapter 3.

This was a startling statement from a late nineteenth-century physicist. Physicists at this time defined an experimental space by its insulation from external disturbances; this insulation alone guaranteed the precision of measurements conducted within.[10] In Vienna, Exner struggled to make his own laboratory measure up to this standard. The three physics institutes of the University of Vienna were a short walk from the university. From the Ring, one headed down Währingerstrasse, one of the largest thoroughfares in Vienna, with heavy electrical streetcar traffic, and turned right half a block down Türkenstrasse. The building itself was an old customs house with little insulation against the electrical disturbances from the street. The rooms assigned to Exner and his students were spread throughout the third, fourth, and fifth floors, which were least insulated from "commotion" in the rest of the building. It had proven too expensive even to heat the room used for the experimental physics class. Compounding the professors' problems, the lecture room lacked podium and benches, so students were forced to write on their knees. Between the state of the building and its congested location, Serafin Exner declared it in the 1890s "absolutely unusable" for the purposes of experimental physics. An 1896 letter to the dean of the Philosophy Faculty signed by the directors of the three Physical Institutes, Exner, Boltzmann, and Viktor Lang, painted a dismal picture of conditions in the laboratories:

> Any precision measurements whatsoever are impossible from the start due to the constant shaking [of the building]; finer measurements cannot be carried out not only during the day but even at night, in part due to the traffic on Währingerstrasse and due to the wind, in part due to the constant currents in the adjacent houses. Moreover, the spaces are extremely impractically arranged and far too small for scientific purposes, such that often due to the presence of the observer in the room the temperature there rises in a troublesome way.[11]

In sum, throughout the three institutes, "not one solid spot, shaft, or floor is to be found where an instrument could be set up; in all three not one room is to be found in which a relatively constant temperature could be produced and none in which magnetic measurements could be carried out . . . [N]one of the provincial universities—not to mention Graz or Prague—is in this respect

10. On the problem of laboratory insulation in late nineteenth-century Berlin, see David Cahan, *An Institute for an Empire: The Physikalisch-Technische Reichsanstalt, 1871–1914* (Cambridge: Cambridge University Press, 1989).

11. Philosophische Dekanat 1895/6, Akt 117, Archiv der Universität Wien, Vienna.

even remotely so poorly provided for as Vienna. Even the local middle schools have far more suitably arranged laboratories than the university."[12] Indeed, Erwin Schrödinger once remarked wryly that even the alpine cottage where he studied atmospheric electricity was better constructed than Exner's Türkenstrasse laboratory.[13] Not until 1913 did Viennese physics obtain a suitable home.

In practice, then, disturbances could not be blocked out. They had to be compensated for in other ways. By the late nineteenth century two approaches to the shortcomings of insulation existed. At this time, much of physical science centered on the measurement of physical constants. Measurements that fluctuated were therefore suspect, and physicists tried to suppress fluctuations *physically* by insulating or adjusting their instruments.[14] A second option was to suppress fluctuations *statistically:* physicists treated deviations from a measured average as random error. In general, physicists attributed variability in their measurements to outside disturbances, a margin of uncertainty they called "isolation error."

In the field, on the other hand, Serafin Exner and his students used the term "disturbances" not as a measure of subjective uncertainty but to identify real objects and phenomena. So, for instance, one student defined "disturbances" by determining the distortion of the earth's local electric field due to hills and valleys, telegraph poles, church towers, even a standing person.[15] In the Alps, Exner deemed insulation as detrimental to measurements as it was essential in the city. The very presence of a building altered the local topology and made it almost impossible to standardize measurements of atmospheric potential. This is why the open design of an "alpine" or "Swiss" cottage was conducive to this research. The very best condition for

12. Ibid.

13. Erwin Schrödinger, "Beiträge zur Kenntnis der atmosphärischen Elektrizität LI: Radium-A-Gehalt der Atmosphäre in Seeham 1913," *Sitzungsberichte der Kaiserlichen Akademie der Wissenschaften zu Wien IIa* 122 (1913): 2023–67, quotation on p. 5; *Gesammelte Abhandlunge, Collected Papers* (Vienna: Verlag der Österreichischen Akademie der Wissenschaften, 1984), vol. 1, p. 79.

14. On compensation methods in radioactivity measurements, see Coen, "Scientists' Errors." On the surprising productivity of disturbances in physics labs, see Christoph Hoffmann, "The Design of Disturbance: Physics Institutes and Physics Research in Germany, 1870–1910," *Perspectives in Science* 9 (2001): 173–92.

15. Hans Benndorf, "Über die Störungen des normalen atmosphärischen Potentialgefälles durch Bodenerhebungen" and "Über gewisse Störungen des Erdfeldes mit Rücksicht auf die Praxis luftelektrischer Messungen" ("Beiträge zur Kenntnis der atmosphärischen Elektricität. VI and XXIII"), *Sitzungsberichte der Kaiserlichen Akademie der Wissenschaften zu Wien IIa* 109 (1900): 923–40 and 115 (1906): 425–56.

measurement was a complete absence of civilization, the only disturbance then being the physicist himself. Exner's program for atmospheric electricity research was the ultimate example of science "in the open."

To realize this ideal Exner designed a portable electroscope, a device for measuring the electrical potential of the air. The gadget fit snugly in a pocket, and he assured his colleagues that the accompanying conducting rod looked just like an alpine hiker's walking stick. But the most important feature of the device was that it did not need to be charged before each measurement. This would allow the physicist to watch the fluctuations of the atmospheric potential as they unfolded over time. As Exner explained, "A sustained observation of the [indicator] gives a clear image of the processes simultaneously occurring in the air."[16] Here the contrast between laboratory and alpine physics becomes clear. In the lab, physicists shielded their instruments to ensure a steady reading and labeled any remaining variability "error." In the field, physicists instead treated variations as further clues to the nature of the phenomena under study. As Exner's favorite writer, Goethe, had counseled, the naturalist must learn to follow the rise and fall of the barometer and to see in it the respiration, the variable pulse, of the earth itself.[17] Exner suspected that what he was seeing in the fluctuations of his electroscope was the vertical movement of the negative charge carried by fluctuating water droplets in the atmosphere.

Equipped with such a portable electroscope, Exner's students carried atmospheric research into the Alps and beyond. The resort area of Seeham, where Exner's student Erwin Schrödinger monitored his electroscope one summer in a lakeside house, was close to Brunnwinkl in the Salzkammergut. There, too, research was shaped by the architecture and culture of the summer retreat, and Schrödinger's "disturbances" were often young women.[18] Other researchers went farther afield. Eager to compare measurements from very different climates, Serafin traveled in the late 1880s to Ceylon and in the following decade to Egypt. Otherwise laconic, on these voyages he wrote letters to his siblings "which made our mouths water and inflamed our imaginations."[19]

16. Franz Serafin Exner, "Über transportable Apparate," p. 660.

17. Cf. Mark Sommerhalder, *Pulsschlag der Erde!: Die Meteorologie in Goethes Naturwissenschaft und Dichtung* (Bern and New York: P. Lang, 1993).

18. Moore, *Schrödinger*, p. 70.

19. Marie Exner von Frisch to Gottfried Keller, 28 March 1889, in Smidt, *Kellers Glücklicher Zeit*, p. 184.

Meanwhile, Serafin sent a student to Siberia—in February. This was truly a venture into the open. The job fell to Hans Benndorf, the son of Serafin's dear friend Otto and a Brunnwinkler since childhood. Benndorf hoped to set up his electroscope in a warm house in a Siberian town. But he soon found that the towns were filled with smoke, and he complained that the local houses were "hermetically sealed."[20] In this case, perfect insulation—the urban physicist's dream—made measurement impossible. So Benndorf carried his instruments out onto the snowy plains, only to find that his own clothing was disturbing his measurements. Furs were out of the question—they produced static electricity that agitated his electroscope, and they restricted his movements. Gloves, too, got in the way, so he trained himself to work with bare hands: "One gets inured to it relatively quickly," he observed in his published report.[21] Benndorf's efforts were rewarded eight years later with the Lieben Prize from the Austrian Academy of Sciences.

Or consider Felix Exner's study of heat transport in the Wolfgangsee, begun in 1899, the season of heavy floods and Hilde's near-drowning.[22] From observations made on the few days of sunshine in July, Felix found that the temperatures at several depths below the eleven-kilometer-long lake's surface displayed curious fluctuations. These had a daily period, but their course did not agree with either of the two most obvious hypotheses—that the lower depths were warmed by conduction of heat from the surface or directly by sunlight.[23] Eight years later, with help from his wife and his cousin Karl von Frisch, Felix measured these fluctuations again, and again he found large fluctuations at 7 and 12 meters below the lake's surface, such that the daily maximum at one depth coincided with the daily minimum at the other. He reasoned that the discontinuity between the temperatures at these depths indicated the presence of *layers* of water of different densities (warmer water above, colder below). This led him to a hypothetical model (fig. 12). He supposed that at the surfaces between these layers there arose a rocking motion, setting up standing waves of a daily period (determined

20. Hans Benndorf, "Messungen des Potentialgefälles in Sibirien" ("Beiträge zur Kenntniss der atmosphärischen Elektricität II"), *Sitzungsberichte der Kaiserlichen Akademie der Wissenschaften zu Wien IIa* 108 (1899): 341–70, quotation on p. 348.

21. Ibid., p. 349.

22. Karlik and Schmid, *Exner und sein Kreis*, p. 150.

23. Felix M. Exner, "Messungen der täglichen Temperaturschwankungen in verschiedenen Tiefen des Wolfgangsees, *Sitzungsberichte der Kaiserlichen Akademie der Wissenschaften zu Wien IIa* 111 (1900): 707–25.

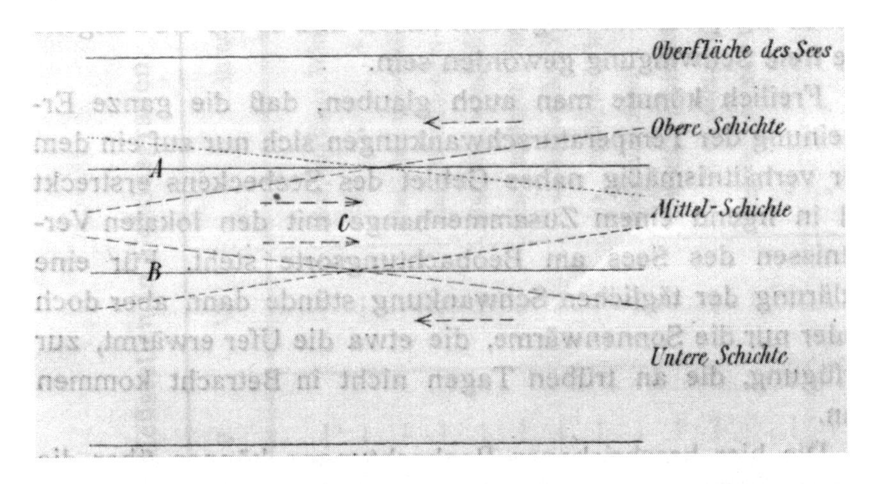

FIGURE 12 Felix Exner's dynamical model of the Wolfgangsee. The arrows indicate the directions of current flow in each layer (see text for explanation). Source: Felix M. Exner, "Über eigentümliche Temperaturschwankungen von eintägiger Periode im Wolfgangsee," *Sitzungsberichte der Kaiserlichen Akademie der Wissenschaften zu Wien* 117 (1908): 9–26, figure on p. 19.

perhaps by winds, perhaps by the shape of the lake). By virtue of these standing waves, for half the day point *A* would lie in the top layer and its temperature would be raised, while *B* would lie in the bottom layer and its temperature would be lowered; for the other half day the relations would be reversed. Between *A* and *B* there would be a layer without temperature fluctuations, and, more mysteriously, there would be nodal points like *C*. Beneath the placid surface of the Wolfgangsee, there lurked motions of an unexpected complexity.[24]

In this, his first geophysical research, Felix Exner had already hit on a problem that would prove critical to the burgeoning field of meteorology. As we will see, he would return throughout his career to the problem of the turbulence produced at discontinuous boundaries. In the 1920s, for example, Felix tried blowing air over a long, narrow box of sand in his laboratory to produce imprints of the waves in the air itself. He found that the length of the waves grew as the distance from the source of wind increased. This effect could be approximated by a solution to the continuity equation of

24. Felix M. Exner, "Über eigentümliche Temperaturschwankungen von eintägiger Periode im Wolfgangsee," *Sitzungsberichte der Kaiserlichen Akademie der Wissenschaften zu Wien IIa* 117 (1908): 9–26.

classical hydrodynamics. But at the boundaries between the layers of air, Felix noted, no truly eddyless motion was possible, since the concavities of one layer were larger than the convexities of the next.[25] Felix was of course familiar with Helmholtz's studies of the frictional motion at the boundaries of fluids, published in the late 1850s.[26] Yet in characteristic Austrian fashion, Felix, unlike Helmholtz, did not seek a deterministic mechanics. As Felix would later write, the ubiquity of "fluctuating" movements in nature signaled the limits of classical hydrodynamics. As we will see, Felix would later argue on this basis against Vilhelm Bjerknes's seminal theory that cyclones were waves at the surface between two air "fronts." Most immediately, however, Felix Exner's study of the Wolfgangsee was a training in alpine physics, an exercise in reading fluctuations as clues to hidden processes.

In their field research, Exner and his students took on the persona of the hunter, fending for themselves *im Freien* while reading nature's signs in the fluctuations of their electroscopes.[27] They fashioned themselves according to an Austrian-liberal ideal of masculinity, following literally in the steps of such venerable Austrian geologists as Friedrich Simony, legendary for his daring ascents in the Austrian Alps, and Eduard Suess, an expert on the Alps who was also a prominent liberal politician.[28] This was physics on the model of natural history: a quest for accidental clues and telltale traces. The goal was not to pare away fluctuations but to interpret them. It was a style of physics seemingly far removed from the precision measurement of the late nineteenth-century laboratory.

FLUCTUATIONS IN THE LABORATORY

One would expect the techniques of alpine physics to be as useless in a laboratory in the heart of Vienna as they were inside the "hermetically

25. Felix M. Exner, "Zur Physik der Dünen," *Sitzungsberichte der Kaiserlichen Akademie der Wissenschaften zu Wien IIa* 129 (1920): 929–52; "Dünen und Mäander, Wellenformen der festen Erdoberfläche, deren Wachstum und Bewegung," *Geografiska Annaler* 3 (1921): 327–35.

26. Helmholtz, "Ueber Wirbelbewegungen," *Journal für die reine und angewandte Mathematik* 55 (1858): 25–55.

27. On the self-fashioning of male scientists as heroes in eighteenth-century France, see Mary Terrall, "Heroic Narratives of Quest and Discovery," *Configurations* 6 (1998): 223–42.

28. On Simony, see Albrecht Penck, *Friedrich Simony: Leben und Wirken eines Alpenforschers. Ein Beitrag zur Geschichte der Geologie in Österreich* (Vienna: Hölzel, 1898); on Suess, see *Eduard Suess, Forscher und Politiker* (Horn: Österreichische Geologische Gesellschaft, 1981). Certain British physicists of the period vaunted their own "heroic" scientific exploits in the Alps; see Bruce Hevly, "The Heroic Science of Glacier Motion," *Osiris* 11 (1996): 66–86.

sealed" huts of Siberia. In the turn-of-the-century physics lab, there should have been no need to analyze fluctuating variables. After all, fluctuations were merely imperfections of the laboratory's insulation.

However, at the turn of the twentieth century, the two familiar methods of suppressing fluctuations—physical and statistical—became problematic. Rapid advances in high-precision measurement made it possible to detect the discontinuities of the microscopic world before they could be accounted for theoretically. Scientists detected microscopic fluctuations before they realized what they were seeing. To their eyes, the noise their new instruments recorded could just as well have been an experimental artifact as a natural irregularity. As instrumentation outpaced theoretical knowledge of the microscopic world after 1900, physicists in Serafin Exner's laboratory had the advantage of a unique perspective. They began to think of "fluctuation phenomena" as a new subfield of physical inquiry, one that drew analogies among phenomena as diverse as radioactivity, Brownian motion, critical opalescence, and imperfect crystals.[29] As we will see, their training in the Alps had taught them to treat fluctuations in their measurements as signal rather than noise.

These Viennese physicists had also been swayed by Exner and Boltzmann to believe in the reality of nature's molecular substructure. In Austria at the turn of the century, as in the 1860s, atomism remained a challenge to orthodox theology and a self-consciously progressive position. Yet atomists circa 1900 also confronted new challenges from radical philosophical movements. Most prominent among these were Ernst Mach's positivism, which reduced physical phenomena to aggregates of sensory impressions, and Wilhelm Ostwald's energetics, which drew broad conclusions from the hypothesis that nature was built of continuous energy, not discrete matter. Molecular research in Austria remained the pursuit of philosophical moderates.

Research on Brownian motion illustrates this continuity. In the late 1860s, when molecular theory was still one of the new fruits of the ascent of Austrian liberalism, Sigmund Exner had made measurements of this phenomenon that remained an undisputed reference for thirty years. Then, in 1900, soon after his study of the Wolfgangsee, Felix Exner took

29. See Reinhold Fürth, *Schwankungserscheinungen in der Physik* (Braunschweig, 1920), pp. 1–11. Important contributors to the field included Einstein, Marian Smoluchowski, and Jean Perrin. In the social sciences, fluctuations became a target of research in the late nineteenth century, as Francis Galton's eugenic theory attributed new significance to deviations in human populations. Mathematically, the mean fluctuation in a variable x is defined as $((x^2) - x^2)^{1/2}$; the mean *relative* fluctuation in x is defined as $((x^2 - x^2)^{1/2}/x$. On Exner's research group see Karlik and Schmid, *Exner und Sein Kreis*.

FIGURE 13 Vienna physicists, 1908. From left to right, seated, Fritz Kohlrausch is second, Marian von Smoluchowski is fourth; from left to right, standing, Egon von Schweidler is second, Karl Przibram is third, Felix Ehrenhaft is fourth, and Felix Exner is last. Erwin Schrödinger is absent. Source: Karlik and Schmid, *Franz Serafin Exner und sein Kreis* (Vienna: Verlag der Österreichischen Akademie der Wissenschaften, 1982), p. 100. Courtesy of the Austrian Academy of Sciences.

up where his father had left off. Felix hoped to reconcile this phenomenon with kinetic theory. If the suspended particles participated in the "molecular motion" of the liquid, he suggested, one would expect their kinetic energy to be proportional to the liquid's temperature.[30] To confirm this would require determining precisely the dependence of the liveliness of the motion on the ambient temperature, making proper heating and insulation crucial.

30. Only one study seemed to Felix irreconcilable with a view of the phenomenon as "molecular motion," that of G. Quincke, who had attributed the motion of suspended particles exclusively to the effects of heat and light on the liquid. Quincke claimed that if the liquid were at the same temperature as the environment, no movement could be observed. Felix M. Exner, "Notiz zu Brown's Molecularbewegung," *Sitzungsberichte der Kaiserlichen Akademie der Wissenschaften zu Wien IIa* 109 (1900): 843–47.

Fortunately, Felix was able to perform the experiment in his father's laboratory at the university's Physiological Institute, opened just the year before. Despite the tribute to "doctrina" in its inscription, this institute had been designed with the needs of experimenters in mind. It was set as far back as possible from the streetcar traffic on Währingerstrasse, and the laboratories were suitably separated from the classrooms.[31] Felix took further care to insulate his instruments. Still, instead of the direct linear relationship he expected between the square of the particles' velocities and the temperature, Felix's results were "very irregular." Worse, the interpolated line implied that the particles would come to rest well before the temperature reached absolute zero. Felix's attitude toward these results was surprising, characteristic not of the turn-of-the-century physics laboratory but of his training in field science. He stressed not the quantitative but the qualitative value of the results, as clues to the character of invisible phenomena—"as measures of the one and only visible [*sichtbar*] motions capable of making apparent to us [*veranschaulichend*] the internal motion of a fluid."[32]

Felix Exner's study of Brownian motion helped to awaken new interest in the phenomenon in Vienna, at a time when Jean Perrin in France and Albert Einstein in Switzerland were also searching for a molecular explanation. Marian von Smoluchowski, a member of the tight circle of Serafin Exner's students that included Benndorf and Schrödinger, mentioned Felix's kinetic perspective in a letter to Perrin.[33] In 1906 Smoluchowski arrived at a molecular account of Brownian motion that he judged "simpler, more direct and therefore more convincing" than the one Einstein had published the year before. The differences between Smoluchowski and Einstein's approaches are revealing. Einstein began by postulating certain conditions of randomness for the particles' motion, on the basis of which he justified applying a Boltzmannian statistical mechanical distribution function to the particles' displacement. Smoluchowski instead described the path of a single particle as it collided with smaller particles, using complex probabilistic considerations in a highly generalized manner.[34] Unlike Einstein's derivation, Smoluchowski's was visualizable. Tracing the crooked path of

31. "Physiologisches Institut," in *Neubauten für Zwecke des naturwissenschaftlichen, medizinischen, technischen und landwirtschaftlichen Unterrichtes an den Hochschulen in Wien 1894–1913* (Vienna: K. K. Hof- und Staatsdruckerei), pp. 16–20.

32. Felix M. Exner, "Brown's Molecularbewegung," pp. 844–47.

33. Letter from Smoluchowski to Perrin, quoted in Nye, *Molecular Reality*, p. 140.

34. Paul Hanle, "Erwin Schrödinger's Statistical Mechanics" (Ph.D. diss., Yale University, 1975), pp. 113-16.

a single particle was an approach that reflected Smoluchowski's training in alpine physics, where he had learned to watch fluctuations patiently as clues to hidden phenomena.

Like many of his Austrian colleagues, Smoluchowski was indeed an alpine physicist at heart. For him the fascination of atmospheric and geological phenomena was inseparable from the lure of the mountain landscape and the challenge of scaling its heights. In his free time, Smoluchowski was an avid mountain climber who tackled some of the most challenging peaks in the Alps. These excursions were research and holiday at once. One, for instance, led to the 3,108-meter-high Sonnblick Observatory, where Smoluchowski and his companions admired the view and the meteorological instruments. Smoluchowski regularly published succinct yet vivid accounts of his hikes in the notices of the German and Austrian Alps Society. These were the reports of a naturalist as well as an adventurer, enriched with descriptions of atmospheric and geological phenomena. Some of these he subsequently pursued in theoretical detail—leading him, for instance, to an explanation of the blue of the sky and to mathematical models of the formation of mountains and glaciers.[35] The naturalist's view he nurtured on his hikes and in his atmospheric research shaped his strategies as a theorist. He proceeded not, as did Einstein, on the basis of general principles, but in the manner of a naturalist observing the course of a process, or a hunter tracking his prey.

SCIENTISTS' ERRORS OR NATURE'S FLUCTUATIONS?

For Serafin Exner's research group, one challenge in particular made clear the relevance of field research to the laboratory. The problem arose with the arrival in 1903 of a new tool for making visible the motion of molecules: the ultramicroscope. In the ultramicroscope, a strong beam of light perpendicular to the viewing angle scatters off the particles under study, which show up as bright dots on a black background. This method made it possible for the first time to resolve particles smaller than the wavelength of light. At this unfamiliar scale of magnification, the paths of particles appeared to be far more irregular than previously imagined, a result that observers struggled to reconcile with kinetic theory.[36] Exner's former doctoral students Karl

35. On Smoluchowski as an alpinist, see Armin Teske, *Marian Smoluchowski, Leben und Werk* (Warsaw: Polish Academy of Sciences, 1977), chapter 6; on the Sonnblick excursion, see p. 18.

36. See Mary Jo Nye, *Molecular Reality: A Perspective on the Scientific Work of Jean Perrin* (New York: American Elsevier, 1972), pp. 99–102.

Przibram and Felix Ehrenhaft were the first physicists to apply this instrument to the study of Brownian motion. More famously, Ehrenhaft also used it to observe the time of fall of charged droplets in an electric field, concluding that the charge on an individual droplet deviated from multiples of Robert Millikan's value for the electron's charge *e*.[37]

This was just one case of Ehrenhaft's notorious iconoclasm. In an anti-dogmatic spirit characteristic of Austrian science, Ehrenhaft laid waste to his students' beliefs in everything from relativity to the law of inertia. Then he provoked them to argue back. "Are you dumb? Are you stupid? Do you really agree with everything I say?" he would demand. One of those students was Paul Feyerabend, who credited Ehrenhaft with stimulating his own irreverent critique of scientific authority. In this sense, Ehrenhaft was part of what Feyerabend portrayed as a distinctively Austrian tradition of science for "a free society."[38]

In their dispute over the electron charge, Ehrenhaft accused Millikan of biasing his own results by treating observed deviations from the average measured charge as mere "error." He posed the problem as follows:

> Nonetheless the large fluctuations in the values found by multiple observers, for example with the Wilson method,[39] remain for the present unexplained. Under such conditions, if one wishes to steer clear of further hypotheses and the interpretations, which one author has very frequently advanced with regard to these measurements, then the conclusion would be unavoidable, that these fluctuations of the elementary quantum or electronic charge are *to be seen as grounded in Nature itself* and the interpretation of the experiment is perhaps to be modified.[40]

By neglecting these "fluctuations," Millikan had proceeded in a manner typical of laboratory physicists of the day. Like "disturbances" or "error," "fluctuations" was a catchall term used by physicists for unexplained

37. Gerald Holton, "Subelectrons, Presuppositions, and the Millikan-Ehrenhaft Dispute," in *The Scientific Imagination: Case Studies* (New York: Cambridge University Press, 1978), pp. 25–83.

38. Paul Feyerabend, *Science in a Free Society* (London: Verso, 1978), pp. 109–11.

39. Ehrenhaft most likely referred to C. T. R. Wilson's water drop counting method, introduced in 1909, rather than the older falling cloud method used by H. A. Wilson and Millikan.

40. Felix Ehrenhaft, "Über die Messung von Elektrizätsmengen, die kleiner zu sein scheinen als die Ladung des einwertigen Wasserstoffions oder Elektrons oder von dessen Vielfachen abweichen," *Sitzungsberichte der Kaiserlichen Akademie der Wissenschaften zu Wien IIa* 119 (1910): 815–66, quotation on p. 826.

variations in measurements of phenomena that were expected to be constant. The Viennese, however, had been trained through field research to treat fluctuations as phenomena in their own right and to hunt for their source.

In 1910, Karl Przibram looked more closely at the distribution of charge values Ehrenhaft had recorded. Przibram found a close fit between Ehrenhaft's results and the distribution that would be expected if the data were characterized by error curves centered on multiples of the fundamental charge. Perhaps Millikan's mean values were significant after all. But Przibram was not content to dismiss the variations as "error"; he wanted to know their source. A solution to the mystery lay close at hand. Had he and Ehrenhaft not recently used the ultramicroscope to study Brownian motion? Przibram suggested that the variability in the time of fall of the ionized droplets could result from just this phenomenon.[41] In this case, were the fluctuations "grounded in Nature?" Were they isolation error? The answer of course depended on whether the topic under study was the electron charge or Brownian motion.

In the lab, as in the field, Exner's group severed terms like "error" and "disturbance" from the vague meanings they carried elsewhere and gave them concrete definitions. Thanks to the interdisciplinary nature of the Exners' science, this procedure resembled contemporary practices in physiology and psychology, such as the quantification of reaction times. It was a departure, however, from existing standards within the international physics community. Most European physicists continued to use "error" in a normative sense to designate a failure on the part of the experimenter.[42] Exner and his students instead used error to designate an ineradicable variability rooted in nature itself. Not until the late 1920s would this objectification of error enter the mainstream, due to the advent of an indeterministic theory of quantum physics and the rise of an operationalist philosophy of science.[43]

41. Karl Przibram, "Ladungsbestimmungen an Nebelteilchen II," *Sitzungsberichte der Kaiserlichen Akademie der Wissenschaften zu Wien IIa* 119 (1910): 1719–53.

42. This is neatly illustrated by Giora Hon's study of a variety of critiques of a single experiment at the turn of the century: "Is the Identification of Experimental Error Contextually Dependent?" in *Scientific Practice*, ed. Jed Buchwald (Chicago: University of Chicago Press, 1995), p. 170–223, and "On the Concept of Experimental Error" (Ph.D. diss., Cambridge University, 1985).

43. For instance, the approach of the Viennese group to the quantification of experimental uncertainty has an affinity with that of Berlin physicists Zernike and Ising in 1926, long after the Viennese had pioneered the study of fluctuations. See Mara Beller, "Experimental Accuracy, Operationalism, and the Limits of Knowledge, 1925–1935," *Science in Context* 2 (1988): 147–62.

It was in their research on radioactivity that Exner's group went farthest toward redefining "error" and with it the entire concept of probability as employed in the physics of the day.[44] The history of research on the rates of radioactive decay is, at first glance, a story of how variability in the field was isolated in the laboratory and rendered predictable. Whether radioactive elements could be expected to emit alpha particles at the same rate in different locations, altitudes, and climates was an open question at the turn of the century. If a gram of radium were delivered from Vienna to Paris, how radically might its output on arrival be expected to deviate from the value certified by the Austrian laboratory? The Viennese were uniquely positioned to study radioactivity, thanks to their access to the uranium mines in St. Joachimsthal, Bohemia, at the time the only source from which to extract radium.

Beyond practical concerns of shipping and pricing was the looming question of the cause of radioactive decay. The most prevalent assumption was that its cause was environmental. Leading researchers tested radium's decay rate against nearly every conceivable change in the environment, but their precision instruments never seemed to produce perfectly constant values.[45] In what would later and almost inadvertently become a much cited paper, a young American researcher named Howard Bronson reported: "The precision of an individual measurement is, however, limited by a slight oscillation of the needle. Repeated attempts have been made to eliminate this by better shielding from external electrostatic action, and by careful attention to contacts, but all to no purpose."[46]

Shortly after Bronson published this complaint, radioactivity researchers from around the world gathered in Liège for the First International Congress of Radiology and Ionization. Discussion centered on the need for international collaboration to locate radioactive materials and secure them for the general use of the scientific community. A paper from Exner's student Egon von Schweidler entitled "Fluctuations of the Radioactive Transformation" appeared to be a theoretical analysis unconnected to these

44. This episode is discussed in more detail in Coen, "Nature's Fluctuations." On the Vienna Radium Institute, see Maria Rentetzi, "Gender, Politics, and Radioactivity Research in Interwar Vienna," *Isis* 95 (2004): 359–93; Wolfgang Reiter, "Stefan Meyer: Pioneer of Radioactivity," *Physics in Perspective* 3 (2001): 106–27.

45. For a full list of these experiments, see Stefan Meyer and Egon Schweidler, *Radioaktivität* (Leipzig and Berlin: Teubner, 1927), pp. 38–41.

46. Howard Bronson, "The Effect of High Temperatures on the Rate of Decay of the Active Deposit from Radium," *Philosophical Magazine* 11 (1906): 143–45, quotation on p. 145.

practical concerns.[47] Schweidler's primary research was on atmospheric electricity, not radioactivity. His approach to fluctuations derived from the field, not the laboratory. He was trained, moreover, in the Austrian tradition of probabilistic reasoning, as made clear by a lecture he delivered to the Philosophical Society of the University of Vienna in 1900, "On the Objective Validity of the Probability Calculus." Although the lecture has been lost, its title evokes the critical perspective he brought to statistical physics.[48]

In his contribution at Liège, Schweidler considered the "law of radioactive decay," as it had been derived by Ernest Rutherford and Frederick Soddy at McGill University in 1902. By weighing a sample of a radioactive element repeatedly over the course of several days, Rutherford and Soddy had arrived at a formula expressing m, the amount of the substance remaining after time t, as a fraction of the total initial amount M; $m = Me^{-\lambda t}$. Here λ is a constant specific to each radioactive element. In essence, physicists thought of each atom as akin to a little bomb set to go off at a certain time: the average of those times was $1/\lambda$.

Schweidler's modest but revolutionary suggestion in 1905 was that this equation expressed no more than a statistical law. In other words, it was valid only on average if one looked at a large enough sample over a long enough period of time. The key to this interpretation was the new meaning that Schweidler assigned to λ: not an average of predetermined lifetimes, but the *probability* of decay for an atom of a given element.[49] Here, Schweidler used probability explicitly in the second sense discussed above, as the frequency of an event. In fact, the decay of a radioactive atom has since become the most familiar textbook example of a probabilistic phenomenon and a standard for definitions of randomness.

Were Schweidler's interpretation correct, it should have been possible to measure deviations from the familiar decay law at small enough scales. Calculations using the latest probabilistic mathematics gave the Austrians a sense of the scale on which they would have to hunt for fluctuations in the decay rate in order to prove Schweidler's hypothesis. For a sample of one million atoms, for instance, the fluctuations would be just 0.1 percent

47. Egon von Schweidler, *Über Schwankungen der radioaktiven Umwandlung* (Liège, 1905).

48. The lecture is cited in *Rückblick auf die ersten 25 Vereinsjahre der Philosophischen Gesellschaft an der Universität Wien* (Vienna: Philosophische Gesellschaft der Universität Wien, 1913).

49. More precisely, $\lambda\Delta t$ equals the *probability* that an atom undergoes a transformation in the time Δt.

of the total alpha particle output; for a sample of ten thousand atoms, the fluctuations would be 1 percent.

For almost three years after the Liège conference, the Viennese seemed to be the only physicists interested in the fluctuations of radioactive decay. Then suddenly in 1908 physicists in Manchester and Berlin rushed to confirm Schweidler's prediction. Almost simultaneously, these same laboratorics announced two innovative methods for detecting individual alpha particles: an electrical detector based on an amplification of an alpha particle's ionizing effect and a screen that scintillated when struck by an alpha particle. This timing was not coincidental. Schweidler's theory offered a means of calibrating these new instruments. According to Schweidler's calculations, physicists would be able to use the size of fluctuations as an estimate of the number of atoms present. More important, were the theory confirmed, experimenters could justifiably claim that the variability of measurements obtained with their sensitive new instruments was "random." They could thus set aside the possibility of systematic error and take the mean value as an accurate time-average rate of decay.

Erwin Schrödinger later complained that too many physicists had misinterpreted the meaning of Schweidler's "discovery." Its importance lay not in the derivation of a handy relation between the mean fluctuation size and the number of atoms but rather in "the fundamental knowledge of the probabilistic character of the decay constant λ."[50] Schrödinger's point applies equally to historians of physics as to his contemporaries.[51] Rather than discrediting this application-oriented reading of Schweidler's theory, we should recognize that it reflects the interpretation of the large majority of physicists outside of Vienna. They understood Schweidler's hypothesis above all as an indispensable guide for calibrating instruments and observers, and only secondarily as a statement about a fundamental randomness in nature.

50. Erwin Schrödinger, "Schweidlersche Schwankungen," *Sitzungsberichte der Akademie der Wissenschaften in Wien IIa* 128 (1919): 177–237, quotation on p. 179.

51. Bruce Wheaton, one of the few historians to have remarked on Schweidler's theory at all, describes Schweidler's innovation as "a means to determine the absolute number of decaying atoms" by measuring the fluctuations from the exponential decay law. Wheaton, *The Tiger and the Shark: Empirical Roots of Wave-Particle Dualism* (Cambridge: Cambridge University Press, 1983), 143. Other discussions of Schweidler's thesis appear in J. van Brakel, "The Possible Influence of the Discovery of Radio-Active Decay on the Concept of Physical Probability," *Archives for the History of the Exact Sciences* 31 (1985): 369–85, quotation on pp. 373–74; E. Amaldi (see note 35); Gerd Gigerenzer et al., *The Empire of Chance: How Probability Changed Science and Everyday Life* (Cambridge: Cambridge University Press, 1989), p. 181.

In Vienna, Exner's group was particularly critical of the seat-of-the-pants error analysis on which early "confirmations" of the Schweidler theory relied. While their colleagues in Britain, Germany, and France collapsed the distinction between experimental "errors" and natural "fluctuations," the Viennese stubbornly insisted on separating fluctuations inherent in the phenomenon from artifacts of the experimental arrangement. They had learned this approach through their field research. At Seeham in 1913, for instance, Schrödinger found it impossible to eliminate "[s]mall, entirely irregular fluctuations" of his electrometer's needle, "despite much effort." He was "therefore disposed to view them as real Schweidler fluctuations of the γ- or secondary-radiation in the tube." Other European physicists would have dismissed the fluctuations as irrelevant "error." Following Exner's example, Schrödinger instead recorded the "behavior" of the needle over the course of five to ten minutes.[52] Rather than masking these fluctuations, he hunted for their source.

A further critique came from another of Exner's most devoted students, Fritz Kohlrausch. Kohlrausch was another product of a scientific dynasty. His father was a chemist and his grandfather and uncle were successful physicists; the latter succeeded Helmholtz as the director of Berlin's Imperial Institute of Physics and Technology. Kohlrausch criticized colleagues elsewhere for failing to distinguish between fluctuations in the rate of radioactive decay and those produced by their experimental setups. Methods of detecting alpha particles individually, including scintillation screens and electrical counters, only registered particles within a small solid angle. Assuming that the directions in which alpha particles are emitted are random, this angular selection introduced spatial fluctuations indistinguishable from the temporal fluctuations predicted by Schweidler's theory. In 1912 Kohlrausch and Schweidler attempted to use an individual counting method without angular selection, but their results were inconclusive. From the perspective of 1928, Kohlrausch concluded that tests performed with individual methods had demonstrated merely that "an element of indeterminism" entered the experiment at some stage, not necessarily in the process of decay itself. Such methods were "*not suitable for deciding the question of the chance nature of atomic decay.*"[53]

52. Schrödinger, "Beiträge zur Kenntnis der atmosphärischen Elektrizität LI: Radium-A-Gehalt der Atmosphäre in Seeham 1913," pp. 5, 7.

53. Kohlrausch, "Die experimentelle Beweis für den statistischen Charakter des radioaktiven Zerfallsgesetzes," *Ergebnisse der Exakten Naturwissenschaften* 5 (1926): 192–212, quotation on 198; emphasis in the original.

The Austrians were critical not simply of the experimental methods employed in early tests of Schweidler's theory but more fundamentally of the poorly defined concept of "error" other physicists had used. What was needed, according to Kohlrausch, was a better measure of the deviation of experiment from theory. Kohlrausch proposed to simulate the alpha-particle experiments with a lottery game, a model that had figured in probability theory since the Enlightenment.[54] In collaboration with Schrödinger, Kohlrausch had first performed lottery drawings to model the *H*-curve of Boltzmann's statistical mechanics. For the alpha-particle problem, Kohlrausch recorded 5,000 drawings from an urn of 100 numbered tickets. In this way, he built up a store of "experimental data that one *knows* to have a chance character."[55] With this data, he could model previous experiments and reassess their results. For example, by varying the range of ticket numbers corresponding to "alpha particle emitted" or "no alpha particle emitted," he could vary the probability of decay. Then he could simulate counting the number of atoms decaying in a given time interval by recording the number of tickets in the prescribed range that appeared in each sequence of a given number of drawings. Ultimately, the aim was to be able to compare, on the one hand, the deviation of radioactive data from the proposed statistical law, and on the other, the deviation of urn drawings from probability calculations. More precisely, Kohlrausch defined the "mean error" as the expected value of the difference between the size of the observed fluctuation about the observed mean and predicted size of the fluctuation. For each experiment, the mean error of the corresponding lottery model could be compared to the mean error of the experiment itself. The mean error thus functioned as a measure of the precision to be expected in a comparison between theory and experiment for a statistical law.

Here was a move characteristic of Viennese physics: to transform "error" from a vague synonym for uncertainty into a precisely defined measure of variation with a concrete interpretation in the physical world. Schrödinger had done much the same thing a few years earlier in evaluating experiments on Brownian motion. The problem was one Ehrenhaft had encountered. Given the fluctuations produced by Brownian motion, how closely could the measured values in the droplet experiments be expected to approach

54. Kohlrausch cited Paul and Tatyana Ehrenfest's suggestion that Boltzmannian statistical mechanics could be made more intuitive by analogy to lottery drawings. Paul and Tatyana Ehrenfest, *Encyclopädie der mathematischen Wissenschaften*, vol. 4 (1911), part 32.

55. Kohlrausch, "Der experimentelle Beweis," p. 192, emphasis in the original.

the true values?[56] Schrödinger objected to the common criterion, which involved calculating the average time of fall from an ever longer series of measurements, until oscillations in the average shrank to an acceptable size. Since the members of the earlier series are contained in each later series, the later ones are not statistically independent of the earlier ones. That this procedure gradually reduced the spread in the data was therefore no reliable sign of the data's consistency. Schrödinger proposed an alternative criterion similar to Kohlrausch's notion of mean error. One would compare the mean fluctuation of the speed-of-fall measurements from their "true values" to the mean fluctuation of the Brownian displacements from their "true values." The problem, of course, was that it was "not practically possible to insert the true values."[57] Schrödinger had come up against the core difficulty of measuring statistical fluctuations: what is the meaning of a "true value" for a quantity known only through its statistical distribution?

Felix Exner faced this same problem of verifying a statistical hypothesis in the context of meteorology. In the 1910s and 1920s he was studying correlations between weather "anomalies," deviations from the long-term average weather in a given location. How, he wondered, could one determine whether suspected correlations, such as that between winters in Europe and in Greenland, were "real"? One option was to use the margin of error for the correlation coefficient that British statisticians had formulated. But Felix rejected their definition of error as arbitrary, because one had to decide a priori which values counted as "real," without further empirical input. Only if data taken from other time intervals gave a similar correlation factor did Felix feel he could judge the result "with some probability as real." Like Schrödinger and Kohlrausch, Felix Exner evinced a Humboldtian impulse to collect ever more empirical data.[58] In the place of vague notions of "error" or "uncertainty," these physicists set concrete physical interpretations of the variability of their measurements.

To do so, Serafin Exner would argue, was to render the concept of probability "objective." It was during the First World War, with his laboratory closed and his students in battle, that Serafin Exner managed to put down

56. Erwin Schrödinger, "Zur Theorie der Fall- und Steigversuche an Teilchen mit Brownscher Bewegung," *Physikalische Zeitschrift* 16 (1915): 289–295; cf. Hanle, "Schrödinger's Statistical Mechanics," pp. 104–8.

57. Schrödinger, "Zur Theorie der Fall- und Steigversuche," p. 294.

58. For the term "Humboldtian Science," see Susan Faye Cannon, *Science in Culture: The Early Victorian Period* (New York: Science History Publications, 1978), pp. 73–110.

on paper his own thoughts on the Viennese physics research of the past decade.

OBJECTIVE CHANCE

Unlike the overwhelming majority of the Austrian population, Serafin Exner viewed the war skeptically.[59] Like other members of the Exners' circle, including Josef Breuer and Marie von Ebner-Eschenbach, he voiced his doubts privately.[60] He urged his students not to enlist, but they went nonetheless. His daughter Hilde, niece Nora, and nephews Alfred Exner and Karl and Otto von Frisch all volunteered for medical service. Meanwhile, Serafin plunged himself into a writing project. "In grave and difficult times one gladly escapes to a field that is entirely removed from human arbitrariness; thus did these lectures come to be written during the war."[61] A manuscript dated 17 September 1914 suggests that he began to draft what would become his best-known work soon after word reached Brunnwinkl of the outbreak of hostilities.[62] The book was a survey of contemporary physics for a general audience, culminating in a long final section on "laws of nature," in which Serafin elaborated on the probabilistic view he had introduced in his 1908 rectoral address. He drew on evidence from his laboratory's extensive research on radioactive decay and Brownian motion. Already in the notes written in 1914 he used emphatically probabilistic language to describe radioactivity: "Chance rules the atom since the decay law is a law of chance that is valid only for large numbers[,] that is to say, the individual events must be random and mutually independent."[63] But this book was to be more than an overview of recent physics. In the same spirit in which he urged

59. Hans Benndorf, "Gedenkrede," p. 9. Serafin expressed his opposition to war in general in his unpublished history of the world, *Vom Chaos zur Gegenwart. Eine kulturhistorische Studie*, Sig. I888.089 MS, Library of the University of Vienna. For the patriotic response that was common among Austrian scholars see Günther Ramhardter, *Geschichtswissenschaft und Patriotismus: Österreichische Historiker im Weltkrieg 1914–1918* (Munich: Oldenbourg, 1973).

60. Breuer to Ebner-Eschenbach, 17 July 1914, 10 Sept. 1914, 24 Oct. 1914, in *Marie von Ebner-Eschenbach—Dr. Josef Breuer: Ein Briefwechsel, 1889–1916*, ed. Robert A. Kann (Vienna: Bergland, 1969).

61. Franz Serafin Exner, *Vorlesungen über die physikalischen Grundlagen der Naturwissenschaften* (Vienna: Franz Deuticke, 1919), p. iii.

62. Franz Serafin Exner, notes (1914), Archiv der Österreichischen Akademie der Wissenschaften, Vienna, Nachlass Franz Exner.

63. Ibid., p. 44.

his students not to go to war, he wrote this inquiry into nature's laws as a moral lesson to the public.

Nowhere in this text did Exner suggest that a doctrine of chance should replace determinism. He wrote instead of "freeing" the mind from its old habits, as relativity theory had accomplished: "[W]hoever cannot make himself free of the idea [*Vorstellung*] of absolute laws finds himself in the same situation as someone who clings to the concepts of absolute time and absolute motion."[64] His discussion of scientific method was a call for critical thinking and open-mindedness, married to an unrelenting attack on dogmatism. By the time the second edition appeared in 1919, however, Exner found himself in the unexpected position of defending his own discipline against charges of dogmatism.

In 1918, after four years of world war, Oswald Spengler famously prophesied the "decline of the West." Spengler saw evidence of this decay all around him: in the savagery of warfare, the mob mentality of socialists, the avarice of capitalists, and the spiritual void of mechanistic science. And the last of these, like the war-torn civilization it served, had already begun to work its own undoing. Indeed, physics seemed to Spengler to have left reason behind altogether. With its turn from causal explanation to statistical description, physics had entered a "subjective" phase, reliant on a notion of "fate," producing "a picture of history and not 'Nature.'"[65]

Spengler's accusations and gloomy predictions would continue to haunt Exner to the end of his life. He would spend his final years trying to prove Spengler wrong by writing an optimistic history of the world. In this never-published manuscript, Exner defended a vision of historical progress rooted not in destiny but in chance. Just as molecular collisions drove the evolution of a gas toward equilibrium, so, in Exner's view, did chance encounters between cultures drive cultural evolution: "It is understandable, that in the course of progress random external influences gain ever more importance and bring a faster pace to evolution."[66] Exner would thus turn Spengler's accusation on its head, making chance the vehicle of progress.

In 1918, Spengler had as yet no inkling of the fundamentally probabilistic principles of quantum physics that would emerge in the following decade. His point of reference was instead "the bizarre hypotheses" that had been

64. Franz Serafin Exner, *Vorlesungen* (1919), p. 711.

65. Spengler, *The Decline of the West* (Oxford: Oxford University Press, 1991), 218.

66. Franz Serafin Exner, *Vom Chaos zur Gegenwart*, typescript, Library of the University of Vienna, p. 119. On this manuscript see Stöltzner, "Exner's Indeterminist Theory."

proposed to explain the energy emitted from radioactive elements. Spengler cited the statistical theory of radioactive decay, a product of Exner's laboratory, as evidence of a "subjective" turn in physics.[67] He charged that "the atomic theory is thus a dogma, not empirical knowledge."

To this accusation Exner would object: "[T]he first claim is thus false, since this theory is not dogma, but rather a hypothesis, and the second is self-evident, since which physicist would ever have seen in it an empirical fact? If the atomic theory were a dogma, how much endless intellectual trouble and work could physicists have saved themselves in their efforts to make the approval of this hypothesis probable; yet had it ever appeared to them as a fact, then this activity would have had no sense at all."[68] As Exner portrayed it, the task of physics was to make hypotheses "probable," leaving no room for dogmatism. As he wrote in the *Lectures,* it was the moral duty of scientists to adapt their "habits of thought" to their experience of the world: "Otherwise instead of laws we erect dogmas and science turns into a religion."[69] A "probabilistic" attitude was thus the surest defense against the threat of dogmatism.

Yet dogmatism was not the only danger. Spengler's principal challenge to modern science was his radical relativism, his claim that "each person has his own world."[70] The challenge Exner faced—one familiar to Austrian liberals—was to advance his probabilistic methodology while avoiding the Spenglerian implication that science was just another form of solipsism.

As it stood, the theory of probability seemed to Exner to risk just this interpretation. Laplace, for instance, had understood chance as the human ignorance of causal factors. This implied that an event could be chancelike for one person but not for another. "[T]he concept of 'chance' [*Zufall*] or of the 'chance event' [*zufälligen Ereignisses*] is such an unstable one in the literature because it frequently involves a personal factor."[71] Chance must

67. "Above all this is manifested in the bizarre hypotheses of atomic disintegration which elucidate the phenomena of radioactivity, and according to which uranium atoms that have kept their essence unaltered, in spite of all external influences, for millions of years, quite suddenly, without assignable cause, explode, scattering their smallest particles over space with velocities of thousands of kilometers per second. Only a few individuals in an aggregate of radioactive atoms are thus struck by Destiny, the neighbors are entirely unaffected. Here, too, then, is a picture of history and not 'Nature,' and although statistical methods here also prove to be necessary, one might almost say that in them mathematical number has been replaced by chronological." Spengler, *Decline*, p. 219.

68. Franz S. Exner, preface to the second edition of *Vorlesungen* (Vienna: Deuticke, 1922), p. xi.

69. Franz S. Exner, *Vorlesungen* (1919), p. 711.

70. See Rudolf Haller's essay in Elisabeth Nemeth and Friedrich Stadler eds., *Encyclopedia and Utopia* (Dordrecht and Boston: Kluwer, 1996).

71. Exner, *Vorlesungen* (1919), p. 667.

be a feature of the world, Exner reasoned, since the laws of probability were confirmed by experience. Here Exner followed Venn's empiricist logic, but took it a step further, for Venn had never entertained the possibility of uncaused events. Unwittingly, perhaps, Exner echoed his father's attacks on the Hegelian psychologists for the murkiness of their concept of chance. The first Franz Exner had sought a transparent language for science that would be a foundation for a utopian liberal community. His son's ambition was no less grand. His goal was to show that "chance" is "something grounded in nature itself . . . and not first introduced into it by man." Serafin sought a definition of chance that would be transparent and universal, much like the definitions he chose for the properties of colors. All these concepts required interpretations that were unambiguous in the sense of being independent of the faculties of an individual observer. As his students had done in their efforts to "prove" through lottery drawings the chance character of radioactive decay, Exner drew his criterion for chance from the empirical distributions of outcomes in cases of equal probability. By setting probability on this empirical foundation, Exner fortified it as a tool with which to dismantle "dogma" without stumbling into the pit of radical relativism.[72]

In Austria during the First World War, "dogma" was not a lightly chosen term. Faced with a church that was one of the most reactionary in Europe, Austrian liberals remained fiercely anticlerical. It was in these years that philosophers like Ernst Mach and Friedrich Adler denounced German physicist Max Planck for invoking religion to support his account of a scientist's ethical duties. Mach accused Planck of trying to turn the physics world into a "community of believers," and he opted out: "Freedom of thought is more precious to me."[73] It is no coincidence, then, that Planck, who had defended determinism as a necessary assumption in science, was one of Exner's foremost opponents. As the Austrians understood it, Planck's argument for the necessity of determinism hinged on his religious beliefs. Exner's attack on determinism as a dogma was thus part of an Austrian strain of stridently secular science.

Beyond refuting Spengler and Planck, Exner's exposition of probabilism also had a constructive goal. This became clear in a review in Friedrich Poske's Vienna-based journal for science instruction. As it had been in the 1880s, this journal remained the voice of Austria's liberal educators defending the "humanistic value" of the study of natural science. Poske gave

72. On Exner's adoption of a frequentist interpretation of probability, see Stöltzner, "Vienna Indeterminism."

73. Heilbron, *Dilemmas of an Upright Man*, p. 54.

Serafin's *Lectures* his highest recommendation for "every teacher of physics." In his words, the book presented the probabilistic viewpoint ultimately to foster a critical perspective: "In this the book does not mean to present a physical worldview, but contents itself instead with a critical investigation of the foundations of our physical knowledge."[74] As Poske made clear, Exner's probabilism was, in the end, a contribution to the humanist program of liberal education in Austria.

THE DOGMA OF CAUSALITY

Exner's students and colleagues read his *Lectures* with excitement. As a clear-headed strike against dogmatism, Exner's critique of the concept of natural laws was exemplary of the goals of the future members of the Vienna Circle and its American incarnation, the Unity of Science Movement. Hans Reichenbach, who favorably reviewed the *Lectures* in *Die Naturwissenschaften* in 1921, would later maintain that Exner had given the first clear critique of the causal principle on the basis of statistical physics.[75] Philipp Frank likewise credited Exner with "drawing attention to the possibility" that causality does not apply to the microscopic world. Writing in 1932, Frank also pointed out the affinity between Exner's position and that of Richard von Mises, who argued in 1922 that deterministic descriptions could not apply to the physics of discrete particles.[76]

Exner's exposition of an acausal physics convinced scientists as well as philosophers. In 1922 Erwin Schrödinger praised Exner as the first to "launch a very acute philosophical criticism against the *taken-for-granted* manner in which the absolute determinism of molecular processes was accepted by everybody. He came to the conclusion that the assertion of determinism was certainly *possible,* yet by no means *necessary,* and when more closely examined *not at all very probable.*" As Schrödinger explained, acausality seemed "unthinkable" simply because people had become habituated to causal reasoning over many generations, as the result of their

74. Poske, review in *Zeitschrift für den Physikalischen und Chemischen Unterricht* 36 (1923): 133–34, quotation on p. 134. A review in a journal for pharmacists likewise praised Exner's "pedagogical ability," his ability to communicate to nonphysicists, and recommended in particular the final section on natural laws. *Pharmaceutische Monatshefte* 3 (1922): 40.

75. Reichenbach, *Philosophische Grundlagen der Quantenmechanik,* p. 11.

76. Frank, *The Law of Causality and Its Limits,* ed. R. S. Cohen, trans. by M. Neurath and R. S. Cohen (Dordrecht: Kluwer, 1998), pp. 70, 72.

experience of regularity in nature. Recently, phenomena like Brownian motion and radioactive decay showed that regularity need not imply determinism. Causality had thus lost "its rational foundation." "Exner's assertion amounts to this: It is quite possible that Nature's laws are of thoroughly statistical character. The demand for an absolute law in the background of the statistical law—a demand which at the present day almost everybody considers imperative—goes beyond the reach of experience. . . . For a doubtful attitude in this respect is today by far the more natural." In language characteristic of Austrian liberalism, Schrödinger concluded that one should not expect "atomic theory to substantiate the dogma of Causality."[77]

Schrödinger's interest in the evidence against determinism had been piqued a few years earlier by the research of Marian von Smoluchowski. After Smoluchowski's death in the dysentery epidemic of 1917, Schrödinger carried on where his friend had left off. In a rash of intellectual excitement late in the war, he fervently scribbled queries, equations, and exclamations until he had filled fifty pages with his notes. Fluctuation phenomena, he wrote, held out the hope of definitively contradicting the supposedly irreversible laws of thermodynamics. In their various forms, fluctuations made "deviations" from the laws of thermodynamics experimentally observable. In this sense, fluctuation phenomena could provide "a new *experimentum crucis,* and once consensus prevails, a new proof of the *relative* validity of the Boltzmannian conception as opposed to [classical] thermodynamics. (*Absolutely* valid theories do not exist.)"[78]

Like Schrödinger, Smoluchowski had welcomed the new statistical perspective as a liberation from stale dogmas. In 1913 he had lectured to the *Deutsche Naturforscher und Ärzte* on "Experimentally Demonstrable Molecular Phenomena Contradicting the Usual Thermodynamics." This title required explanation: "The title of my review sounds somewhat revolutionary, and I think actually that it would have been quite a venture even ten years ago to speak so disrespectfully of the traditional conception of thermodynamics. But today first we have generally less respect for dogma in physics."[79]

Fritz Kohlrausch likewise adopted Exner's unshakeable skepticism toward "absolute" theories. Exner's death in 1926 prompted Kohlrausch to

77. Erwin Schrödinger, "What Is a Law of Nature?" in *Science, Theory, and Man* (1935; New York: Dover, 1957), pp. 133–47, quotations on pp. 142, 147.

78. Schrödinger's notebooks on Smoluchowski's work are transcribed in Hanle, "Schrödinger's Statistical Mechanics." This quote is on p. 268; emphasis in the original.

79. Quoted and translated in Hanle, "Schrödinger's Statistical Mechanics," p. 116.

cite as an epigraph to his work a particularly revealing passage from Exner's 1917 preface to his *Lectures:*

> These laws are however not given to us by Nature like a piece of matter, rather we infer them from the observation of individual cases; and the generalizations thence derived as laws are always a product of men; to test these as to their properties, and especially as to the extent of their validity, remains nonetheless one of our most important tasks.[80]

It was the iconoclastic potential of fluctuations research, its liberal spirit, that inspired Vienna's young physicists. Trained in alpine physics and imbued with the moral ideals of their mentor, Exner's students went on to shape the probabilistic turn in Austrian physics. Foremost among these pioneers were Felix Exner and Erwin Schrödinger.

FELIX EXNER: "FATHER OF A STATISTICAL METEOROLOGY"

His interest in geophysics secured by his youthful study of the Wolfgangsee, Felix Exner spent the last year of the old century in Germany studying with Emil Wiechert, a geophysicist, and Wilhelm von Bezold, one of the founders of dynamical meteorology. When he returned to Vienna in 1901, he took up an assistantship at the Central Institute for Meteorology and Geomagnetism (ZAMG).[81] The ZAMG had been founded in 1851 to coordinate observations from stations under construction throughout the empire's vast territory, which provided Vienna with an inexhaustible supply of raw data. The institute received strong imperial support due to the practical benefits—above all agricultural and military—expected from weather forecasting.[82] Thanks as well to the achievements of individual scientists—Julius Hann and

80. K. W. F. Kohlrausch, *Probleme der Gamma-Strahlung* (Braunschweig, 1927), i; Franz Exner, *Vorlesungen über die Physikalischen Grundlagen der Naturwissenschaften* (Vienna, 1917), p. v.

81. Nonetheless, Felix's university studies had not included the meteorological classes taught by Josef Perntner, the Central Institute's director, and by the former director Julius von Hann. Instead he had acquired a broad grounding in mathematical physics and philosophy, including classes on thermodynamics and on electricity and magnetism with Boltzmann, spherical functions with Hasenhörl, vectors and quaternions with Smoluchowski and with Schweidler, experimental physics in the laboratory of his uncle Serafin, "Psychologie und Logik der Forschung" with Mach, "practical philosophy" with Jodl, and psychophysics with Müller. (Archiv der Universität Wien, phil. PA 1584, box Nr. 70: Felix Exner.)

82. Christa Hammerl et al., eds., *Die Zentralanstalt für Meteorologie und Geodynamik, 1851–2001* (Graz: Leykam, 2001).

Viktor Conrad in climatology, Max Margules in dynamical meteorology—Vienna had become known by 1900 as a leading center for meteorology research, as well as the home of the field's major journal, the *Meteorologische Zeitschrift.* Felix Exner would cement that reputation.

Notwithstanding their dependence on state support, Vienna's meteorologists regarded their forecasting duties with disdain. Forecasting was a strictly empirical operation at the time. Scientists made predictions by finding archived weather maps with patterns similar to current ones. In the eyes of meteorology's elite, such guesswork did not amount to science. Felix set out to change this. Before Norway's Vilhelm Bjerknes could even announce his intention of making meteorology an exact science, Felix Exner showed just how this might be accomplished.

In 1902, Felix made what was apparently the first attempt to model weather patterns from basic physical principles on a large geographic scale.[83] This was a problem of combining the seventeenth-century principles of dynamics with nineteenth-century thermodynamics and setting it all in a rotating coordinate frame. The two basic equations for the thermodynamics of the atmosphere had been derived by William Thomson and others in the 1860s and 1870s.[84] One described the conduction of heat in a layer of air, while the other related barometric pressure to the height above the earth's surface. To derive equations with which to predict changes in air pressure, Felix made a number of simplifying assumptions. He neglected friction, horizontal acceleration, heat transfer, variations in the height of the earth's surface, and motion of air in the vertical direction. This allowed him to pose the problem in terms of coordinates that took account of the rotation of the earth, arriving at a first-order linear differential equation for the change in pressure with time. Helpfully, Felix introduced a diagram to visualize the equation's significance (fig. 14).

Felix found that the model's predictions coincided with observations from European weather stations 70 to 90 percent of the time. In his evaluation,

83. Huw C. Davies, "Vienna and the Founding of Dynamical Meteorology," in Hammerl, *Zentralanstalt*, pp. 301–12. The most thorough discussion of methods in nineteenth-century meteorology to date is Gisela Kutzbach, *The Thermal Theory of Cyclones* (Boston: American Meteorological Society, 1979). For evaluations of Felix's dynamical model, see Friedrich Lauscher and Georg Skoda, "Zum Gedenken an Felix M. Exner," *Wetter und Leben* 33 (1981): 94–102; Heinz Fortak, "Felix Maria Exner und die österreichische Schule der Meteorologie," in Hammerl, *Zentralanstalt*, pp. 354–86. On statistical methods in meteorology, see O. B. Sheynin, "On the History of the Statistical Method in Meteorology," *Archive for the History of the Exact Sciences* 31 (1984/85): 53–95.

84. On the earlier work, see Kutzbach, *Thermal Theory*, chapter 3.

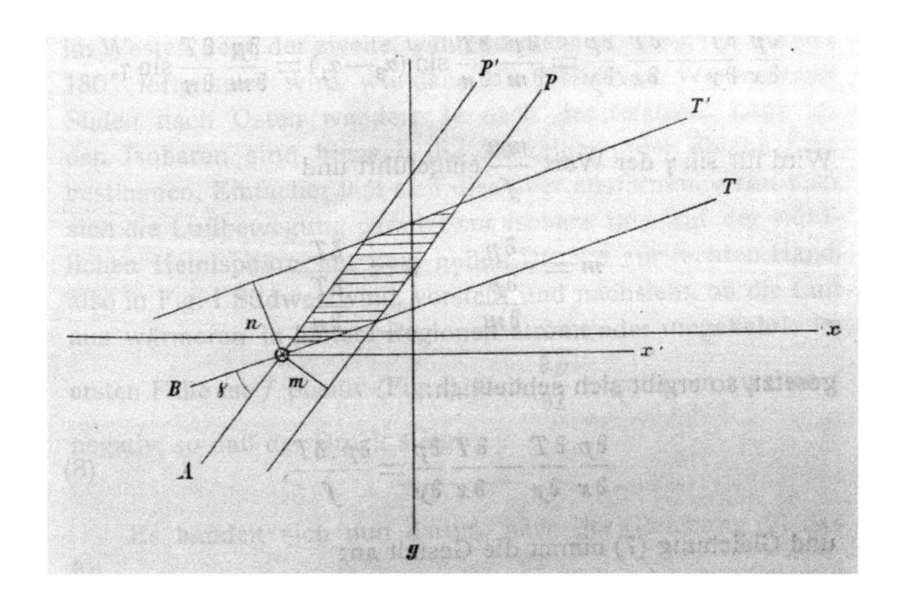

FIGURE 14 Felix Exner's representation of his equation for changes in air pressure. In the figure, the straight lines P and P' are isobars (lines of equal pressure), T and T' are isotherms (lines of equal temperature). In this case the pressure difference $\delta p = p - p'$ is taken to be positive, as is the temperature difference $\delta T = T - T'$. The area f enclosed by the intersection of the isobars and isotherms can be positive or negative (in the figure it is positive, since $T > T'$, $p > p'$). What his final equation said was that this area was inversely proportional to the change in pressure with time at that location. An example showed how to use the diagram. In a frictionless situation in the northern hemisphere, the movement of air is parallel to the isobars with higher pressure to your right if your back is to the wind (due to the combination of the pressure gradient and the Coriolis force from the earth's rotation). Now ask yourself whether the wind flows from warmer to colder regions or vice versa (here, it flows from warmer to colder, since T' is lower than T). In the first case, the area f is positive, and therefore $\delta p/\delta t$ negative, and the pressure falls; in the second case, the pressure rises. The smaller f is, the larger the change in pressure, and vice versa. In other words, by simply overlaying a map of isotherms on a map of isobars and "hold[ing] it to the light," one could quickly judge the expected changes in air pressure. Source: "Grundzüge einer Theorie der synoptischen Luftdruckveränderungen," *Sitzungsberichte der Kaiserlichen Akademie der Wissenschaften zu Wien IIa* 115 (1906): 1171–1246, figure on p. 1185.

even if the sign of the change in air pressure could "only be determined with a 70 percent probability," this could still "be of value for the prognosis of a pressure distribution." Felix introduced here a probabilistic style of reasoning that has since become ubiquitous in meteorology. He calculated from observations that a blind prediction that the weather tomorrow will be the same as today's will be correct 60 percent of the time. Therefore, what determined a theory's value for forecasting was its ability to predict *changes* in the weather, as Felix's model could do—if only 70 percent of the time. Given all the simplifying approximations he had made, Felix reflected with satisfaction, "one cannot deny the method a certain validity." Already during the first flush of his enthusiasm for forecasting, Felix displayed a tolerance for nature's unpredictability that might seem out of character in a man seeking to reduce the weather to classical physics. Perhaps for this reason, Felix has recently been referred to as "the father of statistical [weather] forecasting."[85] As we will see, it was this tolerance for uncertainty that would set Felix—along with the "Austrian school"—apart from their Norwegian competitors in the years to come.

In order to fine-tune his model for a better fit with observations, what Felix now needed was data, lots of data. A voyage around the world in 1904–5 allowed him to visit the American Weather Bureau in Washington, D.C., which compiled observations of an unprecedented uniformity across an immense geographical expanse. Upon his return to Vienna in 1905, Felix used this data to make a qualitative assessment of his model's value. He plotted an observed weather distribution, calculated and plotted the predicted distribution after four hours, and compared this to the actual distribution in the records. Based on this evaluation, Felix was more circumspect when he published again on the dynamical model in the *Meteorologische Zeitschrift* in 1908. Rather than claiming "practical" value for his method, he described his model as "a crude approximation" to the solution of weather forecasting, whose value lay in its power to explain rather than predict. "Yet for the forecaster it is already a great advantage to have any real foundation at all for his prediction and thereby to make his reasoning more precise."[86]

It was thus in 1908, the year of his uncle Serafin's rectoral address, that Felix first began to express doubt about the feasibility of using the differential

85. Lauscher and Skoda, "Felix M. Exner," p. 96.

86. Felix Exner, "Über eine erste Annäherung zur Vorausberechnung synoptischer Wetterkarten," *Meteorologische Zeitschrift* 25 (1908): 57–67, quotation on p. 62.

equations of classical hydrodynamics to model the complex phenomena of the atmosphere. He raised not only a practical question of methodology, but the more fundamental question of the scope of determinism in nature. Five years later, he would write of meteorology as "the prototype of those sciences that, despite strenuous efforts, have not arrived at any, or hardly any, laws."[87] At the very moment when foreign colleagues like Bjerknes and Lewis Richardson were optimistically embarking on a deterministic program for meteorology, Felix Exner was becoming convinced of the limits of meteorological prediction.

Felix soon became a spokesman for the value of probabilistic methods in the physical sciences. In 1912 he introduced the German-speaking scientific world to the method of correlation, developed in Britain by Francis Galton and Karl Pearson. In Central Europe even meteorologists were unfamiliar with correlation, despite the fact that Galton had developed the method in order to solve a problem in meteorology.[88] The essence of the method was to evaluate the degree of dependence of two series of measurements by comparing not the measurements themselves but their deviations from the mean of the series. Since nineteenth-century meteorologists were chiefly concerned with tracking variables like temperature, wind, and rainfall, they might have been expected to incorporate Galton's innovation. But meteorologists of the day generally believed that if the phenomena they studied could be freed entirely of observational errors and "local influences," they would reveal simple periodicities—the ebb and flow of a Newtonian universe. As one French scientist wrote in the 1860s, "the irregular atmospheric variations must be viewed not as random, but rather as oscillations of unequal periodicity."[89] In the eyes of typical nineteenth-century meteorologists, irregular fluctuations were not real phenomena but errors. There was thus no demand for a method like Galton's that would enable the study of fluctuations themselves.[90]

Felix Exner was able to recognize the potential of the correlation method for meteorology because he did not expect nature to display Newtonian regularities. The value of the method in his eyes was that it was equally valid for probabilistic as for deterministic phenomena. This he made clear when

87. Felix Exner *Dynamische Meteorologie*, 2nd edition (1917; Vienna: Springer, 1925), p. 1.

88. Porter, *Statistical Thinking*, pp. 272–296.

89. Quoted in O. B. Sheynin, "Statistical Method in Meteorology," p. 71.

90. Cf. Porter, *Statistical Thinking*, p. 296.

he introduced the method to his colleagues at the Natural Scientific and Medical Society in Innsbruck in 1912.[91] "Strict" correlations, he explained, were exemplified by the laws of mathematics. But in the everyday practice of science, one often encountered correlations "which are not as strict and exact as physical laws." These were better called "rules" or "regularities," since they admitted of "exceptions."[92] The method of correlation could be used to analyze such regularities, which appeared frequently in science as in daily life. Phenomena described by such regularities were to some degree unpredictable, but as the number of observed instances of a given phenomenon increased, so did its predictability, in accordance with the law of large numbers. Thus many physical "laws" were merely the result of this principle of the emergence of regularity out of randomness.[93] In structure and language, Felix's argument for the value of correlation methods mirrored his uncle Serafin's case for a probabilistic physics at his rectorial inauguration in 1908.

Felix likewise presented meteorology as an inherently probabilistic discipline in his 1917 textbook—but this was modesty on his part. He was in fact largely responsible for bringing probability to bear on the science of weather. And the value of some of his work is just now being recognized by meteorologists today. Building on the work of Gilbert Walker, Felix helped pioneer the use of correlation to analyze what he referred to as weather "anomalies," deviations from the long-term average weather in a given location. His studies of the North Atlantic Oscillation have been rediscovered by meteorologists of the twenty-first century, who acknowledge that his orientation, for better or worse, was fundamentally "statistical."[94]

Anomalous as Felix's statistical approach might seem among his contemporaries, in Austria he was not alone. An interesting parallel is his colleague and successor as the director of the ZAMG, Wilhelm Schmidt. Just six years younger than Exner, Schmidt had likewise studied physics in Vienna with Boltzmann and Serafin Exner. Throughout his research, Schmidt worked with an analogy that spoke of his training in the Austrian tradition of statistical physics. He modeled his mathematical treatment of

91. Published as *Über die Korrelationsmethode* (Jena: Verlag Gustav Fischer, 1913), based on a lecture delivered to the Naturwissenschaftlich-medizinischen Verein in Innsbruck, 26 Nov. 1912.

92. F. M. Exner, *Korrelationsmethode*, p. 5.

93. Ibid., p. 6.

94. Heinz Wanner et al., "North Atlantic Oscillation—Concepts and Studies," *Surveys in Geophysics* 22 (2001): 321–82, quotation on pp. 331–32.

the turbulent currents of the atmosphere on Boltzmann's approach to the random molecular motions in gases. In his comprehensive survey of energy exchange between air masses (*Austausch*), Schmidt developed this analogy in detail, introducing *Austausch* correlates of the thermodynamic concepts of "conduction," "diffusion," and "friction." Following Boltzmann's reasoning, Schmidt explained that even if we could observe the trajectories of each particle in the atmosphere, we would be no closer to understanding weather phenomena. "One progresses far more quickly, as in the 'kinetic gas theory,' if one tries instead to conceive of and combine the effect of the totality of all the particles." The goal for Schmidt, as for Boltzmann, was thus "a general view of the mean states, of the average effects."[95]

By the 1920s, then, Felix Exner and Wilhelm Schmidt were each convinced that the study of turbulence was essential to progress in meteorology. In the lyrical examples they offered, the inherent complexity of the natural world was visible in such effects as the swaying of tree branches or the paths of falling leaves.[96] To capture the effects of such complex phenomena, both Schmidt and Exner relied on statistical methods, but they also made creative use of what Peter Galison has called mimetic experimentation.[97] Consisting of direct imitations of natural effects in the laboratory, mimetic experimentation was a recourse suited to the reduced means of Viennese meteorology during and after the First World War, when Vienna lost its empire, its economic security, and, of course, its vast network of observing stations. It was through laboratory experiments that Schmidt and Exner became familiar with the turbulent motions at the surfaces between air masses, with far-reaching consequences for Austrian meteorology in the years to come.

This focus on turbulence and the limits of predictability set Exner and Schmidt on a course of research that diverged sharply from what historians have considered the agenda of the day: deterministic weather prediction.[98] To Exner and Schmidt, Lewis Richardson's forecasting factory and Vilhelm Bjerknes's simplified "polar front" looked naïve at best. As Exner wrote in a cool review of Richardson's 1922 book, "It does not seem probable to the reviewer that one will ever achieve fundamental progress along the

95. Wilhelm Schmidt, *Der Massenaustausch in freier Luft und verwandte Erscheinungen* (Hamburg: Henri Grand, 1925), p. 8.

96. Schmidt, *Massenaustausch*, 7; Exner, "Ungeordneter Bewegung."

97. Galison, *Image and Logic*, chapter 2.

98. Frederik Nebeker, *Calculating the Weather* (San Diego: Academic Press, 1995).

path indicated. For that the processes are too complicated and burdened by too many details." Richardson's key failing in Exner's eyes was that his deterministic model could not capture the complexity of atmospheric phenomena. As he explained, "Theoretical physics has trained in small-scale processes, and it is to be expected, that for large-scale phenomena other theoretical foundations will have to be found which exclude all the details."[99] The solution, it seems, would have to be statistical.

Exner's reception of Richardson's work was warm, however, compared to what Robert Marc Friedman has described as Exner's "troublesome" and "ugly" personal response to Bjerknes's polar front theory, the most famous meteorological innovation of the twentieth century.[100] Regardless of how he behaved in person, on paper Exner's criticism was tempered but sharp. His objection was quite specific. By no means did he reject the concept of discontinuous air masses in the polar region. To the contrary, as his distinguished Austrian colleague Heinrich von Ficker noted, Exner deserved a portion of the credit for this very concept.[101] Indeed, Ficker pointed out that Bjerknes had failed entirely to cite the work of Austrian meteorologists, including himself and the late Max Margules. This was an especially egregious omission given the availability of these references in Exner's 1917 *Dynamical Meteorology*. As Ficker stressed, however, the Austrians were not interested in a "priority dispute."

According to Ficker, Bjerknes had convinced the scientific world of his polar front theory through the "weight of his personality" as well as through the intentionally schematic character of his presentation. By contrast, his Austrian forerunners had presented their research in its full mathematical complexity, while their extensive use of observational data made their publications even more opaque to the casual reader. Thus, Bjerknes had achieved a more heuristically useful presentation. But was it a fair representation of the natural phenomena in question?

99. Felix Exner, review of Lewis F. Richardson, *Weather Prediction by Numerical Process* (Cambridge: Cambridge University Press, 1922), in *Meteorologische Zeitschrift* 40 (1923): 189–90, quotation on p. 189. It undoubtedly did not help Richardson's case that he had included in the description of his weather farm a disparaging caricature in which Exner may well have recognized himself: "an enthusiast" in the basement "observing eddies in the liquid lining of a huge spinning bowl" (Richardson, *Weather Prediction*, p. 220).

100. Robert Marc Friedman, *Appropriating the Weather: Vilhelm Bjerknes and the Construction of a Modern Meteorology* (Ithaca: Cornell University Press, 1989), p. 199.

101. H. Ficker, "Polarfront, Aufbau, Entstehung, und Lebensgeschichte der Zyklonen," *Meteorologische Zeitschrift* 40 (1923): 65–79.

Exner doubted Bjerknes's theory on several grounds.[102] It seemed futile to him, for instance, to try to explain cyclone formation as a local effect, rather than as a product of the global air circulation. To Exner, Bjerknes's concept of the polar front failed precisely because it did not take into account the factors that had emerged as a focal point of research in Austria: instability and turbulence. If cyclones really were waves in the surface between warmer and colder air masses, as Bjerknes claimed, then this boundary must be stable. Yet, as Exner's own experiments with spinning basins and miniature sand dunes had convinced him, no truly eddyless motion was possible at the surface between two layers of air. Schmidt's work on *Austausch* reinforced this point. Significantly, Schmidt's 1925 discussion of discontinuity surfaces did not cite Bjerknes once. To the Austrians, then, the Norwegian version of the polar front was neither original nor wholly credible.

Still, what truly piqued Exner's hostility to the new Oslo meteorology was more methodological than conceptual. The first issue was the theory's simplification of the complex and perhaps fundamentally unpredictable forces of the atmosphere. The second was the aggressive manner of Bjerknes's campaign for his theory. This two-pronged critique was succinctly encapsulated in the 1925 edition of Exner's *Dynamical Meteorology,* where Exner wrote that the polar front "is to be regarded merely as a schematic hypothesis."[103] As a schematic theory, it ignored the true nature of atmospheric phenomena, and as a hypothesis it deserved to be advanced gingerly, not as the battle cry it had become to the Norwegians. As Friedman justly notes, Bjerknes denied evidence that the front might be a more complicated structure than his theory suggested, and he silenced even collaborators if they threatened "the hegemony" of his polar front.[104]

In short, Bjerknes's manner struck his Austrian colleagues as *dogmatic.* To men like Exner and Schmidt, trained in the liberal Austrian tradition of probabilistic reasoning, nothing was more threatening to the scientific community than dogmatism, whether it came from the clergy or from scientists themselves. Compounding this effect was Bjerknes's campaign for a *deterministic* model of weather. Given the alignment in Austria of determinism with dogmatism, and probabilism with intellectual freedom,

102. See Felix Exner, "Sind die Zyklonen Wellen in der Polarfront oder Durchbrüche derselben?" *Meteorologische Zeitschrift* 38 (1921): 21–23; *Dynamische Meteorologie,* p. 355; "Über die Zirkulation zwischen Rossbreiten und Pol," *Meteorologische Zeitschrift* 44 (1927): 46–53.

103. Felix Exner, *Dynamische Meteorologie,* p. 355

104. Friedman, *Appropriating,* p. 200

the self-presentation of Norwegian meteorology in the 1920s must have struck the Austrians as antithetical to their own style of science.

Even the language in which these theories were couched was telling. Bjerknes, of course, seized on the military metaphor of the front, relying on images of "struggle" and "victory."[105] This choice reflected meteorology's increasing relevance to warfare, but it also accentuated Bjerknes's use of the polar front theory as part of a nationalistic campaign. Exner's use of metaphor was quite different. Even in 1918, while responsible for making meteorology serve Austria's war aims, Exner favored images of freedom, not battle. In his words, air masses were "free" and "independent."[106] If Bjerknes's air masses were constrained by their position in a hemispheric "battle," Exner's were individuals whose trajectories were highly contingent. Exner imagined the air masses as semiautonomous actors: "[H]ow will they spread, where will they move?"[107] Reflecting his identification with the tradition of liberal science in Austria, Exner chose metaphors of freedom, contingency, and individuality over the dominant language of warfare.

In such ways, the dispute over the polar front became an opportunity for Felix Exner and his Austrian colleagues to carve a new identity for themselves. With the loss of the empire and of their advantages of scale, Austrian meteorologists in the 1920s desperately needed to define themselves anew with respect to the international scientific community.[108] They needed to give a clear meaning to the phrase "the Austrian school of meteorology." Given Felix Exner's role in shaping this identity through his clashes with Bjerknes, it is fitting that one vivid exposition of "Austrian" meteorology appeared in the obituary for Exner published by the Austrian Academy of Sciences in 1930:

> In general Exner's book—and his entire research orientation—stands in conscious opposition to the so-called "Nordic School" and its manner of presentation. It cannot not be denied that they were born of the complex circumstances of a mountain nation, which on one hand imposes a more sharply critical view,

105. Friedman, *Appropriating*, p. 188.

106. Felix Exner, "Studien über die Ausbreitung kalter Luft auf der Erdoberfläche," *Sitzungsberichte der Kaiserlichen Akademie der Wissenschaften zu Wien IIa* 127 (1918): 795–847, quotation on p. 847.

107. Felix Exner, "Ausbreitung," p. 800.

108. See, e.g., Christa Hammerl's account of the 1922 international meteorology conference hosted by the Austrians at the Sonnblick Observatory (Hammerl, *Zentralanstalt*, p. 139).

> but on the other also affords more profound insights into weather phenomena. Here it will not be nearly as easy to find in nature the desired simplified conditions for a theoretical approach; only gradually do ideas accommodate the ever more complicated facts; and so was Exner of necessity led from his theoretical works to experiments in the laboratory and to the organization of special observations, above all in the mountains. Such a method indeed forms an important part of what abroad is designated the "Vienna School" and has a decisive, corrective influence on the other schools, including that of Bjerknes. And if today, through Exner's work, our ideas of cold air masses in high latitudes, from their origins to their penetration in the temperate zones, the manner of their propagation, are far less simple, indeed cannot be integrated into a single picture—yet they certainly approach closer to the truth.[109]

Hailing Felix Exner as their leader, the Austrians' defined their own contribution to meteorology in terms of the antidogmatic and antideterministic values of alpine physics and Austrian liberalism.

ERWIN SCHRÖDINGER: A MORPHOLOGICAL STATISTICS

It may seem like an accidental irony of history that Felix Exner, who devoted his career to the science of weather, chose university classes in molecular physics and epistemology over the chance to study with the master of Austrian meteorology, Julius Hann; while Erwin Schrödinger, who was drawn to abstract theories of the microscopic world, did enroll in this most practical of physics courses, a complement to his research on atmospheric electricity under Serafin Exner's guidance. Schrödinger exploited this background during the First World War, when he was assigned to teach meteorology to pilots. His wartime lecture notes clearly reflect his training in a specifically Austrian school of meteorology.[110] He addressed topics associated with Hann, such as the fluctuation of meteorological factors and the significance of mean values. He also included such typically Austrian concerns as the "influence of mountains on the weather," and the "Föhn," the distinctive Alpine winds.

During the war Schrödinger also pursued theoretical research on atmospheric physics. In 1917 he published a paper on shock waves in the atmos-

109. *Almanach für das Jahr 1930*, Akademie der Wissenschaften in Wien, pp. 257–61, quotation on pp. 259–60.

110. Archive for the History of Quantum Mechanics, reel 39, section 2; the notebook comprises forty pages.

phere that bridged his earlier wartime experience in the artillery with the theoretical methods of Austrian meteorology. The motivation for this study came from research by Wilhelm Schmidt, the Viennese author of the *Austausch* concept, and Schrödinger also cited the work of Austrians Alfred Wegener and Felix Exner.[111] His early training had given him the background in hydrodynamics necessary to work comfortably with the methods of dynamical meteorology.

If Felix Exner has been neglected by historians, Erwin Schrödinger has been amply scrutinized; yet he continues to present a mystery. In 1926 Schrödinger revolutionized atomic physics by expressing the state of a hydrogen atom in the mathematical form of a standing wave. Shortly after the publication of his wave equation, German physicists Max Born and Werner Heisenberg claimed that its interpretation forced the abandonment of space-time continuity and determinism at the atomic level. Schrödinger disputed their interpretation. Had he undergone a sudden transformation from an avid probabilist into a reactionary determinist?

A clue lies in Schrödinger's early training in the Austrian style of alpine physics. Schrödinger began in late 1924 to work with a startling hypothesis introduced the year before by Louis de Broglie. De Broglie had proposed that wave and particle descriptions were dualistic, such that any particle could be assigned a wavelength, equal to Planck's constant h divided by its momentum. Schrödinger was interested in using this description to make his calculations of gas statistics more visualizable. Drawing on his background in atmospheric physics, Schrödinger pictured the atom as the "'whitecap' on the wave radiation that forms the basis of the world."[112] Suggestively, Schrödinger also assigned a problem on atmospheric shock waves to a new doctoral student just weeks before he first wrote down his wave equation.[113]

In the months after publishing his first paper on wave mechanics in early 1926, Schrödinger continued to work with the hydrodynamic analogy. This is best seen from his correspondence with the Frankfurt physicist Erwin Madelung. Madelung had trained in engineering and electrodynamics, so he was familiar with hydrodynamics as the language of the electrical theory

111. Erwin Schrödinger, "Zur Akustik der Atmosphäre," *Physikalische Zeitschrift* 18 (1917): 445–53; *Gesammelte Abhandlungen*, vol. 4, p. 3.

112. Moore, *Schrödinger*, p. 188. In a similarly visual vein, he wrote in 1925, "I have tried in vain to make for myself a picture of the phase wave of the electron in the Kepler orbit" (ibid., p. 192).

113. Moore, *Schrödinger*, p. 192.

of the late nineteenth century. His 1922 handbook for mathematical physics stressed the need for fluency in a variety of mathematical disciplines in order to learn how to think "visually" and "physically."[114] On 9 October 1926 Madelung wrote to Schrödinger of a new interpretation of his wave equation, apparently continuing an ongoing dialogue of which there is no further record.[115] Two weeks later, Madelung sent this interpretation to the *Zeitschrift für Physik,* where Max Born's probabilistic interpretation had been published three issues earlier.[116] Madelung's motivation was to show that the essential aspects of the new quantum theory could be "modeled" (*modellmäßig dargestellt*) through the equations of classical hydrodynamics in a manner that was "immediately intuitive" (*unmittelbar anschaulich*). In this representation, "quantum jumps" would be replaced by "a slow transition to the state of nonstationarity." In other words, a hydrodynamic model would restore space-time continuity. Madelung intended to represent the quantum mechanical behavior of a single electron as a *continuous current.* He showed that solutions to the time-independent and time-dependent Schrödinger equations could be transformed into stationary states of a hydrodynamic current and states with periodically varying current density, respectively.[117]

Madelung's inspiration for this strategy came directly from Schrödinger. In his second paper on wave mechanics, published in the spring of 1926, Schrödinger had separated the time-dependent equation into real and imaginary parts, just as Madelung would later do, and had suggested that the real part of $\psi \delta \psi / \delta t$ gave the space density of the electric charge. Over the course of 1926, then, Schrödinger and Madelung elaborated in communication with each other a picture of the electron spread out in space as a wave. This argument was revived by David Bohm in the 1950s and has remained a reference point for new interpretations of quantum mechanics ever since.[118]

114. Jungnickel and McCormmach, *Intellectual Mastery,* vol. 2, p. 344.

115. Erwin Madelung to Erwin Schrödinger, 9 Oct. 1926, Schrödinger Nachlass, Zentralbibliothek für Physik, Universität Wien.

116. Erwin Madelung, "Quantentheorie in hydrodynamischer Form," *Zeitschrift für Physik* 40 (1926): 322–26.

117. He demonstrated this by separating an alternative form of the time-dependent equation into real and imaginary parts and making appropriate transformations of variables, such that the imaginary part expressed hydrodynamic continuity, and the real part gave the equation for a rotation-free current subjected to a conservative force.

118. David Bohm, *Causality and Chance in Modern Physics* (Philadelphia: University of Pennsylvania Press, 1957). Cf. Louis de Broglie, *Une tentative d'interpretation causale et non-linéare de la mécanique ondulatoire* (Paris: Gauthier-Villars, 1956).

Atmospheric physics played an unappreciated role in Schrödinger's development and interpretation of the wave equation. This helps explain why Schrödinger rejected Born's and Heisenberg's interpretations. In particular, it illuminates his famous commitment to the goal of visualizability (*Anschaulichkeit*) for atomic physics.[119] Schrödinger proposed the hydrodynamic analogy in 1926 explicitly as a means of recovering *Anschaulichkeit*. In the debates that followed, he stressed that waves were fundamental to nature, as he had suggested earlier with the image of a particle as the foam on a wave's crest.

Compare this to the image of nature Felix Exner had popularized just a few years earlier: a world made up of waves, oscillations, fluctuations.[120] As Felix made clear, this was a view gleaned from experience with the physics of the atmosphere, as well as the geomorphology of mountains, rivers, and lakes. During their summers in the foothills of the Alps, both Felix Exner and Erwin Schrödinger had been trained to follow nature's fluctuations continuously in time.

Schrödinger's concern was to reconcile quantum theory with this task. While one might think that space-time continuity implied determinism, for Schrödinger this in no way precluded a probabilistic standpoint. On the contrary, Felix Exner had explained nature's affinity for fluctuations probabilistically, as a consequence of statistical mechanics: "It is thus a direct consequence of probability that ordered movements are not preserved."[121] Schrödinger likewise sought to understand the unpredictability of the quantum world on the model of statistical mechanics.[122] For instance, he emphasized the similarities between the mathematics of wave mechanics and the Fokker-Planck equation for the time-development of a stochastic system, which he had previously applied to fluctuation phenomena like radioactivity and Brownian motion. The wave equation might describe "the behavior of a statistical ensemble, just like the so-called Fokker partial differential equation,"

119. Cf. Jan von Plato, *Creating Modern Probability* (Cambridge: Cambridge University Press, 1994).

120. Felix Exner, "Über natürliche Bewegungen in geraden und gewellten Linien," *Naturwissenschaftliche Wochenschrift*, n.s. 19 (1920): 385–90; "Über ein Bewegungsprinzip in der Natur," lecture delivered at the Akademie der Wissenschaften in Vienna on 30 May 1923 (Vienna: Österreichsiche Akademie der Wissenschaften, 1923).

121. Felix Exner, "Bewegungsprinzip," p. 8.

122. Admittedly, there were moments when his confidence in probability faltered, as in a letter to Willy Wien in the summer of 1926 (see Moore, *Schrödinger*, p. 225). It seems safe to say, nonetheless, that the statistical mechanical analogy represented his mature view.

he wrote to Planck in 1927.[123] The time-dependent wave equation and the Fokker-Planck equation have *nearly* the same form. Yet their differences had important implications: the Fokker-Planck equation is irreversible in time, the wave equation reversible. Schrödinger could not see past this discrepancy. However, his colleague Reinhold Fürth, the successor of Mach and Lampa to the experimental physics chair in Prague, plowed ahead with this classical analogy. Inspired by Schrödinger's suggestion, Fürth showed that a classical analogue to Heisenberg's uncertainty relations could be derived from the Fokker-Planck equation for diffusion in one dimension, such that the product of the uncertainty in position and in momentum would be greater than or equal to the diffusion constant.[124] Schrödinger had written of this statistical mechanical analogy that it had "probably impressed itself on everyone who has long been intimate with both fields of ideas," statistical and quantum.[125] Thanks to his situation in the former Habsburg lands, Fürth was one of the few actually in this position in the 1920s. He was the author of the first survey of "fluctuation phenomena," the field pioneered in Vienna in 1905.[126]

Schrödinger himself harbored high aspirations for a statistical mechanical reformulation of quantum mechanics. In an undated manuscript he set out a plan for a book on statistical thermodynamics, evidently more ambitious than the one he actually published:

> For I am becoming convinced that quantum mechanics is statistical-thermodynamical not only in its origins, but that it will return to this source at the end of its course . . . Statistical mechanics, originally nothing but an "applied field" of the old mechanics, has taken on a life of its own, so strongly, that the foundations of the old mechanics must be shifted. The "We cannot know the initial conditions precisely" has gained independence [*hat Leben gewonnen*]. We cannot determine them precisely in principle. What was originally a pragmatically drawn boundary has become a theoretical one.[127]

123. Quoted in Karl Przibram, ed. and trans., *Letters on Wave Mechanics: Schrödinger, Planck, Einstein, Lorentz* (New York: Philosophical Library, 1967), p. 19.

124. Reinhold Fürth, "Über einige Beziehungen zwischen klassischer Statistik und Quantenmechanik," *Zeitschrift für Physik* 81 (1933): 143–162.

125. Erwin Schrödinger, Über die Umkehrung der Naturgesetze (Berlin: Akademie der Wissenschaften, 1931), p. 148; *Gesammelte Abhandlungen*, vol. 1, p. 412.

126. Reinhold Fürth, *Schwankungserscheinungen in der Physik* (Braunschweig, 1920).

127. Erwin Schrödinger, "Schreiben," one-page typescript, W33-318, Schrödinger Nachlass, Zentralbibliothek für Physik, Universität Wien.

Schrödinger's ambition, then, was a fluctuation physics of the atomic world. That nothing else satisfied him seems natural when we picture him back at Seeham before the First World War, observing "nature's pulse" in the fluctuations of indicator needles.

Indeed, the image of these jittery pointers was deeply engraved in Schrödinger's memory. They were a reminder of the scope of chance not merely in the physical world, but in all facets of human life. This, of course, was the broader lesson that Serafin Exner had pressed on his students. Through encounters with natural contingency, Exner intended these young men to cultivate antidogmatic characters. They would learn to manage uncertainty in physics and beyond. Years later, Schrödinger ruminated on this lesson in a poem entitled "Parable." Combining his experience of alpine physics with his studies of Eastern philosophy, Schrödinger described the human attitude best suited to an unpredictable world. How should one face life's endless battery of accidents? Schrödinger chose an undogmatic detachment, a spirit of *laisser-aller* that bespoke the liberal goals of his mentor Exner:

PARABLE
Friend, what in life seems
Weighty and meaningful,
Whether it bends us low
Or cheers and delights us,
Acts, wishes, and thoughts,
Believe me, mean no more
Than the dial's random fluctuation
In the experiment that we design
To fathom nature:
Mere molecular collisions.
You can't sniff out the law from the light spot's crazy
tremors.
Your joy and trembling
Is not the point of this life.
The world spirit, if he sets about it,
May, after thousands of runs,
Finally enter a result.—
Is it really our concern?
[PARABEL
Was in unserem leben, freund,
wichtig und bedeutend scheint,
ob es tief zu boden drücke

oder freue und beglücke,
taten, wünsche und gedanken,
glaube mir, nicht mehr bedeuten
als des zeigers zufallschwanken
im Versuch, den wir bereiten
zu ergründen die natur:
sind molekelstösse nur.
Nicht des lichtflecks irres zittern lässt dich das gesetz
erwittern.
Nicht dein jubeln und erbeben
ist der sinn von diesem leben.
Erst der weltgeist, wenn er drangeht,
mag aus tausenden versuchen
schliesslich ein ergebnis buchen.—
Ob das freilich uns noch angeht?][128]

Felix Exner and Erwin Schrödinger followed parallel paths to probabilism. As young physicists in the Salzkammergut, patiently watching their instruments trace the evolution of the atmospheric elements, they became convinced of the ubiquity of fluctuating movements in nature. At Serafin Exner's Vienna institute, they trained in the theory and practice of fluctuation physics, learning to observe and calculate the statistical significance of microscopic disorder. Serafin Exner himself showed them the implications of these observations. Statistics alone could account for the laws of nature. The statistical view liberated physics from the dogma of determinism, and it did so without opening a skeptical abyss. Fluctuation physics thus became a symbol of the struggle against dogmatism in a city rife with anticlericalism.

Just as fluctuation physics was a resource for liberal self-fashioning, so did the private life of the *Bildungsbürgertum* shape Austrian physics. The peculiar geography of the alpine summer retreat, real and imagined, the design of the cottages and the myth of isolation, set the conditions of possibility for alpine physics. And alpine physics in turn shaped the study of fluctuations and the probabilistic turn in Austrian physics. As Serafin Exner made clear, it was necessary to leave civilization behind in order to see the phenomena in this way.

128. Reprinted in Moore, *Schrödinger*, p. 6. Moore's translation was helpful to my own.

CHAPTER NINE

The Irreplaceable Eye

Visual Statistics, 1914–1926

IN THE deceptive calm of June 1914, Marie Exner von Frisch's son Ernst was revising his history of the Wolfgangsee. In the new introduction he apologized for presenting the turn-of-the-century Austrian reader with yet another celebratory memorial to a past age. "We live in an age of jubilees, monument unveilings, and memorial crosses. A time will come when one will laugh over the eagerness of this generation to dig up commemoration days, in order to comfort itself for the apparent uneventfulness of the present."[1] Indeed, that time was closer than Ernst suspected.

Just two months later, the illusion of uneventfulness was shattered. With the outbreak of the world war, Ernst temporarily lost interest in historical research. Still, he retained his post as the librarian of the Vienna parliament for the duration of the hostilities, and within a year he returned to his local history. "One had grown accustomed to the war," he noted; surely peace was imminent. In a new preface, he observed that 1915 was the five hundredth anniversary of the settlement of St. Gilgen. During the Thirty Years' War, Brunnwinkl had been "a place of peace in the midst of a tumultuous world."[2] Three centuries later, Ernst observed, its role had not changed.

In the autumn of 1914 the imperial government in Vienna gave jurist Hans Frisch orders to flee his post at the University of Czernowitz, the capital

1. Ernst quoted his own remark in the 1915 preface to his *Geschichte der Brunnwinklmühle 1615–1882* (Vienna: self-published, 1918).

2. Ibid.

of Bukovina and the easternmost outpost of Habsburg education (now Chernivtsi, Ukraine). With a caravan of library books, Hans headed for the Wolfgangsee, accompanied by several of his now homeless colleagues. Brunnwinkl also provided brief sanctuary during the war for several men and women of the third generation who worked for the military medical service. Karl Frisch, for one, spent his honeymoon there in the summer of 1917. By then the tennis courts were overgrown with grass, and local authorities were threatening to build a road past the cottages with labor provided by Russian prisoners of war. Nonetheless, as far as the Brunnwinklers were concerned, not much had changed in St. Gilgen.[3]

With Vienna overcome by a food shortage and political unrest, several women and children of the Exner-Frisch circle moved year-round to Brunnwinkl, where they could take advantage of the garden and the plentiful supply of firewood.[4] They left behind a window onto their experiences in the form of a handwritten and illustrated pamphlet, wryly titled *Der Suppentopf: Ein geistiges Erbauungsblatt* (The Soup Pot: A Journal of Moral and Intellectual Edification). In sketches, riddles, stories, and poems, the women found ironic humor in their situation. Under "new pictures from the theater of war," for instance, were illustrations of daily life at Brunnwinkl. A drawing of two women playing chess by candlelight bore the caption "Even in the darkest nights the battles rage." As they explained in a "note to our subscribers," they intended to "help you bear the long war's effects through humor."

The women's irony reflected the social inversions of the home front—a "fatherless society," as one contemporary called it. As historian Maureen Healy explains, this label fit in two ways. With the soldier supplanting the "diligens pater familias" as a figure of authority, men unfit for duty lost their traditional sphere of power. At the same time, members of the "imperial family," as the peoples of Austria-Hungary were known, lost their trust in the omnipotent figure of the emperor-father. During the war, hard-hit citizens addressed deeply personal requests for aid to Franz Josef and his successor Karl, all in vain.[5] This crisis of patriarchal authority motivated questions about the nature of family, state, and the metaphorical family-state—about what made these communities cohere or drove them to dissolution.

At Brunnwinkl, the Exners struggled to ensure the survival of their own

3. Karl Frisch, *Fünf Häuser am See*, pp. 81–82; Hans Frisch, *50 Jahre Brunnwinkl*, pp. 15, 36–37.

4. Hans Frisch, *50 Jahre Brunnwinkl*, p. 36.

5. Maureen Healy, *Vienna and the Fall of the Habsburg Empire: Total War and Everyday Life in World War I* (Cambridge: Cambridge University Press, 2004), pp. 258–99.

community. This effort was punctuated by their celebration in May 1917 of the thirtieth anniversary of the planting of the linden tree, the communicative hub of their hamlet. Serafin Exner composed a poem for the occasion in which he gave the tree the power of speech. For thirty years it had "recorded what occurred in your realm," and now it wished to "refresh . . . the memory images [*Erinnerungsbilder*]."[6] Behind this fantasy lay Serafin's conviction that visual memories were the foundation of aesthetic realism, the building blocks of a universal language. "Memory images" were to be Brunnwinkl's source of continuity and cohesion, as Adolf Exner had foreseen at the colony's founding. Yet the linden's appeal was also a reminder of the tenuousness of this utopian ideal in the cataclysm of the world war.

WAR AND THE SCIENCE OF PERCEPTION

If the home front was literally and figuratively a "fatherless" world, then Erwin Schrödinger's experience was typical. By war's end, his father's business had failed and the old man's health was deteriorating rapidly. Within two years, Schrödinger would lose his father, mother, and grandmother. In the midst of this personal chaos and the larger breakdown of social order, Serafin Exner's laboratory became Erwin's refuge. He spent day after day there, throwing himself into his research.[7] Schrödinger later recalled the "warm and in the finest sense familial bond that unites all Austrian physicists." He went so far as to speculate that Vienna's next generation of physicists, "who had long revered the dear Father Exner as their spokesman [*Senior*]," had inherited from him their best qualities.[8] Exner's students consistently referred to this group as a "family" and to Exner as "Väterchen" (Papa).[9] What Exner's lab offered was a surrogate family, a model of scientific collaboration imported from Brunnwinkl.

Characteristic of these collaborations was Schrödinger and Kohlrausch's foray into color theory. In the course of their measurements in Exner's lab,

6. Hans Frisch, *50 Jahre Brunnwinkl*, p. 36.

7. Moore, *Schrödinger*, p. 115.

8. Letter to Stefan Meyer, quoted in Gabriele Kerber et al., eds., *Dokumente, Materialien, und Bilder zur 100. Wiederkehr des Geburtstages von Erwin Schrödinger* (Vienna: Fassbaender, 1987), pp. 28–29.

9. Hans Benndorf, "Worte der Erinnerung" and "Gedenkrede"; Egon von Schweidler, obituary for Exner in *Almanach der Akademie der Wissenschaften* (1927), pp. 179–84; see quotations from other obituaries in Karlik and Schmid, *Exner und sein Kreis*, p. 66.

this pair discovered remarkable differences in their perceptions of color. "The relationship of the objections is entirely reciprocal. If F. K. declares the mixing field of E. S. to be 'far too green,' then inversely E. S. finds the mixing field of F. K. 'far too red.' The mixing fields that one observer has adjusted to be identical differ for the other observer almost as much as a lemon from an orange!"[10] As this teasing banter suggests, these experiments became games with the goal of seeing as if through a friend's eyes.

In this sense, research on color was a response to wartime conditions. It was a playful part of the cameraderie among Exner's "sons." It was also an antidote to the sensory deprivation and solipsistic anxieties of life in the trenches. We know from a rich body of historical research on the First World War that soldiers' experiences in the trenches were marked by disturbances of communication and sensory perception.[11] "Sensory irregularities" were not just a symptom of war but also a principal cause for exclusion from military service. Kohlrausch, for instance, had been released from the artillery due to an ear injury in 1917. Schrödinger himself was sensitive about the imputation of a visual defect. As he would discover in the course of their research, he was what was known as an anomalous trichromat. This was a condition that would once have kept him from work in the navy or railways, where recognizing the colors of flags or signal lights was vital. Schrödinger countered that his vision was as accurate as anyone's, just qualitatively different.[12]

Schrödinger's military service impressed on him the irreducibly personal conditions of sensory experience. Assigned to the artillery service for the first two years of the war, he found himself becoming a "wreck, physically and mentally." "I am no longer accustomed to work or to think for half an hour. Every rational thought is entangled with another one: what's the use of it all if the war is not finally finished." During the "utterly boring" days in the trenches, he reported, "I fill my mind with the psychology of the fundamentals of consciousness (memory, association, the concept of time)."[13] Schrödinger's psychological musings reflected his anxiety about

10. Schrödinger, "Die Gesichtsempfindungen" (1926), in *Gesammelte Abhandlungen*, vol. 4, pp. 183–288, quotation on p. 219.

11. The most pertinent studies are Eric J. Leed, *No Man's Land: Combat and Identity in World War I* (Cambridge: Cambridge University Press, 1979); Stephen Kern, *The Culture of Time and Space: 1880–1918* (Cambridge, Mass.: Harvard University Press, 1983); Modris Eksteins, *Rites of Spring: The Great War and the Birth of the Modern Age* (New York: Anchor, 1989).

12. Schrödinger, "Gesichtsempfindungen," p. 221.

13. Schrödinger, wartime diary, quoted in Moore, *Schrödinger*, pp. 90–91.

his own mental state. He was intent on observing himself not simply for symptoms of the loss of "rationality" but also to understand how it was that one could become, as he put it, "accustomed to the war." During long watches, he studiously analyzed his own sensory perceptions. One night, for instance, a guard alerted Schrödinger that he had seen lights advancing up a hill toward their unit. Schrödinger realized that the lights were really due to the luminescence of electrical discharge from the tips of barbed wire just a few feet away, a phenomenon known as "Saint Elmo's Fire." This was an effect of atmospheric electricity that Schrödinger would have known from his studies with Serafin Exner. The movements of the observers themselves produced a parallax effect that made the lights appear to be in the distance.[14] The effect thus existed partly in the world, partly in the observers.

The First World War was a watershed for the study of sensory illusions and malfunctions. Gestalt psychology, for instance, was shaped by Kurt Goldstein's experiences with the visual incapacities of brain-injured soldiers in Germany.[15] Eric Leed has argued that when vision failed soldiers in the First World War, aural experience gained in significance. For Max Wertheimer, who had studied neurophysiology with Sigmund Exner in Prague, a stint in the artillery led to investigations of the psychological phenomena of sound localization.[16] Erwin Schrödinger's artillery service likewise prompted him to reflect on his own hearing. It was a well-known fact that an explosion not audible to a nearby observer may be audible from a distance of 50–100 kilometers. Later in the war, after his transfer to the meteorological service, Schrödinger treated this "abnormal audibility" as a physical rather than a physiological anomaly, an effect of atmospheric acoustics.[17] Given these experiences, it was hardly surprising that Schrödinger would turn his attention to visual "abnormalities" once peace was signed.

For psychiatrist Robert Exner, Karl's only child, the war pressed the question of the conditions under which communication about visual experiences became possible. In 1914, on the eastern front at Lemberg (now L'viv,

14. Ibid., p. 82.

15. Harrington, *Reenchanted Science: Holism in German Culture from Wilhelm II to Hitler* (Princeton: Princeton University Press, 1996), pp. 145–51. See too Kurt Lewin, "Kriegslandschaft," in *Kurt-Lewin-Werkausgabe*, vol. 4, *Feldtheorie*, ed. Carl-Friedrich Graumann (Bern: Hans Huber, 1982), pp. 315–26.

16. Ash, *Gestalt Psychology*, p. 188.

17. Schrödinger, "Zur Akustik der Atmosphäre," *Physikalische Zeitschrift* 18 (1917): 445–53, *Gesammelte Abhandlungen*, vol. 4, p. 3.

Ukraine), Robert was captured by Russian soldiers. From December 1915 until November 1920 Robert served as a physician in prisoner-of-war camps, first in an Ottoman region and then in Siberia.[18] His experiences in these camps became material for his postwar studies, inspiring publications both on the psychopathology of prisoners of war and on visual communication.[19] In one memorable anecdote, a fellow physician who was teaching at a Mongolian village school told Robert that he had been dismissed for declaring that the sun and moon were spherical. Robert, who had learned the Tatar language since his capture, recounted this anecdote to a group of Tatar nomads and smugglers. They only laughed. Europeans believed the sun and moon to be spherical, they explained, but one could tell just by looking that these were really round "surfaces." In vain, Robert tried to convince them otherwise by drawing their attention to the shadows on the moon and by using a ball to demonstrate the transformation of a two-dimensional image into three dimensions. "But no one can go there and make sure that the moon is really a sphere," replied the Tatars.[20] This impasse evidently set Robert thinking about the cultural specificity of visual experience.

Leading Robert's thoughts further in this direction were two of his fellow prisoners of war in Siberia, professional artists whom he described as Expressionists. The paintings they made during their imprisonment contrasted dramatically with their prewar work. The first painter, "a Jew," began to use exclusively blue and red pigments, "which were brought into the most strikingly harsh contrasts for a futuristic representation [*Darstellung*] of despair." The second, "a German," turned to blues and greens to paint "hazy" imagined landscapes.[21] Both painters showed signs of depression and complained of the "endless barbed wire" of the prison camp. Robert termed the painters' condition "barbed-wire depression," suggesting not just the experience of confinement but also the visual monotony of the camp. (Schrödinger likewise diagnosed his wartime fascination with lights dancing on the ubiquitous barbed wire as a symptom of visual boredom.) Robert was struck by the fact that in both painters a purely emotional

18. Robert Exner, "Lebenslauf," Anlage I. Dekanat des medizinischen Fakultäts Z. 1652, May–June 1939, Archiv der Universität Wien, Vienna.

19. Ibid., Anlage III.

20. Robert Exner, "Die Genesis der vier Kardinalfarben," *Psychiatrisch-Neurologische Wochenschrift* 40 (1938): 3–4.

21. Ibid., p. 15. On color concepts and race, see too R. Exner and R. Routil, "Studien zur Farbentheorie," *Annalen des Naturhistorischen Museums in Wien* 57 (1949/50): 6–11.

disturbance seemed to have produced a clear shift in color valuations. Robert approached the painters' case quite differently than he did the Tatars. He treated the Eastern and Western views of the moon as incommensurable, an ultimate barrier to dialogue. He approached the artists' landscapes, on the other hand, as representations that were idiosyncratic yet intelligible within the idiom of Expressionism. Together, these encounters struck Robert as evidence that vision must be profoundly culture-bound.

THE PROBLEM OF COLOR NAMING

Stimulated by the perceptual disruptions of wartime, Schrödinger and Kohlrausch took up the science of color. They soon found that comparing "my red" to "your red" was a challenge. Accounting for individual differences in perception turned out to necessitate major revisions to the empiricist theory of vision. Could a theory that began with physics rather than psychology account for the variability of individual observers? Could two people's "reds" be measured against an objective standard? Drawn by the promise of comparing Schrödinger's "lemon" to Kohlrausch's "orange," the pair set out to extend the three-color theory in just this way.

Their starting point was the problem of quantifying brightness, over which Serafin Exner and Franz Hillebrand were battling at this time. In certain experimental situations, it seemed to Schrödinger that brightness could be easily defined. This was the case if the observer was simply asked to judge when two color fields were no longer distinguishable, that is, completely equivalent. Attaching an objective measure to these colors was straightforward. In the language of projective geometry, this case corresponded to a "lower metric." But this form of measurement was limited. In the lower metric, the brightness of two color fields could be compared only if they were of the same hue, just as in projective geometry only collinear or parallel lines could be compared in length.[22] Something more was needed to quantify situations in which the observer was asked to *compare* two color fields. Here, individual variations in color perception could no longer be ignored. The question was how to represent this dependence on the observer within an objective framework of measurement.

The first step was to derive a "higher" metric from the lower. The "bridge" between the two color spaces was what Helmholtz had defined as the

22. Schrödinger, "Grundlinien einer Theorie der Farbenmetrik im Tagessehen" (1920), *Gesammelte Abhandlungen*, vol. 4, 33–132, quotation on p. 36.

fundamental sensations. Geometrically, the fundamental sensations were linearly independent, meaning that any color could be matched by a combination of these three primaries. Schrödinger could therefore express the brightness of a color as a function of the fraction of each primary it contained, weighted by the specific brightness coefficients, whose values Serafin Exner had recently measured. He then sought a measure of the "nonidenticalness or separation" between any two colors, "which must be a minimum for maximally similar colors." On this assumption there exist in the color space surfaces of constant brightness. But these surfaces had the unusual property that any two points on them were closer to each other than to any other point on a ray from the origin. The geometry of Schrödinger's "higher" color space was of the counter-intuitive kind called non-Euclidean, where straight lines were not the shortest distance between two points.

In this space, the natural choice for the "line element," the measure of distance, was the "just noticeable difference" (*Ebenunterschiedlichkeit*), a concept introduced by Fechner in the mid-nineteenth century. Defining the line element in this way implied that "the colors we see as equally spaced from a given color are those small changes in this color which, under constant observing conditions, are *just noticeable*."[23] Those colors would form an ellipsoid (an elongated sphere) in the color space. Following Fechner, Schrödinger approximated the threshold perceptible difference as constant. Any color could then be uniquely located on this surface given three values, such as hue, saturation, and brightness.

But this geometrical exercise did not yet take into account the fact that different people see colors differently. Where would the observer figure in this system of measurement? Schrödinger's solution was to make the axes of the coordinate system correspond to a given observer's primary sensations. This meant that the coordinate system would have to be empirically determined for each individual. "Admittedly, the manner in which we lay down coordinates, divide up and measure it, is an artificial, mathematical construction . . . ; but this holds in entirely the same way for the usual position space."[24] What Schrödinger seems to have had in mind was an analogy to Einstein's general theory of relativity. It was thanks to his studies of relativity that Schrödinger had the mathematical tools to tackle non-Euclidean geometry. He seems to have adopted as well Einstein's understanding of "relativity." Measurements of color qualities would be "relative to the observer,"

23. Schrödinger, "Die Gesichtsempfindungen," 280.

24. Schrödinger, "Grundlinien," 44.

much as measurements of length and time in Einstein's theory depended on the motion of the observer. A properly chosen measure, analogous to the space-time interval, would remain invariant in all coordinate frames, grounding what Schrödinger called the "reality" of the color space.

Schrödinger acknowledged that this geometrical representation was still a mere approximation, because it assumed that the least perceptible difference between colors was constant. As he noted, "The threshold is not in general a sharply defined value; rather, for each stimulus difference there exists a probability for a correct judgment of difference that quickly grows to one as the difference in stimulus increases."[25] This added a random source of individual variability.

Having set out to rescue the three-color theory, Schrödinger found that the implications of his analysis were radical. Certain color measurements, such as the relative brightness of the two fields composed of the same spectral wavelengths, would yield the same result for any observer, "normal" or otherwise, or if replaced by a photometer.[26] For other comparisons, however—such as the indistinguishability of two fields, one of sunlight and one of white light composed of two complementary spectral colors—the individual human eye is irreplaceable:

> The two seemingly equal lights are, taken in and for themselves, thoroughly different, they have nothing to do with each other, only that they appear equal to *this* eye, which in its color judgment cannot not be appealed, nor controlled nor replaced by any measuring instrument.[27]

For Sigmund and Serafin Exner, the simple fact that people could converse about colors underwrote the conviction that individual variations could be collapsed into an average or "normal" eye, producing a unique correlation of stimuli and sensations. Schrödinger disagreed. Since it was by "custom and agreement" that people learned to apply color names, it seemed clear "that the attribution of the same name to the same color by different people provides no criterion for the equality of their sensations."[28] In Schrödinger's new framework, the notion of the "normal eye," so central to the Exners' research, became a mere abstraction corresponding to the negotiated norms of language.

25. Schrödinger, "Die Gesichtsempfindungen," p. 266. On the problem of fluctuating sensory thresholds in the history of psychology see Gigerenzer and Murray, *Cognition*, chapter 2.

26. Schrödinger, "Grundlinien," p. 42.

27. Ibid., p. 43.

28. Schrödinger, "Gesichtsempfindungen," p. 217.

Robert Exner took this argument farther. Citing this and other research from his uncle Serafin's laboratory, he argued in the 1930s that the system of primary colors, the "primary sensations" of the Helmholtz theory, was neither a physical nor a physiological fact. Where his uncles had taken the primary sensations to be the biological arbiters of beauty, Robert explained them as mere abstractions produced through social interactions. The primary colors "represent[ed] highly conceptual abstractions for use in everyday life," and they varied from culture to culture. Linguistic usage thus reflected the "average capability" of individual observers, but such an average was a social construction, not a norm incarnated: "The model of the four cardinal color concepts is thus to be thought of as having developed under the pressure of the necessary *phasis* and *praxis* of the cultures and not for instance by pioneers and eccentrics! Certainly in every age and in every culture this development must have reflected the average capacities of the individual members of the culture."[29] Drawing on anthropology and linguistics, Robert explained how color names evolved to produce "the same effect in many differently disposed brains!" That color concepts remained "free of individual idiosyncrasies" could be attributed to linguistic usage, not the existence of "normal" eyes.[30]

It is illuminating to compare the conclusions of these young members of the Exner circle with the reflections of their Viennese contemporary Ludwig Wittgenstein. Color naming was one of Wittgenstein's primary examples of a "language game," as a posthumously published manuscript on color makes clear.[31] For Wittgenstein, colors had no meaning outside of the rules of linguistic usage; a private language of color was impossible. Like Wittgenstein, and unlike scientists of the previous generation, Schrödinger and Robert Exner each acknowledged the role of linguistic usage in overcoming individual differences in the perception of colors. Unlike the philosopher of language, however, both scientists retained a physicalist and realist view of colors. They maintained the possibility of a direct, if not unique, correlation between physical stimuli and subjective sensations.

In their postwar work, Schrödinger, Robert Exner, and other Viennese collaborators speculated on the process through which a language of color might evolve in a natural environment. They began by considering the earliest phase of the evolution of human color perception, the adaptation of early

29. R. Exner, "Kardinalfarben," p. 2.

30. Ibid., p. 16.

31. Ludwig Wittgenstein, *Remarks on Colour*, ed. G. E. M. Anscombe, trans. Linda L. McAlister and Margarete Schättle (Berkeley: University of California Press, 1977).

ocean-dwelling vertebrates to their surroundings. In the blue-green light below the ocean surface, they theorized, it would have been an evolutionary advantage to respond to red light as to an alarm signal. Red was thus the original "color," the opposite of which was simply "pale." Robert Exner took this research in an anthropological direction, suggesting, for instance, that red had acquired a "magical" significance for early humans, evident in its use for painting and ornamentation. Each culture adapted its color concepts to its local needs. In Arab cultures, for instance, the sacred status of the color green might have derived from its importance in regions with little vegetation. Blue, by contrast, was a relatively young concept, which Robert speculated had developed in northern lands in step with perspectival painting. The color scheme of a given community thus evolved through verbal interactions and the production and exchange of visual representations.

In the new framework to which Schrödinger, Kohlrausch, and Robert Exner contributed, color concepts were abstractions with a realist foundation. They corresponded to local efforts to communicate about the environment. Under normal conditions, then, humans managed to use language to overcome individual differences in the perception and valuation of colors. In the process, local groups developed highly differentiated visual cultures.

In wartime Vienna, it turns out, the concept of a local visual culture was not merely academic. It was a practical problem of pedagogy, design, and politics.

A NATIONAL STYLE

In 1917 Viennese color theory found a new audience. Fritz Kohlrausch was hired to introduce classes in color theory to the Vienna Arts and Crafts School.[32] This innovation was part of a series of reforms the school undertook during the war. The institute faced shortages of money, materials, and personnel. It was even forced to move next door to the Museum for Art and Industry when medical workers took over its building. Work at the school now focused on military memorials, posters, and related projects.

For the school's predominantly female population at the time, such projects were a means of participating in the collective experience of the Great War, with its fever of patriotism and self-sacrifice.[33] The war memorial that

32. Kohlrausch also taught during this period at the Graphische Lehr- und Versuchsanstalt in Vienna, where he dealt with optics. Despite his injury, he also retained an honorary position at the Akademie der Musik that he had held since 1912.

33. Healy, *Vienna and the Fall*, pp. 163–210. On the war experience see Ekstein, *Rites of Spring*; Paul Fussell, *The Great War and Modern Memory* (New York: Oxford University Press, 1975).

Hilde Exner sculpted for a town in the Salzkammergut invites such an interpretation. In her creation, David stands motionless after his defeat of Goliath, his head bent low, as if overcome by the sense of a prodigious power within. Hilde intended the sculpture to convey "the sheer superhuman accomplishments of our heroes."[34] This tribute to the transcendence of the personal echoed Hilde's ambition to be known as more than a mere "*female* artist."

For much of the war, though, Hilde and her cousin Nora did not live as artists at all. Nora returned to Vienna from Munich in 1914 to work day and night organizing the loading of trains with medical supplies and the unloading of wounded soldiers. She also traveled to the front, where she distributed snow caps and helped design a mobile medical station.[35] Hilde too worked on a medical train, traveling for months at a time without a bed to sleep in. In 1915, Hilde was in Hungary with imminent plans for a respite in Vienna, where Nora lay ill in an overcrowded hospital. Hilde was in the midst of writing to her cousin of her eagerness to see her when she received word of Nora's death.[36] Hilde survived her cousin by just seven years, dying apparently of an illness contracted during the war. Her David stood as a memorial to her own brief life.

Had Hilde and Nora been born a decade later, they might instead have spent the war as students at the Arts and Crafts School. Director Alfred Roller refused to let the war distract the school from its mission of developing well-rounded, autonomous artists, rather than mere craftsmen.[37] As he insisted in 1919, "the advancement of crafts and production must not be a political matter."[38] Even as Roller championed the ideal of the independent artist, however, the school felt more strongly than ever the need to define a national style against competitors in Paris and London. As the war followed on the heels of the first of several financial collapses at the Wiener Werkstätte, marketing a distinctive Austrian style became an urgent need. As Viennese critic Bertha Zuckerkandl observed in 1916, "The primary stipulation, however, if Viennese fashion is to achieve a leading position abroad, is that it must be artistic, inventive, and true to its materials—the distinctive characteristics of Austrian style. . . . What we should be aiming at, therefore,

34. "In memoriam, Hilde Exner," Collection of P. Dijkgraaf, Den Helder.

35. "Nora v. Zumbusch," *Neue Freie Presse* 3 March 1915, p. 13.

36. Hilde Exner to Nora Exner von Zumbusch, undated, Collection of P. Dijkgraaf, Den Helder.

37. Manfred Wagner, *Alfred Roller in seiner Zeit* (Salzburg and Vienna: Residenz Verlag, 1996), pp. 204, 212, 219.

38. Wagner, *Roller*, p. 220.

is an international fashion boasting our national quality."[39] While the influx of women artisans alarmed many of her male colleagues, Zuckerkandl saw it as an opportunity to define an Austrian style in new fields like fashion.

With the collapse of the empire in 1918, craft productions became a tool for forging a national identity for the new Austrian Republic. This Austria was a rump state, a "leftover" as Clemenceau put it. German Austrians had long identified with the Habsburg Empire, with the culture of the German-speaking lands, and with their home province, but they did not identify with the new political entity known as Austria.[40] Vienna's agricultural wealth had been lost to Hungary and its industrial resources to Czechoslovakia. Inflation could not be controlled until 1924. The nation was, moreover, torn by political strife. While Socialists held power in Vienna, clerical conservatives and paramilitary organizations gained strength in the provinces. It was in this fragmented environment that the problem of constructing a national identity would have to be worked out. Hugo von Hofmannsthal, for instance, looked to Austrian crafts to revitalize the defeated nation's reputation abroad. Austria would be known not for its Baroque palaces but for "the villa-esque, the cottage and that element of highly cultivated rusticity."[41] Consistent with the ideology of the Arts and Crafts movement before the war, a national style was to be found in the unity of form and function—the infusion of art into all aspects of daily life—and of the *Gesamtkunstwerk*, the all-encompassing work of art. An often cited example was a country home in northern Moravia designed by Josef Hoffmann and the Wiener Werkstätte for the Primavesi family. The design was brightly colorful because, as Zuckerkandl explained, "Traditional Moravian popular culture is rooted in color." The goal was "a conscious continuation of surrounding nature."[42] Such celebrations of landscapes and folk traditions became common after the war, culminating, for instance, in the ubiquitous slogan "Schönes Österreich."[43]

39. Bertha Zuckerkandl, "Die Wiener Mode-Ausstellung" (1916), quoted and translated in Gabriele Fahr-Becker, *Wiener Werkstaette, 1903–1932* (Cologne: Taschen, 1995), quotation on p. 188.

40. See, for instance, Istvan Deak's essay in *After Empire: Multiethnic Societies and Nation-Building*, ed. K. Barkey and M. von Hagen (Boulder, Colo.: Westview, 1997).

41. Hofmannsthal, "Die Bedeutung unseres Kunstgewerbes für den Wiederaufbau" (address to the members of the Austrian Werkbund, 1919), in *Gesammelte Werke* (Frankfurt: Fischer Taschenbuch, 1979), vol. 9, pp. 55–68, quotation on p. 58.

42. Quoted in Fahr-Becker, *Wiener Werkstaette*, p. 85.

43. *Schönes Österreich: Heimatschutz zwischen Ästhetik und Ideologie* (Vienna: Österreichisches Museum für Volkskunde, 1995).

It soon became a truism that a trademark of the "Austrian style" was its use of color. Zuckerkandl identified Austrian crafts' "distinctive palette of colors." Ever since, art historians have considered a "subjective" and "individualistic" approach to color to be a defining feature of Austrian modernism.[44] A contrast is typically drawn between Vienna's "subjective" approach to color and the "objective," "scientific" color theory taught at the German Bauhaus and based on Wilhelm Ostwald's *Farbenlehre:* "In contrast to the Bauhaus, where color received a scientific exposition, which ruled out all subjectivity, Roller's color theory permitted individuality and self-reliance."[45] This opposition fits well with the traditional view of modernist Vienna as a playground of pure subjectivity.

In fact, little research has been done on the actual teaching of color theory at the Arts and Crafts School.[46] Much of the weight of claims like Oberhuber's rests on the poorly documented association of the school's directors with a German painter, Johannes Itten, who lived and taught privately in Vienna from 1917 to 1919 before being invited to teach at the Bauhaus. As Itten later recounted, it was during this stay in Vienna that he began to notice the "unique and characteristic features" of the compositions of certain of his students. "They were self-interpretations in a graphic medium.[47] In subsequent years, Itten increasingly encouraged his students to use color and form as means of self-expression. He honed techniques for identifying his students' personalities and moods on the basis of their drawings—much as

44. "Color theory, which had been an important element since the eighteenth century, became, under Roller, joined to its relation to materials and the resulting contrast of color values, in order to make comprehensible in this way the individuality of processes relating to color. Since every relationship to colors is subjective, these became better understood through Roller's principle. The question of color thus became related to the individual personality of the artist." Oberhuber, "Die Professoren und ihre schulischen Methoden," in *Kunst und Lehre am Beginn der Moderne: Die Wiener Kunstgewerbeschule, 1867–1918,* ed. Gottfried Fliedl (Salzburg and Vienna: Residenz Verlag, 1986), p. 374.

45. Oberhuber, "Die Professoren," p. 374. On the Bauhaus's relationship with the Vienna Circle, see Peter Galison, "Aufbau/Bauhaus: Logical Positivism and Architectural Modernism," *Critical Inquiry* 17 (1990): 709–52.

46. To date Oberhuber's is the only account of color theory as taught at the Vienna Arts and Crafts School in the early twentieth century. The two major studies of the school in this period do not discuss the teaching of color theory: Erika Patka, ed., *Kunst: Anspruch und Gegenstand: von der Kunstgewerbeschule zur Hochschule für angewandte Kunst in Wien, 1918–1991* (Vienna: Residenz Verlag, 1991); Gerlinde Sagbauer, "Strukturen des Weimarer Bauhauses und der Wiener Kunstgewerbeschule 1919–1925," Hausarbeit für die Studienrichtung "Bildnerische Erziehung" (ÖNB 1,290.000-C).

47. Johannes Itten, *The Art of Color: The Subjective Experience and Objective Rationale of Color,* trans. Ernst van Haagen (New York: Reinhold, 1966), p. 176.

psychiatrists like Robert Exner were beginning to do. Itten's precise relationship with the Arts and Crafts School during his years in Vienna remains a matter of debate among art historians, but it seems likely that Roller unsuccessfully courted Itten to join the faculty. The most authoritative history of the school in these years argues that Itten's teaching philosophy was inspired largely by his contact with the school, in keeping with the traditional view of the school as a breeding ground for subjective aesthetic theories.[48] Yet Itten himself complicated this picture by describing the artistic process in the terms of experimental psychology, writing of "stimuli," "sensations," "visualizations," and "reactions."[49] As this language suggests, such "subjective" experiences might be objectifiable and even measurable. Familiar accounts of Vienna's turn to subjectivism fail because they ignore the subtleties of meaning that "subjective" and "objective" carried in the interstices between Viennese art and science.

BETWEEN THE LABORATORY AND THE WORKSHOP

In pursuit of the goal of producing autonomous artists, Roller considered introducing a number of new fields of study during the war years. As the minutes of a 1916 faculty meeting explained, "At this time it is hardly possible to keep certain categories of students fully occupied. However, since the period of study will be shortened due to military service, the moment calls for a more intensive program." Color theory was one of the few proposed fields to be implemented. The faculty report went on to explain the motivation for the new class and to describe how it was to be taught:

> The subject of color theory, which is now missing from the curriculum, would begin with experiment and proceed to theoretical considerations, making the student familiar with the laws of vision, with the effects of light, of contrast, of color. The field is gaining ever greater practical significance and is not easily accessible for those without scientific preparation. To obtain a healthy basis for answering the pertinent questions, which every practicing craftsman will incessantly face, is a recognized, pressing desire of all involved.[50]

48. Gabrielle Koller, "Die Kunstgewerbeschule des K.K. Österreichischen Museum für Kunst und Industrie, Wien, 1899–1905" (Ph.D. diss., University of Vienna, 1983), esp. p. 45.

49. See Dieter Bogner and Eva Badura-Triska, *Johannes Itten* (Vienna: Museum moderner Kunst, 1988).

50. Kunstgewerbeschule Studienordnung 139, 11 July 1917, Archiv der Universität für angewandte Kunst, Vienna.

The school was evidently intent on promoting the symbiosis of art and science that had characterized Viennese modernism since the turn of the century.

Consistent with this aim, Fritz Kohlrausch built a seamless bridge between his physical research in Exner's lab and his teaching duties at the art school. As Kohlrausch noted in his annual reports for the school from 1917–18, Exner provided the space and equipment for Kohlrausch's art students to carry out experimental exercises on color. In Kohlrausch's description these were "independent experiments . . . on light-intensity measurements, color mixture on the color wheel and the three-color apparatus, spectroscopic experiments on additive and subtractive mixtures of colors . . . , photographic exposures, etc." As Kohlrausch explained in his first-year report, the goal of the course was to study color in its "three-fold mechanism," physical, physiological, and psychological:

> The lectures are to give a picture, based on scientific knowledge, of the operation of the "external and internal eye" as well as of the development of a visual perception. Corresponding to the three-fold mechanism the material is divided into three parts:
>
> 1. Beginning with the physical basis of light, the optical structure of the eye and the formation of the retinal image are explained.
> 2. This stimulus to the retina is the objective cause of the visual sensation; this qualitative and quantitative dependence on the properties of the stimulus, on the location of the effected position on the retina, as well as on the "state" [*Stimmung*] of the organ form the principle content of the second part.
> 3. The (mental) processing of this visual sensation leads to the formation of visual perceptions, which contain a judgment of the seen object and depend on experience, on remembered thoughts and such.[51]

Kohlrausch's description of his course reflected Exner's belief in the usefulness of a clear distinction between the "objective" or physical aspects of color perception and its physiological and psychological aspects.

In the two years before Kohlrausch left Vienna for the Technical University in Graz, he had the good fortune to teach talented young artists like Vally Wieselthier and Hertha von Bücher. These women went on to successful careers and were eager to experiment with color in their work. The ceramics of Wieselthier, in particular, embodied the school's ideals of

51. Jahresbericht 1917/18, 1918, Akt 27, Archiv der Universität für angewandte Kunst, Vienna. In the Jahresbericht for the following year (1919, Akt 18), Kohlrausch again summarizes this three-part organization, "objective," "physiological," and "psychological."

craft production. She produced vibrantly colored and often witty designs, either singly or in small series, which she aimed to make unique yet affordable. As she stressed, even reproductions were shaped individually by hand and thus became originals.[52] In a similar style and palette, Wieselthier contributed to the new movement in Viennese fashion. With objects like her beaded purses, Wieselthier celebrated women's new opportunities for self-expression in the wake of the war. As a member of the Wiener Werkstätte and later as the director of her own studio, Wieselthier epitomized the dual ambition of Roller's school: producing independent artists while furthering a national movement.

The introduction of color theory to the Arts and Crafts School in 1917 was part of an effort to resolve a tense political predicament. A state-funded institution charged with bolstering Austrian crafts on the international market, the school's director nonetheless clung to the ambition of turning out "autonomous" artists. Kohlrausch's approach to color turned out to be ideally suited to this situation. Ernst Brücke's 1866 color theory textbook for the Arts and Crafts School had treated subjective factors as unavoidable hurdles to the achievement of aesthetic ideals.[53] Kohlrausch instead gave Vienna's arts and crafts community a language with which to recognize color as both a personal and an intersubjective experience. By combining a physicalist realism with due attention to individual differences, Kohlrausch showed art students how a color scheme could simultaneously express an individual personality and define a national style.

COMPETING MODERNISMS

If Vienna's color culture was not primarily "subjective" or "individualistic," we might wonder why it has so often been described as the antithesis of the "scientific" approach to color at the Bauhaus. This account is, in effect, a misunderstanding of the efforts of Viennese scientists and artists to distance themselves from the German model. The Viennese defined their own approach to color in part as an attack on Wilhelm Ostwald, whose plan to reform the teaching of color in Germany eventually reached from the Bauhaus to the elementary schools of the Weimar Republic. Ostwald's color theory bore signs of its origins in international conflict. His first insight

52. Quoted in Marianne Hörmann, *Vally Wieselthier*, p. 115.

53. Brücke, *Die Physiologie der Farben für die Zwecke der Kunstgewerbe* (Leipzig: Hirzel, 1866), cf. Lenoir, "Politics of Vision," p. 165.

into the rationalization of color came on a steamer returning home from London just months before the outbreak of the war. Ostwald conceived of color theory explicitly as a scientific solution to the problem of social order. During the war he declared that the Germans alone had solved "the problem of organization" by replacing individualism with a precisely ordered collectivism, and he launched his massive project of color standardization to this same end.

Ostwald's program was nothing short of "a new color culture for Germany." Through articles and textbooks, both academic and popular, through factory-scale production of color samples and user-friendly atlases, Ostwald aimed to disseminate his standards into every German classroom, workshop, and factory. He made his goals explicit at the first German "Color Day," an event he organized in conjunction with the German Werkbund in 1919. In his address he called for the introduction of color theory in kindergarten, in order to teach children to apply colors "harmoniously," with "ease and certainty." In just a few years it would be possible to produce all German "appliances, clothing, rooms, houses, cities" in harmonious colors. The effect of this "would be felt in the general aspect of our *Volk* and environment and would give a uniquely appealing touch to the total appearance of our life."[54] Moreover, the production of such harmoniously colored objects would give Germany a decisive advantage against competition on the world market. As Ostwald saw it, an international race had begun to rationalize and harmonize color production. An undated manuscript shows Ostwald keeping tabs on Germany's progress compared to its neighbors.[55] Russia appears to have been furthest behind and Germany in the lead. "This gives only a very incomplete overview of the actual dissemination [of color theory]," Ostwald cautioned, "since the industries avoid making reports to keep the competition in the dark." To Ostwald, color was a national problem requiring a national solution.[56]

In order to standardize Germany's use of color, Ostwald began to manufacture color samples and devised a system for quantifying pigments. The

54. Ostwald, "Die Grundlagen der Farbkunde u. d. Farbkunst," manuscript G1 1890 (21), Handschriftenabteilung, Staatsbibliothek zu Berlin—Preussischer Kulturbesitz.

55. Ostwald, untitled one-page manuscript, 514/29-4, Handschriftensammlung, Österreichische Nationalbibliothek.

56. Ostwald again linked science to nationalism when he announced in 1921 that he would only permit translations of his articles to be published when German science regained its full status as a member of the international community. Karl Hansel and Ingeborg Mauer, "Paul Krais, Wilhelm Ostwald und die Werkstelle für Farbkunde in Dresden," *Wissenschaftliche Zeitschrift der Technischen Universität Dresden* 49 (2000): 41–44, quotation on p. 43.

rationalization of color, in his view, demanded a measurement system that was entirely independent of human perception. This ambition was consistent with his monist philosophy, according to which measurements in the physical and in the mental realms were unproblematically equivalent. (The Exners and their students, on the other hand, insisted that physical measurements and psychological sensations could be correlated, but never equated.) Ostwald thus assumed that individual differences in color perception were negligible. In his color atlases, for instance, he claimed to have grouped color samples in such a way that "everyone will order them in the following series, and there has never been a person whose color sense would have required a different order."[57]

The key to Ostwald's method of quantification was his definition of the "remission function" of a pigment. This expressed the fraction of incident light reflected as a function of the wavelength of the light. For each pigment color, he used the wavelength of the corresponding spectral color to define two regions of remission. A large fraction of light of wavelengths longer than that of the corresponding spectral color would be reflected, while a large fraction of light of shorter wavelengths would be absorbed. Call the first region the "remission region" (RR), the second the "absorption region" (AR). In Ostwald's definition, an "ideal" pigment is one that in the RR reflects the same fraction of light as a white paper chosen as a standard "white"; and correspondingly in the AR the ideal pigment absorbs the same fraction of light as a standard black body. Thus for an ideal pigment the remission function is 1 in the RR and 0 in the AR. If h_r is the height of a nonideal pigment's remission function in the RR, then the deviation of the remission function from that of an ideal pigment, $1 - h_r = b$, gives the fraction of particles in the pigment that are completely black. In this sense Ostwald defined b as the "black content" of the pigment. Similarly, the fraction of white particles is given by $1 - h_a = w$. Having determined h_a and h_r, one can conclude that the remaining particles in the pigment are neither white nor black, that is, they are colored. From this line of thought Ostwald derived the fundamental equation of his color theory: $b + w + c = 1$. Not inclined to understatement, Ostwald compared the status of this equation as an empirical law of color theory to that of Ohm's Law in electrical theory.

For Ostwald's purposes, it was crucial that this measurement system was independent of the observer. Only in this way could he justify imposing his laws of color harmony on the nation at large. Unlike the Exner's "normal eye," a product of individual self-discipline, Ostwald's color

57. Ostwald, *Farbenfibel* (1928), p. 2.

standards were to be externally imposed. His color norms expressed his technocratic ambitions just as surely as the Exners' displayed their liberal pretensions.

It was Ostwald's effacement of the individual observer from his equations that drew attacks from Vienna. While teaching color classes at the Arts and Crafts school, Fritz Kohlrausch obtained an Ostwald color atlas and began making his own measurements on the samples. His measurements indicated, first, that the form of the remission curve varied significantly among pigment colors. Ostwald had simply assumed it was constant, tracing each curve with only two points drawn from observation. Pigment colors from yellow to orange behaved almost as Ostwald had posited, but colors outside of this region did not fulfill his assumption at all. Still, the crux of Kohlrausch's critique lay deeper. As Kohlrausch repeatedly pointed out, Ostwald's theory failed to allow for *individual variations* in color perception.[58] In the forty years since Hering had shown that the location of the fundamental sensations in the spectrum varied continuously among individuals, only the Helmholtz three-color theory had successfully accounted for these variations (thanks to Schrödinger). Hering had thought this variability would doom Helmholtz's theory, but in fact it gave it new life. A theory that ignored this variability, in Kohlrausch's estimation, could be no more than an occasionally useful "rule of thumb."[59] Schrödinger made the point more forcefully: it was time for Ostwald to concede that deviations of individual observers from the average response curves were not "errors" of the three-color theory but rather a real range of variation in normal vision, "which cannot be done away with [*welcher sich nicht aus der Welt schaffen lässt*]."[60]

Kohlrausch and Schrödinger's accomplishment was to have created a concrete framework in which to separate the universal aspects of color perception from the strictly individual factors. They proceeded in this case much as they had in their research on microscopic fluctuations. Indeed, they worked with a subtle analogy between human eyes and mechanical detectors. A detector of either kind, human or mechanical, displayed random variations in its sensitivity. When Fritz Kohlrausch performed null

58. Kohlrausch, "Beiträge zur Farbenlehre I," *Physikalische Zeitschrift* 21 (1920), part 1, pp. 397, 398, 401–3, part 2, p. 425, 426, part 3, p. 477.

59. Ibid., part 3, p. 477.

60. Schrödinger, "Über Farbenmessung" (1925), *Gesammelte Abhandlungen* 4, pp. 153–56, quotation on p. 156.

measurements of low-intensity radioactive emission, for instance, he factored in the random fluctuations that plagued the measuring apparatus as it operated at its threshold sensitivity. Similarly, in the case of human detectors, Erwin Schrödinger argued that an observer's threshold for distinguishing between colors fluctuated stochastically. An even larger source of uncertainty lay in variations among detectors. In order to account for such variations in their studies of radioactivity, Schrödinger and Kohlrausch inserted parameters that described features of a particular measuring instrument, such as the inertia of its indicator, then considered ensembles of such instruments. In studies of color theory, they likewise added parameters to account for differences between human observers. Whether dealing with colors, microscopic processes, or atmospheric conditions, they consistently treated fluctuations as phenomena, not mere error. This framework allowed them to attend to the average response of their detectors alongside deviations in individual responses.

By neglecting such deviations, Ostwald set out to make color into a "science of order." He intended to free color of its contingent character as an element of human language. He would replace color as an artifact of social interaction with color as a foundation for social order.

THE LANGUAGE OF THE BEES

At stake in Ostwald's clash with the Viennese color theorists was the issue of reforming the language of science. Ostwald, an advocate of Esperanto, sought to construct a scientific language of absolute laws divorced from the contingencies of everyday grammars. His Viennese opponents, whether the young members of Exner's lab circa 1920 or the aging defenders of the humanist *Gymnasium* circa 1907, never imagined that science could transcend the limitations of human language. Indeed, they turned their humanist methods against Ostwald himself, tripping him up on his own linguistic blunders. Karl Frisch's old friend Fritz Paneth, a physical chemist who had studied with Serafin Exner, defended the humanities against Ostwald in a reflexive war of words. Taking a critical lens to Ostwald's own often sloppy use of language, Paneth made Ostwald out to be guilty of neglecting the "scientific method," that is of making unfounded claims.[61] When Ostwald's

61. Fritz Paneth, undated manuscript review of Wilhelm Ostwald's *Ritter der Vergangenheit und Schmiede der Zukunft*, F. Paneth Nachlaß, Abt. III, Rep. 45, Mappe 84, Archiv der Max-Planck-Gesellschaft, Dahlem, Germany.

son wrote to Paneth on behalf of his ailing father and claimed that Paneth had "misunderstood" his father's words, Paneth refused to accept this excuse. Paneth made Ostwald's argument dependent on the form in which he had expressed it.[62] By the same token, the Viennese denied that the scientific language of color could be fully detached from its everyday incarnation. Color was part of what Karl Frisch began in 1920 to call "the language of the bees."[63] "Color," as he put it, "constitutes a sign [*Merkzeichen*] . . . ";[64] not a universal sign imposed by reason but a local product of social interactions.

It was in part during his honeymoon at Brunnwinkl in 1917 that Frisch came to this conclusion. In the summers of 1912 and 1913 he had launched a series of experiments inspired by his uncle Sigmund Exner's speculations on the coevolution of flowers and the senses of insects.[65] So eager was he to continue this research that only an intervention from his brother Hans convinced him to interrupt his work once a day to take a walk with his bride.[66] He had begun before the war with the question of the bees' capacity to distinguish colors, which continued to divide zoologists in the early twentieth century. Some claimed that insects saw no colors at all, while others reported that they had a distinct preference for blue. Karl designed an experiment to decide the matter. He arranged squares of colored papers on a large board; during a "training" period, he fed the bees a sugar solution exclusively on squares of a certain color, say red. He then presented the bees with a multicolored checkerboard that had not been in contact with sugar, and he counted the fraction of the bees on the red squares after given intervals of time. These statistics could then be used to judge the bees' ability to distinguish red.

Working with multiple "training sites" for the bees was too much for one researcher—and, as Karl later noted, there were no "technische Assistentinnen" (female technical assistants) available at Brunnwinkl.[67] So he recruited

62. Wolfgang Ostwald to Fritz Paneth, 22 Feb. 1931, Fritz Paneth to Wolfgang Ostwald, 4 March 1931. F. Paneth Nachlaß, Abt. III, Rep. 45, Mappe 84, Archiv der Max-Planck-Gesellschaft, Dahlem, Germany.

63. K. Frisch, "Über die Sprache der Bienen I," *Münchener Medizinische Wochenschrift* 20 (1920).

64. K. Frisch, "Farbensinn," p. 56.

65. Ibid., and Karl von Frisch, "Der Farbensinn und Formensinn der Biene," *Zoologischer Jahrbücher* 35 (1914): 1–105. On this research see Tania Munz, "The Bee Battles: Karl von Frisch, Adrian Wenner, and the Honey Bee Dance Language Controversy," Journal of the History of Biology 38 (2005): 535–70.

66. Karl Frisch, *Fünf Häuser am See*, p. 83.

67. Karl Frisch, *Fünf Häuser am See*, p. 78.

his relatives, whose free time and cooperative manner made them ideal for the job of counting bees. As the saying went, wherever someone was doing something at Brunnwinkl, he had spectators. Karl demanded not merely that his assistants display "willingness" and forego other "amusements;" they had, moreover, to "believe" in his experiments.[68] In general, the assistants proved enthusiastic and reliable (each had to pass a test for accuracy). Karl also turned regularly to his uncle Sigmund for advice. In Karl's account, then, the cooperation of the Brunnwinkl community was the precondition for the analysis of communication among the bees.

With the aid of these assistants, Karl determined the range of colors bees could distinguish. The colors they recognized seemed to be those of the most common flowers. For instance, bees seemed unable to distinguish a pure red from black; in fact, most "red" flowers were actually purple-red, containing a good measure of blue. This convergence suggested that flower colors and insect vision had coevolved. Karl next rejected the nineteenth-century hypothesis that insects had "favorite colors." He inferred that the insects were consistently attracted to the color to which they had been "trained." In the wild this would correspond to a preference for the color of the flowers that they frequented.

Karl next tested the bees' capacity to distinguish forms and patterns. This research was, as Frisch put it, a "contribution to the psychology of the bee," evidence of the bees' powers of memory and association. Frisch thereby found "new respect for the little bee brain."[69] For these experiments Karl had six boxes built, each with a hole on the front; over each hole he pasted a shape cut out of identically colored paper, using two each of three forms. During the "training" period, he placed a bowl of sugar solution inside only the boxes displaying one of the shapes. He fed the bees in this way several times, rearranging the order of the boxes at each feeding to ensure that the bees were not relying on their sense of orientation to find the food. He then replaced the boxes with identical ones which had never come in contact with sugar and recorded the number of bees flying to each shape. He found that the bees were able to distinguish stars and clovers and related patterns, but not squares, triangles, or ellipses, nor most patterns of checks (fig. 15). Of the forms they recognized, the more abstract ones, such as the pinwheel or the concentric rings, seemed to Frisch to resemble abstract flowers. The patterns that the bees failed to recognize, on the other hand,

68. Karl Frisch, *Erinnerungen eines Biologen,* p. 46.

69. Karl Frisch, "Farbensinn und Formensinn," p. 76.

were those which were "by nature entirely foreign to them." He concluded: "even when they are capable of the most complicated instincts, insects do not easily leave the narrow circle of what is familiar and has been inherited through generations."[70] Frisch did not treat the bees' limited capacity for pattern recognition as a limitation. By bounding their sphere of activity, the bees' taste for the familiar helped their community cohere. From this point, Frisch began to treat the visual capacities of bees explicitly as a problem of insect *sociology*. Insects had served since antiquity as models of "natural" societies, and Frisch continued in this vein.[71]

Working with two hives, located at B1 and B2 in figure 16, Frisch had noticed that a given feeding site was usually visited exclusively by bees from a single hive. He found that it was not even necessary to "train" the bees from each hive to feed only from the site he had assigned to it; the bees took care of this segregation themselves. If he underfed them, some of the bees from one hive would find their way to the feeding site of another hive. But Karl soon noticed something odd in such cases: the area around the feeding site would be littered with dead and injured bees. "One often saw . . . one or two bees suddenly fall on another and drag it fighting violently over the table. They usually grabbed each other's wings with their jaws and tried to sting each other." Now the reason for the segregation at the feeding sites became clear to him: the bees had fought off the intruders from another hive, protecting their food for their own community. With unabashed anthropomorphism, Frisch concluded that the bees that fed from one site "more or less knew each other personally."[72]

Frisch was interested in the possibility that the hive's appearance in particular stuck in the bee's memory. He tried painting colored patterns on the hives and found that this indeed influenced their sense of orientation. But he was curious to learn whether the bees' sense of color could overcome even their aversion to integrating with bees from other communities. To test this, he imported a second stock of bees from Italy, one that was subtly distinguishable from his "German" bees. He ran the experiment as follows: the hive of the "Italian bee *Volk*" was colored blue with yellow stripes, while the hive of the German bees was colored yellow with blue stripes. Each community was given a few days to acclimate to the colors. Then

70. Ibid., p. 79.

71. See the essays by Danielle Allen and Abigail Lustig in *The Moral Authority of Nature*, ed. Lorraine Daston (Chicago: University of Chicago Press, 2004).

72. Ibid., p. 86.

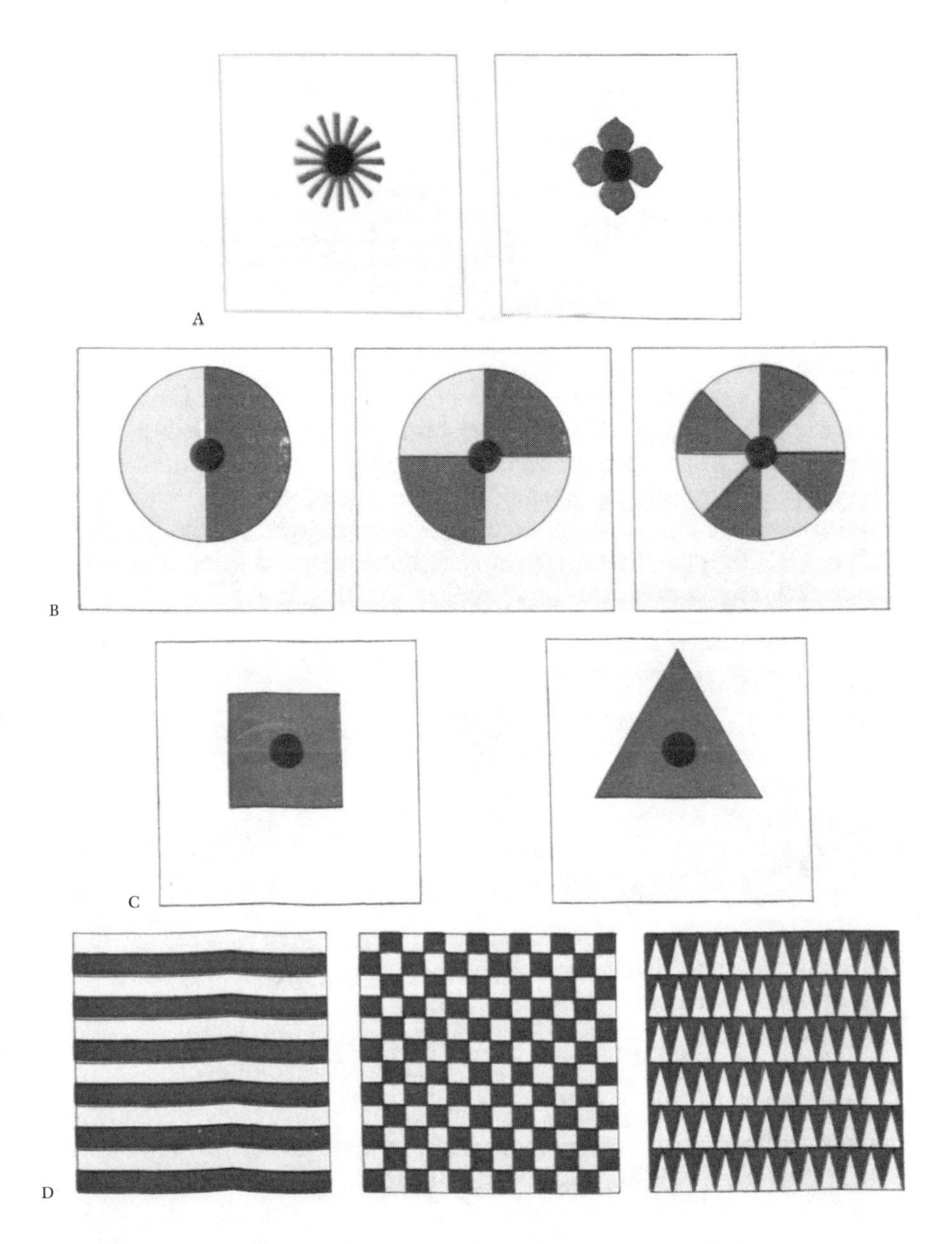

FIGURE 15a–d Karl von Frisch's test of the bee's shape recognition. His bees were able to recognize the patterns in a, but only the third pattern in b, and not those in c or d. Source: Frisch, "Farbensinn und Formensinn," *Zoologischer Jahrbücher* 35 (1914): 66–77.

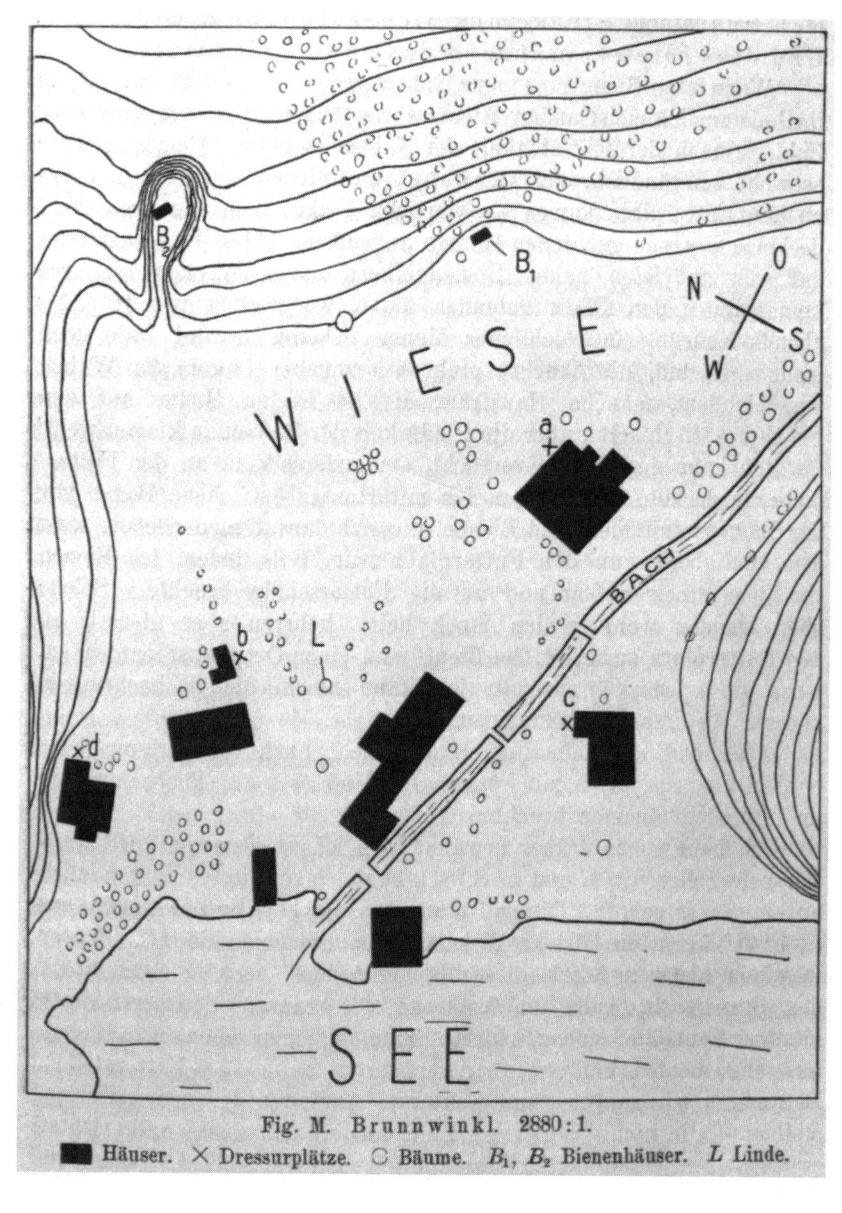

FIGURE 16 Map of Karl von Frisch's bee experiments at Brunnwinkl. Source: Frisch, "Farbensinn und Formensinn," *Zoologischer Jahrbücher* 35 (1914): 81.

the population of each hive was counted, after which the coloring of the two hives was interchanged. Now German bees flew into the Italian hive and vice versa. It was at this point that the different "temperaments" of the two races manifested themselves to Frisch. "While the German *Volk* simply let the Italians in, on the edge of the Italian hive violent fights developed; the majority of the Germans, who stubbornly tried to push their way in, were driven away with bites and stings."[73] In a scene that recalled the violent clashes of nationalist student groups at Habsburg universities, Frisch contrasted the peaceful German hive that welcomed foreigners with the belligerent Italians who defended their turf to the death. Reporting this research just after the outbreak of war, Frisch used racial and territorial language without any hint of irony.

Frisch was intrigued not just by the fact of the fierce patriotism of his bees, but also by the process through which they acquired it. The young bee on its first venture outside the hive went through a ritual Frisch called a "rehearsal."[74] It flew out tentatively, with frequent glances back towards the hive. Frisch supposed it was adjusting its color sense in order to be able to recognize its home. As in the human colony in which these experiments unfolded, visual learning helped the insect colony cohere.

The reflexivity of his investigations came through clearly on the map of Brunnwinkl that accompanied his published article. Nearly at the center is a circle labeled "L," marking the location of the linden tree, the Exners' literal and symbolic center of communication. The linden sustained the Exners' ideal of transparent communication, yet the riddles posted on its trunk celebrated a highly exclusive form of dialogue. As the crossroads of the Exner community, the linden was a fitting reference point for Frisch's study. These experiments were part of a collaborative tradition of visual research at Brunnwinkl. At the heart of the Exners' various projects on the physics and physiology of vision lay the effort to translate among subjective experiences, to see through the eyes of another. Still, their divergences are revealing. Karl Exner, a physicist, located the basis for the equivalence of perceptions outside the viewer, in the geometrical optics of the atmosphere. For the physiologist Sigmund Exner, the universality of visual experiences rested instead on shared features of the sensory apparatus. Karl Frisch, meanwhile, reined in this universalist dream, emphasizing instead that visual experiences bounded (and bound together) a community.

73. Ibid., p. 100.

74. Ibid., p. 87.

A final layer of reflexivity remained implicit. The bees had evolved a sensory apparatus that allowed them to recognize local flowers. They had become naturalists in their own right. Their dances, in so far as they communicated this natural knowledge, could even be compared to a scientific language. That the bee hive figured in Frisch's study simultaneously as a family, a nation, and a scientific community suggests the scope of his motivation. Like so many in a country dismembered and half-starved by war, he was probing the nature of these communities and asking what, if anything, could sustain them.

THE LANGUAGE OF SCIENCE

For Frisch, as for Robert Exner, Schrödinger, and Kohlrausch, color research consistently turned back on itself to become a critique of the language of science. Under the First Republic, young Viennese scientists from a variety of political backgrounds took as their model not Ostwald's technocratic anticlericalism but Serafin Exner's probabilistic humanism. With the Christian Social party holding the largest electoral base in the nation and leading every coalition government, scientists continued to express opposition to religious dogmatism through the language of probability. Yet this language took new statistical forms.

Wartime color research had shown the younger generation that it was meaningless to describe a population exclusively in terms of its mean. These fledgling scientists now faulted Serafin Exner for using his own aging eye as a visual standard.[75] They refused to treat qualitative individual differences in color perception as "abnormalities" or "weaknesses."[76] Young Viennese scientists therefore worked with a variety of statistical distributions beyond the familiar bell curve. In these more exotic distributions, the mean was not always the most probable value and interactions between individuals could not be neglected.

Erwin Schrödinger, for one, observed that statistical independence was not necessarily the most common condition in physics or in society. On one hand, there existed physical and social phenomena in which individuals appeared completely independent, such as the number of alpha particles emitted by a radioactive substance or the relative number of female births. Yet many other effects revealed a subtle interdependence. There was, for

75. Eduard Haschek, "Quantitative Beziehungen in der Farbenlehre," *Sitzungsberichte der Kaiserlichen Akademie der Wissenschaften zu Wien IIa* 136 (1927): 461–68, quotation on pp. 466–67.

76. Schrödinger, "Die Gesichtsempfindungen," p. 494.

instance, "a kind of 'van der Waals cohesion,'" that is, an intermolecular affinity, in the number of people ill with a contagious disease in a given area.[77] In 1924, when Niels Bohr and his collaborators interpreted energy conservation as a statistical effect, Schrödinger welcomed their theory as another example of the insights to be gained from a more varied statistical vocabulary. To Schrödinger, Bohr's hypothesis of a statistical dependence among atoms highlighted "the connection of each individual system with the whole rest of the world . . . Is it an idle speculation, to find in this a similarity to social, ethical and cultural phenomena?"[78] The alignment of particles in such a system arose not from an external field but from internal interactions between individuals. As his statistical description made clear, social order arose analogously to the alignment of color concepts—through interactions between individuals, not through external forces. Watching the rise of Austrian fascism from afar in the mid-1930s, Schrödinger insisted on viewing the nation in just these terms, as a population balanced between independence and cohesion.[79] A generation earlier Serafin Exner had pointed to the statistics of the molecules of an ideal gas as a model of individual freedom in a liberal society. Then, the burden had fallen on the probabilistic language of Austria's scientists to demonstrate how to skirt the twin dangers of dogmatism and solipsism. Now the calculations were even more subtle. Schrödinger held the language of statistics responsible for characterizing both the scope of individual variation and the degree of cohesion of the whole.

Unlike the ubiquitous bell curve of the nineteenth century, these novel statistical distributions could not be defined by just one pair of parameters. For purposes of communication, they required a visual and qualitative form, a morphological statistics. Serafin Exner's students had already spent years practicing a morphological approach to statistics for their measurements of atmospheric fluctuations in the Alps. Now they advanced a new, visual form of statistics as a language appropriate to a modern scientific community. Morphological statistics allowed the viewer to shift between alternate perspectives—the microscopic and the macroscopic, the discrete and the continuous, the individual and the whole.

Such an oscillatory play of attention was related, in part, to new visual technologies. Early filmgoers, for instance, were fascinated by the transition

77. Schrödinger, "Notiz über die Ordnung in Zufallsreihen" (1918), *Gesammelte Abhandlungen*, vol. 1, pp. 214–16.

78. Quoted and translated in Moore, *Schrödinger*, p. 163.

79. Schrödinger, "Gleichheit und Relativität der Freiheit" (1935), in *Gesammelte Abhandlungen*, vol. 4, pp. 356–58.

between the perception of discrete images and of continuous motion, an effect produced by varying the speed of the projector. Viennese scientists employed instruments and effects related to the cinema in their experimental work. In their color research, physicists at Serafin Exner's lab used the "rotating sector," in which portions of a disk were covered with colored papers. When the disk rotated rapidly, an observer perceived a single color, a combination of the colors of the sectors. Schrödinger pointed out that this effect was closely tied to that of the cinematograph.[80] Both depended on the lingering impression of visual images in the mind. Indeed, even in viewing natural motion, Schrödinger noted, an observer perceived several instantaneous stages simultaneously. Likewise, high-speed photography had made it possible to see natural processes such as the fall of a drop of water or oil at a pace that allowed for their full analysis. This application too was relevant to Viennese physics in this period. Felix Ehrenhaft's claim to have observed fractional electron charges rested on his ability to follow the course of a charged droplet, including any Brownian fluctuations in its trajectory.

These new visual technologies made possible new observational practices, but they left open multiple alternatives. Consider for instance the strategies that Karl Frisch developed for communicating his findings on the language of bees. As a youthful letter indicates, Frisch had been interested in the possibility of using a statistical language in biology at least since his university days, inspired by his uncle Serafin's rectoral address.[81] In the war period, he supplemented his written reports with photographs displaying the concentrations of bees on the colored squares or the patterned feeding holes. These photographs were a visual form of statistics. At a distance they could be viewed as evidence of average behavior; close up they could be examined for deviations at the individual scale.

In the 1920s, Frisch turned to film to communicate his research. These were highly reflexive studies of a community's dependence on visual language. The captions of *The Language of Bees* tell the story of a "messenger" bee who uses a "dance" to tell his "comrades" in the hive that he has found food. Frisch's interpretation of the bee's frenetic movements as a "lively circle dance" imposed a peculiarly Viennese-bourgeois sociability on the insects.

80. Schrödinger, "Gesichtsempfindungen," p. 255.

81. In the fall of 1908 Karl apparently asked Serafin for further clarification of the applicability of the probability calculus to biological evolution, and he received a long letter in response. Serafin Exner to Karl von Frisch, 2 Nov. 1908, Collection of Dr. P. C. Dijkgraaf, Den Helder, The Netherlands.

Not only did Frisch's films illustrate the social function of the bees' visual language, they also instantiated reforms to the language of science. In early films such as *The Sense of Color in Bees*, Frisch shifted back and forth between the scale of the individual bee and that of the hive. To this end, he used a few recognizable backgrounds: a close-up of a feeding dish, of a patchwork-colored board, and of a swarm inside a hive. In the first two cases, the evidence consisted entirely of the statistics of bees (the number feeding at a given place and time). The third case was the most difficult to film. Frisch relied on a pointer to follow the movements of the "dancing" bee, momentarily turning the swarm into background noise. In this way, cinematic techniques allowed him to convey large-scale patterns without suppressing small-scale deviations. His messenger bee allegorized the need for a statistical language that did not ignore the individual scale.[82]

In the 1920s Vienna's physicists likewise sought to represent their results in forms that allowed for the play of the viewer's attention between the micro- and macroscale. Schrödinger and Kohlrausch stressed that the lottery simulations discussed in chapter 8 allowed the experimenter to "direct his attention" alternately towards the microscale of individual lottery tickets (molecules) or the macroscale of entire urns (containers of gas/radioactive samples).[83] We might describe this property in cinematic terms as a choice between two modes of viewing: on one hand, close-up, stop-action, single frames; on the other, wide-angle, real-time, moving pictures. This cinematic description is in fact suggested by Schrödinger's unpublished notes. In an effort to make the equation for Brownian motion "visualizable," Schrödinger imagined its representation in film:

> Think of a series of cinema images, which are presented to the eye successively at very short time intervals τ. Each shows a line on which the [particle] shall lie, and the particle as a point. On the first the particle has the position x_0. Each successive image differs from the preceding only in that on it the particle is displaced by $\pm\delta$. For the production of the cinematogram the sign will be decided by the roll of a fair die (for instance $[-1]^n$, where n is the number

82. Both films mentioned here were produced in 1926 and released in 1936 by the Imperial Office for Educational Film (Reichstelle für den Unterrichtsfilm) in Berlin; they are distributed today by IWF Wissen und Medien, Göttingen.

83. See their use of phrases such as "Wir richten unser Augenmerk auf einen individuellen Stein. . ."; "von allen 'A-Urnen' (auf die wir das Augenmerk beschränken können). . .". Kohlrausch and Schrödinger, "Das Ehrenfestsche Modell der H-Kurve, *Physikalische Zeitschrift* 27, 1926: 306–13, quotations on pp. 309, 310, in Schrödinger, *Gesammelte Abhandlungen*, vol. 1, pp. 349–64.

> rolled. If one chooses τ and δ "very small," one obtains the simple image of Brownian motion in one dimension.[84]

Schrödinger's imagined film was an interactive exercise in representing the relationship between aggregate order and individual contingency. A simple change in parameters produced a shift in perspective. Like the lottery game with which Schrödinger and Kohlrausch modeled randomness, such a film of Brownian motion would have been a stochastic simulation for the precomputer age. Schrödinger and Kohlrausch thus embraced visual statistics as a means of displaying macroscopic regularities *and* microscopic fluctuations. Visual statistics was a language that avoided both dogmatic absolutes and the confines of a single perspective.

It was in these same years that the Austro-Marxist philosopher Otto Neurath conceived the ambition of constructing a universal language of visual statistics to facilitate democratic education. In his picture language, the meanings of colors were permanently fixed: green for hope, red for industry, blue for business. These were not, in his view, mere conventions. Neurath asserted that there was "a certain general feeling about the sense of the colours." He was confident that his color assignments would find "a great measure of agreement."[85]

In Neurath's exposition of the language of color we can see the distance between him and the third generation of the Exner circle. Neurath clung to the universalist view of perception which the young Exners abandoned. He warned, for instance, that "snapshots" and films should be used for communication and education only with the utmost caution because they presented a unique, individual perspective, and perspective should not normally enter a scientific argument. To Neurath, moreover, films were particularly dangerous because they relied so heavily on the viewer's memory: it was likely that the last frame in a sequence would make the strongest impression.[86] In place of Sigmund Exner's "*intuitive* statistics," then, Neurath demanded explicit visual statistics in order to discipline a biased memory. Frisch, in making his films, seems likewise to have recognized that he could not rely on a viewer's capacity for intuitive statistics. Frisch made his early films

84. Schrödinger, "Besprechung der letzten Arbeiten Smoluchowskis I," manuscript transcribed in Hanle, "Schrödinger's Statistical Physics," on pp. 274–75.

85. Otto Neurath, *International Picture Language: The Rules of Isotype* (London: Kegan Paul, 1936), p. 34.

86. Neurath, "Visual Education: Humanisation versus Popularisation," ed. Juha Manninen, in E. Nemeth and F. Stadler, *Encyclopedia and Utopia*, pp. 245–335, quotation on pp. 294–95.

"diagrammatic" in much the way Neurath recommended, but they still relied heavily on captions. The films fell short of the goal of transparent communication through pure visuality—even as they allegorized the use of visual communication for building a cohesive community. Theirs was a local language, insisting on the narrative of a harmonious society. They removed any potential ambiguity about the relationship between the micro- and macroscales, between the individual and the hive.

IN 1870 Marie Exner wrote of a family vacation on the Wolfgangsee: "*Our little house by the lake was like a bee hive.*"[87] By the close of the First World War, this was no longer a metaphor. With Marie's son Karl and his relatives tending the insects, the utopian vision implicit in her comparison became clear. In the midst of social breakdown, Brunnwinkl and its hives stood as models of both a harmonious family and a unified state. Yet they did not restore the Habsburg fantasy of the family-state. In the place of the nineteenth-century ambition of fostering universal, transparent communication, the new models of visual communication mirrored a hardening sense of the boundaries between those who belonged and those who did not.

For this very reason, the new science of color proved well suited to the needs of Austria's arts and crafts movement. The history of this alliance again forces us to revise the Schorskean picture of Viennese modernism as a culture of subjectivity. What subjectivity meant to these artists was an amalgam of realist physics, empiricist psychology, nature study, and the mores of the bourgeois home. Individuality and subjective expression were hardly the sole aims of the Arts and Crafts School and the Werkstätte. More immediately, they struggled to promote a distinctive Austrian style on the international market. As Debora Silverman has shown, the art nouveau movement in fin-de-siècle France was characterized by a similar convergence of the ideals of self-expression and national unity. French artisans discussed the former goal in the language of psychology and the latter in the language of history and politics, without attempting to resolve the tension between the two.[88] In Vienna, however, the modernist movement found such a resolution in the local science of color theory. The city's physicists gave its artisans a scientific vocabulary in which color was both a mode of self-expression and a resource for building a national style.

87. Marie Exner von Frisch, *Aus meinen Jugendtagen*, typescript, collection of Dr. Herwig Frisch, Vienna.

88. Silverman, *Art Nouveau*; see too Auslander, *Taste and Power*.

CONCLUSION

A Family's Legacy

AMONG urban visitors to the Wolfgangsee there had long endured the myth that this region was naturally insulated from historical change. As early as 1836, Karl Friedrich Schinkel, the classical master of German architecture, found it "revolting" to see "crude" new villas erected "next to the noble, naïve antique Alpine cottages."[1] The first villa in close proximity to Brunnwinkl went up in 1888,[2] and by 1893 the rail link from Salzburg to St. Gilgen was complete. Still, this same fantasy of isolation held sway over the budding historian of the Exner family Ernst Frisch as late as 1910, when he dug into the local archives to write a cultural history of the Wolfgangsee. He was disappointed to find no records of earthquakes, fires, or wars. "In their seclusion," he wrote, the local peasants "were rarely reminded that behind the steep cliff of the Schafberg the world was not yet over."[3]

This myth did not survive long into the twentieth century. By 1913 a restaurant and lakeside hotel had sprung up in St. Gilgen. Development accelerated rapidly after the First World War, as Austria's economy came to rely on tourism as never before. Bavarian weekenders arrived to take advantage of the weak Austrian currency, followed by international visitors to the new

1. Quoted in Géza Hajós, "Die 'Verhüttelung' der Landschaft," in Klaus Eggert, Géza Hajós et al., *Landhaus und Villa in Niederösterreich, 1840–1914* (Vienna: Böhlau, 1982), pp. 9–56, quotation on p. 16.

2. Hans Frisch, *50 Jahre Brunnwinkl*, p. 22.

3. Ernst Frisch, *Kulturgeschichtliche Bilder vom Abersee: Ein Beitrag zur salzburgischen Landeskunde* (Vienna and Leipzig: Alfred Hölder, 1910), p. iii.

Salzburg festival. By 1938, the Baedeker guide listed St. Gilgen as a *Sommerfrische* destination with five hotels in town.

Certain members of the Exner family complained bitterly of these transformations. Already in 1905 Emilie Exner lamented, "Gilgen is growing ever bigger, the society ever worse; that has the advantage that one meets fewer people in the woods than before."[4] Social undesirables, in her view, were those who preferred to view nature from the distance of a café terrace. Her caricature of the newcomers may have reflected some real differences in manner. Based on a survey of memoirs, historian Hannes Haas concludes that for many of the new vacationers in the Salzkammergut "nature was hardly more than a backdrop, and the mountains were too high for more than platonic love." Children were even forbidden from climbing the Schafberg—a constraint on youthful freedom that the Exners would not have imagined.[5] Two years later Emilie noted that the new houses springing up around the lake were far more lavish than the old ones. "Simply horrid," she declared of the villa built by the Jewish industrialist Max Feilchenfeld after tearing down the old Billroth home; "the garden a rubbish-heap, the house a monstrosity in this charming area . . . Thank God Brunnwinkl has stayed true to its old traditions and does not deign to imitate the *Feilchenfeldisch* elegance . . . And then there are people who want to hear nothing of an aristocracy of birth or of intellect!"[6]

The judgment Emilie pronounced on this new house recalls the description of a modern summer home in the final pages of a short story published the previous year by Ferdinand von Saar, an acquaintance of Emilie's from the Wertheimsteins' salon. Saar's pessimistic tale of the Kostenitz castle implicitly traces the rise and fall of Austria's liberal elite.[7] It closes with the castle's purchase by a wealthy industrialist bereft of reverence for the past, who razes it to make way for a "spacious summer house with all the modern comforts."[8] On the lawn, Saar writes, where once the castle's owners had conversed in sensitive and solemn tones, the new proprietor's many guests laugh and gossip loudly in colorful clothes and tennis shoes. Saar imagines

4. Emilie Exner to Marie Ebner von Eschenbach, 29 July 1905, 9 July 1907, 22 July 1907, 81082 Ja, Stadtarchiv, Vienna. Emilie again disparaged the Feilchenfelds in *Der Brunnwinkl*, p. 19.

5. Haas, "Traum vom Dazugehören," 43.

6. Emilie Exner to Marie Ebner von Eschenbach, 29 July 1905, 9 July 1907, 22 July 1907, 81082 Ja, Stadtarchiv, Vienna. Emilie again disparaged the Feilchenfelds in *Der Brunnwinkl*, p. 19.

7. On this story, see Norbert Miller, *Camera Obscura: Schloss Kostenitz* (Stuttgart: Mayer, 2000).

8. Ferdinand von Saar, "Schloss Kostenitz," in Miller, *Camera Obscura*, 138.

them talking of divorces, socialism, and hypnotism, all themes that seemed to toll the death of the way of life of the old liberal elite.

Emilie was not alone, then, in experiencing the popularization of the summer retreat as an invasion and a fundamental shift. The perceived boundaries of "public" and "private" life were in flux. In earlier years the Exners had never felt a need to cordon off the space within or around Brunnwinkl. The summer home had been defined by its lack of social boundaries, made possible by internalized standards of independence and intimacy. Guests and neighbors adapted to the rhythms of the colony and were absorbed into family life. The flip side of this fluidity was that moral authority won in the domain of the family carried over into "public" life. The Exners' reputations as "many-sided," as "Lebenkünstler"—in short as ideal representatives of the liberal Austrian *Bildungsbürgertum*—depended on the personae they cultivated in their domestic lives.

With the commercialization of the Wolfgangsee, the Exners began to see the need for boundaries. Their successful fight in 1917 to block the construction of a major thoroughfare past Brunnwinkl was a physical manifestation of this recognition. They also sought to mark their territory by articulating what separated them culturally from the newcomers. Ernst Frisch did so in a little guide to the Wolfgangsee published in 1938. He made clear that the book was not intended for the "fashionable lady" who came to the lake simply because it had become fashionable; she was already a lost cause. It was instead for those with the potential to "adapt" to local ways. In this vein he stressed the gulf that divided "locals" from "foreigners":

> Once, about thirty, forty years ago, the residents, who had their own property there or who rented the same house every year, were almost in the majority. There was a good regular crowd [*Stammpublikum*] of summer guests, who enjoyed above all the beauty of nature, who were well acquainted with mountain and lake and tended to socialize with each other. It was an old-fashioned life, just heavenly, until the world war brought an end to it. Then the prosperous middle class was largely wiped out, the rich had become poor and the poor rich, the new means of transportation did the rest to change the scene completely. Since distances no longer exist, an international horde comes and goes the whole summer long in a motley stream. What there still is of the old regulars disappears in this whirl. And when you ask the innkeepers at the end of the summer how business is, you hear to your surprise almost unanimous complaints, which ring of longing for the old days of the solid, steadfast regulars.[9]

9. Ernst v. Frisch, *Sommer am Abersee* (Salzburg: self-published, 1938), p. 9.

Ernst's harangue, like Emilie's caricature of the Feilchenfelds, implicitly invoked Jewish stereotypes. The new summer guests were "noisy, pushy, foreign in the worst sense. But they're tolerated, even encouraged, as an unavoidable symptom of the tourist business."[10] St. Gilgen had always received an unusually high proportion of Jewish vacationers. The social fluidity that the Exners had once valued so highly made it easier for Jews to assimilate at the summer retreat than anywhere in Vienna, except perhaps the Prater park. Still, anti-Semitism had been common in the Salzkammergut since the 1870s, often perpetuated by Jews themselves. In the late 1910s and early 1920s, with the onset of war anxiety and food shortages, anti-Semitism began to swell throughout the region. Jews came to be tolerated by locals only out of economic necessity.[11] Against this background, Emilie Exner and Ernst Frisch's caricatures of newcomers to the Wolfgangsee can be seen as attempts to distance themselves from the family's own infusion of "Jewish blood." Where they had once drawn no distinction between "public" and "private" space, now, as outsiders encroached on Brunnwinkl, the Exners threw up walls, literally and figuratively.

THESE transformations in the configuration of "public" and "private" life at the Wolfgangsee were symptomatic of broader changes. As the Exners' history has shown us, not just the diffusion of liberal values but their very construction depended on a peculiar social geography. The development of a liberal character rested equally on the cultivation of attentiveness and self-discipline at *Gymnasium* and on the acquisition of flexibility and many-sidedness within the circle of the family. Moreover, liberals often made "private" life into a standard for "public" practices. Thus jurist Adolf Exner called on parliament to model the laws of inheritance on the customs and values of the bourgeois family, and he likewise translated the bourgeois ideal of the diligent "pater familias" into a measure of civic rectitude. In the eyes of these liberals, moral standards could legitimately be transferred from the private to the public sphere, but not vice versa. In this vein, historian John Boyer has stressed that liberals in Austria, unlike socialists and progressives, refused to accord legislators the right to intervene in family life in a way that

10. Ibid, p. 10.

11. Albert Lichtblau, "'Ein Stück Paradies': Jüdische Sommerfrischler in St. Gilgen" (pp. 281–309), and Günter Fellner, "Judenfreundlichkeit, Judenfeindlichkeit. Spielarten in einem Fremdenverkehrsland" (pp. 59–126), in Kriechbaumer, *Geschmack der Vergänglichkeit*; Rossbacher, *Literatur und Bürgertum*, p. 306.

could lead to "radical social change."[12] By the first years of the twentieth century, however, it had become impossible to defend the geography of public and private on which Austrian liberalism rested. The movement to legalize divorce was just one example of this trend. The Middle School Inquiry analyzed in chapter 6 offered other clear evidence that Vienna's social topography was changing, as the nature of childhood itself became a matter of political conflict. Ultimately, what looked to Schorske like the middle class's retreat from the public sphere was really a symptom of a larger transformation in the relation of state, society, and family, one that liberals found themselves powerless to stop.

It has often been said that liberals were their own undoing in Central Europe, since their reform program gave votes to their future political opponents. But the question remains why they failed to make liberals of these new voters. In the Austrian case, this might seem particularly surprising. As the franchise widened, so did access to academic education, the focal point of the liberals' bid to reproduce themselves. The *Gymnasium* student's path from certainty to doubt to probability was meant to function as a uniform tool for producing liberal minds in classrooms throughout the empire, from Trieste to Czernowitz, independent of variations in language and culture.

Nonetheless, as Adolf Exner put it, the true sign of political *Bildung* among liberals lay not in the mastery of a calculus but in a *feeling* for political possibilities, a tacit sense for the latitude of deviations from the normal course of events. The Exners took as their models of rationality the farmer, the hunter—men whose learned abilities to navigate an unpredictable environment could not be captured in words, much less mathematical symbols. If liberalism was a character wrought of such experiences, how could liberals ever be produced within the confines of a classroom? A liberal's education was meant to be an adventure, but Austrians were far more likely to recall their *Gymnasium* days as mind-numbingly dull. By the first decade of the twentieth century, the Exners and their allies began to admit that the lessons crucial to the development of a liberal personality could never be replicated on the imperial scale. Indeed, they began to argue that critical reasoning could be cultivated in public schools only by those whose private lives resembled those of the educated middle class.

This transformation entailed not the death of political liberalism in Austria but a complex metamorphosis fraught with paradoxes. With the dissolution

12. Boyer, "Freud, Marriage," p. 96.

of the liberal party in the 1890s and the Christian Socials' rise to power in Vienna, it became difficult to say who best represented the liberal principles of the *Gründerzeit*. The liberals had been dominated from the start by German speakers, and the German nationalists were often viewed as the liberals' successor party. The German nationalists did indeed maintain continuities with the liberals—through their anticlericalism, their opposition to cartels and tariffs, and, in some cases, through the persons of their representatives.[13] Yet they now stressed their German identity in ethnic and anti-Semitic, rather than political, terms.

On the other hand, the Austro-Marxist movement also showed strong continuities with the liberalism of the 1860s. The socialists indeed viewed themselves as "heir to the liberal tradition in the Habsburg domain."[14] Liberalism had shaped Vienna's first cultural organizations for workers, in keeping with the liberal slogan "Knowledge is power, education makes you free."[15] A similarly paternalistic emphasis on *Bildung* was at the heart of socialist reforms in Red Vienna in the 1920s. The socialists' approach to education echoed that of the old liberals in content as well as intent. Like their predecessors, the socialists tried to wrest control of education from clerical conservatives and made the teaching of empirical science a weapon in that battle. Otto Glöckel, the architect of the socialists' education program, stressed the pedagogical value of "observation, the inductive method, skepticism regarding hypotheses, criticism and acceptance of the empirically quantifiable."[16]

Although these goals were touted as reforms, they replicated the decades-old struggles of liberal educators against religious dogmatism. Indeed, Otto Neurath identified the antidogmatic spirit of his program of "democratic education" as a long-standing Austrian tradition.[17] The same goals that had given probabilistic reasoning its place in the liberal curriculum of the post-1848 era made probability and statistics central to Neurath's program of worker education in the 1930s. His invented language of visual statistics was a solution to the very problem that had exercised nineteenth-century liberal educators. It was a pedagogical method that would convey the provisional character of all knowledge while providing a basis for universal, transparent communication.

13. Hobelt, *Kornblume*, pp. 352–53.

14. Rabinbach, *The Crisis*, p. 16.

15. Gruber, *Red Vienna*, p. 83.

16. Michael J. Zeps, *Education and the Crisis of the First Republic* (Boulder: East European Monographs, 1987), p. 65.

17. Neurath, "The Scientific Conception of the World," in Neurath, *Empiricism and Sociology* (Dordrecht: Reidel, 1973), pp. 299–318.

For liberals and socialists alike in the 1920s, programs of adult education were an important means of popularizing science in order to maintain "transparency" and thus defend academic freedom in the face of resurgent religious dogmatism.[18] Yet there are revealing contrasts between the educational strategies of liberals and socialists in general and their uses of visual statistics in particular. The liberal paradigm of public education, illustrated by public lectures at the University of Vienna, forbade explicit political content, aiming in principle to be "value-free." Felix Exner, for instance, took part in such a lecture series at the university in 1926–27 on the subject of the "Austrian Alps." The presentations spanned the Alps' geology and vegetation, its music, costumes, and art. As a whole the lectures conveyed to an urban audience the message that Austria's culture and mountain landscape were intimately linked. Felix Exner's contribution on "The Climate of the Alps" made this point directly. Exner asked his audience to imagine how the climate of Central Europe would look if the Alps ran not west-east but north-south, like the Rocky Mountains. On the east side, Austria's side, the west wind would come down dry and produce a dusty plain like that east of the Rockies. "So we may thank God that the Alps run from west to east; otherwise Central Europe would not be a fertile region that for millennia has sustained us and allowed us to advance."[19] Later published in a single volume, this lecture series painted an ostensibly apolitical vision of the new nation. By contrast, the socialist model of worker education was explicitly political, as seen, for instance, in the "workers' university" run by the Social Democrats.

Statistical languages also functioned differently in the two educational paradigms. Like the third generation of the Exner family, Otto Neurath worked with an associationist model of the process of visual learning. But he seemed to have caught hold of this model from the opposite end. The research of Karl Frisch and Robert Exner emphasized the limited range of visual communication, the extent to which the judgment of abstract patterns depended on visual memories specific to a given community. Neurath distrusted such strategies. While Karl Frisch, for instance, made film into a new language of scientific communication, Neurath argued that film was compromised as a communicative medium by its reliance on the viewer's memory. Neurath sought to eliminate the tacit, intuitive elements of probabilistic reasoning. In place of intuitive statistics, he demanded explicit visual statistics as an unambiguous language for democratic education.

18. Felt, "Öffentliche Wissenschaft."

19. Felix Exner, *Die österreichische Alpen* (Vienna, 1927), 167.

As a learning environment, the museum of visual statistics that opened in central Vienna in the 1930s was as undogmatic and open-ended as it was universally accessible. By contrast, consider the public incarnation that Karl Frisch gave his natural history collection, which can still be visited today. Few travelers to the Wolfgangsee come expressly to see the local museum in the quiet town of St. Gilgen. Yet a rainy day or sheer curiosity draws visitors into the low-ceilinged rooms of antique furniture and historical portraits. The displays lead upstairs into a small addition, its walls lined with glass jars of formaldehyde holding fish, snakes, lizards, and toads. Stuffed birds perch near the ceiling along with a squirrel, while another wall holds hundreds of insects pinned behind glass. There are approximately 5,000 specimens, all native to the lake and its surroundings. Yet the room has no schematic arrangement or informative panels such as one finds in grand museums of natural history like Vienna's Naturhistorisches Museum. As a didactic exhibit the collection feels incomplete, lacking structure and elucidation. It seems more a game than a lesson, yet there is nothing obviously playful about it. Only when we imagine a flurry of young children chasing these creatures around a summer cottage do we begin to understand.

Brunnwinkl was for the Exners a place to play, but playing is a highly exclusive activity. Children take pleasure in cloaking their games in secrecy.[20] The very security that made Brunnwinkl conducive to creative play depended, in the Exners' eyes, on the colony's isolation. The subsequent transition to democratic education in interwar Vienna involved a renegotiation of the boundary between the public and private spheres. Where the nineteenth-century *Bildungsbürgertum* had built public education on a domestic model, early twentieth-century Austrian socialists attempted an inversion, opening up private life to public reform.

ALTHOUGH the socialists' programs of public education, health care, and housing in Red Vienna were models of efficiency and social justice, the party failed spectacularly to safeguard democracy in Austria. Their leader Otto Bauer has been compared to Hamlet for his inability to take decisive action in the face of growing threats from the radical right and left. Still, the roots of the socialists' failure lay less in Bauer's personality than in the paradoxes at the core of Austro-Marxism. Identifying with their liberal predecessors, the Social Democrats insisted on building power through parliamentary means and rejected the use of force. Bauer and his followers focused above

20. Loewenberg, "Creation of a Scientific Community."

all on their intellectual goals for Austria's workers, on their programs of *Arbeiterbildung*. But they were blind to the gathering might of paramilitary groups on the right and left.[21] Paradoxically, the very workers' institutions on which the Social Democrats prided themselves became breeding grounds for left-wing opposition to Bauer's leadership. In March 1933, after Hitler's assumption of power in Germany in January, the Christian Social chancellor Engelbert Dollfuss indefinitely suspended parliament to govern according to the wartime emergency powers decree. The leaders of the Social Democrats failed to respond to this challenge, and Austria's democratic experiment came to an end.

By the time of this crisis, few of the third-generation Exners still lived in Vienna. The family member most closely tied to the government of the First Republic and to the municipal administration in Vienna was Felix Exner, director of the Central Institute for Meteorology, but he had died of a heart attack in 1930 at the age of fifty-four. Premature deaths had likewise taken Hilde and Nora Exner, whose unconventional art and strong patriotism might have made them supporters of the socialist experiments of the 1920s. Others had emigrated but retained close contact with Brunnwinkl. These included Hilde's sister Priska, who had moved to the Netherlands with her Dutch husband, and Karl Frisch and his cousin Franz Exner, who had taken teaching posts in the Weimar Republic.

Among those who remained in Vienna was Marie Exner's eldest son Hans Frisch. By 1933 Frisch had become a full professor of law at the Technical University in Vienna. Yet he was an outspoken critic of Dollfuss and his successor Schuschnigg. Forced into retirement for a publication critical of the regime, Frisch found time to write a book-length critique of the Austrian dictatorship. In it he attacked Dollfuss and his followers for destroying democracy and the rule of law. Striking an anticlerical position characteristic of his liberal forebears, he accused his opponents of hypocrisy, scorning the "intolerant Catholic spirit that emanated from Dollfuß and his successors."[22] He pointed to the damage the Catholic influence had wrought on the Austrian schools, where, for instance, history books had been purged of discussions of the Reformation. The regime's opponents, he attested, felt that their hands were tied, that all weapons were in the control of the state.

21. Rabinbach, *The Crisis*.

22. Hans Frisch, *Die Gewaltherrschaft in Österreich, 1933 bis 1938. Eine Staatsrechtliche Untersuchung* (Leipzig and Vienna: Johannes Günther Verlag, 1938), p. 118.

Yet who were these opponents, among whom Frisch numbered himself? He pointedly called them "the Germans in Austria." Surprisingly, Frisch pinned his own hopes on Hitler's Reich. His tract, completed just after Hitler's annexation of Austria in 1938, celebrated the German nation and its *Führer*. It is hard to imagine that Frisch truly believed that the National Socialists would restore the democracy that Dollfuss had shattered, that their respect for law would surpass that of the Austrian dictatorship. But he clearly miscalculated the Nazis' intentions in one respect. As his book of 1938 makes clear, Frisch expected that the *Anschluss* would clear his political record and guarantee his reinstatement at the Technical University. To his shock, however, in 1940 came a second order for his early retirement, this time on grounds of racial impurity. A year later Hans Frisch was dead of a heart attack at sixty-five.

The application of the Nazi racial laws to the "mixed-blood" Exners was strikingly uneven. Hans's brother Ernst, an archivist in Salzburg, escaped his brother's fate because the local authorities were fond of him and simply sat on the order for his retirement. But their brother Otto fared worse. He was fired in 1942 from his professorship at the medical faculty of the University of Vienna, as well as from his posts with the Red Cross and as chief physician at the Rudolfiner Hospital. Despite a trip to Berlin to plead his case, he was judged unfit to practice medicine due to his Jewish descent. Although he managed to find work at a non–Red Cross hospital, his own health soon failed, and he retired to Brunnwinkl to recuperate.

Otto's youngest brother Karl would join him there in 1944, after his Munich home and zoological laboratory were destroyed by bombs. Karl Frisch too had been threatened with early retirement, but the interventions of powerful colleagues had halted the process. Still, when the Dutch biologist Nikolaas Tinbergen visited Frisch at his Munich lab in 1936, he felt "anxiety on his behalf when I saw that he refused to reply to a student's aggressive *Heil Hitler* by anything but a quiet *Grüss Gott*."[23] Politically conflicted as Frisch must have been, the *Anschluss* nonetheless brought him some benefits. Upon taking his first post in the Weimar Republic in 1921 he had been forced to renounce his Austrian citizenship in favor of German. After 1933, however, tensions between the two nations made it increasingly difficult for him to travel to Brunnwinkl without an Austrian passport—and often with equipment, assistants, and beehives in tow. The imposition of a 1,000 mark border toll was the last straw, prompting Karl to request and receive dual

23. *Les Prix Nobel en 1973*, ed. Wilhelm Odelberg (Stockholm: Nobel Prize Foundation, 1974).

citizenship for him and his family. Given such obstacles, he later admitted that Austria's annexation in 1938 came in some ways as a "relief."[24]

Why did the Nazis allow Karl Frisch to continue his research, while his brother Hans was removed despite his explicit support for Hitler's regime? In his autobiography Karl Frisch explained that his research on perception and communication among bees was deemed "militarily significant," but it seems that this is not to be taken literally. More likely, the implications of Karl's research appealed to the Nazi authorities. The infant science of ethology, pioneered by Frisch, Tinbergen, and Frisch's fellow Austrian Konrad Lorenz, demonstrated physiological causes of animal behavior. Such research could be used to argue that social behavior was racially determined, as Frisch's comparison of German and Italian bees had suggested.[25] Moreover, Frisch openly supported eugenics, and continued to do so in qualified terms until his death in 1982.[26]

Karl Frisch was not the only member of the family whose research fared well under National Socialism. At Vienna's Natural History Museum, a center of Nazi anthropological research during the Second World War, psychiatrist Robert Exner continued his studies of color perception and brain structure on prisoners of war and patients with brain injuries. Looking back on the research he had conducted in Siberia during the First World War, he was led to new conclusions. In the intervening years, he wrote, it had become possible to recognize as "racially" determined perceptual traits that would once have been explained as the result of individual abnormalities.[27] Drawing on subjects of eight different races, he constructed a table relating characteristics of color perception to race and gender.[28] As in the case of Karl Frisch, Robert's Nazi-funded research shared certain continuities with the empiricist tradition in Austrian physiology. Problems of perception remained in the foreground, as did questions of heredity. Now, however, those questions were more likely to be interpreted at the level of the race than of the individual.

24. Karl Frisch, *Fünf Häuser am See*, p. 99.

25. Benedikt Föger and Klaus Taschwer, *Die Andere Seite des Spiegels: Konrad Lorenz und der Nationalsozialismus* (Vienna: Czernin, 2001).

26. On eugenics, see Karl Frisch, *Du and Das Leben* (Berlin: Deutsche Verlag, 1936) and later editions.

27. Robert Exner, "Die Genesis der vier Kardinalfarben," *Psychiatrisch-Neurologische Wochenschrift* 40 (1938): 1–18, quotation on p. 14.

28. R. Exner and R. Routil, "Studien zur Farbentheorie," 6–11, quotation on p. 9.

Unlike Karl Frisch, however, Robert Exner left the political alignment of his research unambiguous. He had secretly joined the Nazi party in 1933, when it was illegal in Austria.[29] When he applied for research funding to the Nazi government in 1939, his list of "political activities" in the previous years included hiding prohibited newspapers from the Dollfuss regime and providing information to the Nazis about researchers with Jewish connections or sympathies. Whether his family knew of these activities is impossible to say. He was still a bachelor in these years and seems to have had little contact with his cousins or with Brunnwinkl.

Not so his cousin Franz Exner, who maintained ties to the Wolfgangsee while teaching law in Munich throughout the 1930s. Like Karl Frisch and Robert Exner, Franz was interested in the biological determinants of psychology. In Vienna during the First World War he had begun empirical research in the field then known as criminal biology. His treatise on the subject published under the Third Reich perpetuated the racial stereotype that the Jewish race showed a marked tendency toward crimes of deceit (though, being physically unimposing, they rarely perpetrated crimes of violence). In other ways, though, his theories echoed liberal wisdom of his parents' generation. He stressed that criminal tendencies resulted from a complex interaction of biology and environment. The development of innate potentials depended on the environment, but the environments to which people chose to expose themselves also depended on innate traits. Frisch stressed that these interactions were not subject to full causal analysis. They involved "mass phenomena," which were appropriately treated statistically. When repeated observations confirmed the hypothesis of a relationship between a crime and an aspect of social life, the confirmation merely "raises the degree of probability of the interpretation, the 'truth' of which remains unprovable."[30] Reading his critique of causal reasoning, it is hard not to be reminded of his father Adolf's notorious rectorial address on the dangers of determinism, which Franz described admiringly in a memoir written in 1944.[31] Yet their intentions diverged radically. Adolf had sought a methodology that would avoid dogmatism and do justice to individual freedom. His son set out to serve the criminal justice system of the Nazi state. He would continue to do so after the war as a defense attorney at the Nuremberg trials.

29. Robert Exner, "Lebenslauf," Anlage II.

30. Franz Exner, *Kriminalbiologie in ihren Grundzügen*, 2nd edition (Hamburg: Hanseatische Verlagsanstalt, 1944), p. 19.

31. Franz Exner, *Einiges über die Exnerei* (Staltach: self-published, 1944), p. 9.

How are we to understand the Nazi sympathies of Hans Frisch, Robert Exner, and Franz Exner? How could individuals brought up to identify with the liberal tradition in Austria have forsaken those liberal values so blatantly? If members of the third generation of this family could welcome Hitler's Reich, how committed could their parents and grandparents have been to liberalism?

To expect that each generation of the Exner family would conform to the ideas of their parents underestimates the complexity of Central European history and of family dynamics. In fact, the familiar strand of Central European historiography that periodizes by generations singles out the generation of the fin de siècle, to which Hans Frisch, Robert Exner, and Franz Exner belonged, for their rebelliousness. This generation "turned their backs" on their parents, writes Karl-Heinz Rossbacher, "in the arts, in literature, in reflecting on the psyche, in the new valuation of dreams, in politics."[32] Take just two of the abundant examples of such conflicts. Theodor Gomperz's daughter Bettina, a sculptor, poet, and progressive social thinker, rejected her father's classically liberal views. Viktor Adler, the founder of Austria's socialist party, discovered his most troublesome left-wing opponent in his son Friedrich. In neither case should the children's divergences call into question the sincerity of the parents' views.

On the other hand, to overlook the continuities in these intergenerational relationships is to fall into a Schorskean trap. Thus Friedrich Adler's radicalism was a rebellion against his father's authority but it was also a reinvigoration of his socialist ideals. Likewise, the same critical, intellectually idealistic, and freedom loving spirit that drove Theodor Gomperz seems to have driven his daughter, even when she turned her critical eye to nineteenth-century liberalism. To understand such dynamics we should consider that the children came of age in a different world than had their parents. The First World War and the unsettled republic in its wake presented political, social, and economic opportunities utterly unlike those of late imperial Austria. Historians have rightly emphasized the political paradoxes that marked these years, particularly the ambiguous incarnations of liberal principles in new guises. In many ways, the politics of the Republic hinged on unforeseeable contingencies like the stock market's crash. No party or individual emerged from these experiences unchanged. In short, it would be historically inaccurate and ethically suspect to infer from an individual's political choices in the 1930s the atmosphere she grew up in.

32. Rossbacher, *Literatur und Bürgertum*, p. 542.

In the case of the Exner family in particular, it is crucial to recognize that Brunnwinkl was not a totalitarian regime. Its young people were explicitly encouraged to think for themselves; blind conformity was shunned. We should recall that by 1933 five of the thirteen members of the third generation were no longer alive, and one was living in the Netherlands. It is an accident of history that those who remained and not others shaped the family's legacy. Who crafted that legacy, and how?

IN the summer of 1931 the Exners celebrated their fiftieth anniversary at Brunnwinkl. From Brunnwinkl's "daughter colony" in Litzlberg came the third Franz Exner with his family, along with the Conrads and Benndorfs, family friends for generations. The highlight of the festivities was a dramatic reenactment of the hamlet's three-hundred-year history. The players spoke verses penned—partly in local dialect—by Ernst Frisch, the family's authority on regional history. Hans Frisch's sons played seventeenth-century millers, Karl Frisch's daughters were an eighteenth-century miller and fisherman, and Hans's daughter closed the show in the role of the debt-ridden miller's wife who had sold her home to Marie Exner von Frisch in 1882.

The function of this performance was to write the Exners seamlessly into a lost world of alpine folk traditions. In the Salzkammergut in the early 1930s, displays of folk costumes and local dialects were commonplace demonstrations of national pride. This version of Brunnwinkl's history cemented the Exners' identities as locals, not mere urban visitors. At the same time, it invited the celebrants to forget the transformations of the wider world, to bury themselves once again in the myth of Brunnwinkl's isolation from historical change.

The "charm" of owning a summer home, as Adolf Exner had explained to his sister Marie back in 1882, lay in the continuity it lent to family life from year to year and from generation to generation. Austrian liberals had always craved stability and feared revolutionary change. This desire for constancy only intensified towards the turn of the century, as they saw new political movements shaking the foundations of the empire. The summer retreat was their bid for continuity. In Theodor Gomperz's words, its charm lay in the absence of novelty.[33] The summer house played a paradoxically conservative role in the history of Austrian liberalism, nurturing the way of life of the old liberal elite while all around their supports were crumbling.

33. Quoted in Rossbacher, *Literatur und Bürgertum*, p. 242.

In the 1930s the Wolfgangsee showed all too many signs of the encroachment of the outside world. From 1935 to 1937 the Austro-fascist chancellor Kurt Schuschnigg summered near Brunnwinkl on the road to St. Gilgen, closely guarded by police and detectives. "All night long the searchlights played over the lake and illuminated the forest on the opposite bank," Karl Frisch recalled.[34] The summers passed without incident, but the atmosphere had undoubtedly been tainted. Then, with the onset of war, streams of refugees from western Germany arrived at the Wolfgangsee. Bombs, however, did not come close.

Brunnwinkl's survival in these years was not fortuitous; nor was the endurance of the Exner family as a coherent unit. The work that went into preserving the family is visible, for instance, in a history that Ernst Frisch drew up in 1936, apparently in response to official inquiries. Ernst dug deep into municipal archives to trace the family back to the parents of the first Franz Exner. He recorded Franz Exner's accomplishments, detailing his academic and administrative service. Briefer sketches of Franz Exner's children and grandchildren followed, emphasizing the men's roles as civil servants and the women's marriages to men similarly well connected to the state. Ernst disposed of his grandmother's Jewish origins with the fewest words possible: "She was born in Vienna, daughter of a merchant. Baptismal register St. Stephan."[35] Although the immediate purpose of this document is unclear, Ernst undoubtedly recognized the threat that Charlotte's Jewish blood posed to him and his relatives in proximity to Nazi Germany. In trying to protect the family, he was effectively rewriting their history.

The work of self-preservation involved not only manipulating the Exners' public face but also shaping the family's self-image. Franz Exner took on this responsibility in 1944 when he printed a memoir "for his grandchildren and great-grandchildren." This included a celebration of Brunnwinkl and of the Exners' community in Vienna from a new perspective. Franz recalled with great affection his childhood home on Pelikangasse in the capital's university district. It neighbored the homes of the Conrad and Benndorf families, and the little community of friends was known as Pelikanwinkl. Here, he attested, "politics played no significant role, aside from the time of the Boer war, at least compared to what it would be today."[36] The Pelikanwinkl was "truly a *Gemeinschaft* (community)." This false nostalgia for an

34. Karl Frisch, *Fünf Häuser am See*, p. 99.

35. Typescript marked "Erhebungen," Collection of Dr. Herwig Frisch, Vienna.

apolitical *Gemeinschaft*, ubiquitous in the interwar years, was just the longing to which National Socialism played. In his scholarly writings, Franz Exner had identified the singular importance of the *Erziehungsgemeinschaft* (child-rearing community) as a key factor in preventing criminality, and he praised the Nazi state for fostering such conditions.[37] Franz, who had in fact made only brief visits to the Wolfgangsee until his wife's death in 1921, conjured Brunnwinkl in similar terms. In verses penned for the fiftieth anniversary celebration in 1931 he described the colony as "an organism" in spite of "individualism." Amidst Brunnwinkl's "unity," however, he singled out in his memoir one individual who had failed to integrate. Emilie Exner "was less popular than her husband. Very clever, ambitious, a writer fighting for the career woman, she was somewhat out of place among the Exners."[38] In his effort to portray Brunnwinkl as apolitical, Franz shifted Emilie Exner to the margins of the family.

It may seem as if these individual members of the family were simply bending to the prevailing political winds in order to weather the storm. This is not just a case of opportunism, however. In these examples we see individuals struggling to preserve the family against the centrifugal force of political upheaval. The Exners' craving for political stability was characteristic of this academic dynasty's unusually close ties to the state. Compared to scientific families elsewhere in Europe at the time—the Darwins or De Broglies, for instance—the Exners' rise depended to a rare degree on state patronage. While publications or aristocratic lineage might boost scientists in nineteenth-century England or France, the Exners' authority in imperial Austria was made possible by academic and administrative appointments. This unique symbiosis of the family and the state in Austria grew out of the state's involvement in education and out of a class structure that facilitated the *Bildungsbürgertum*'s merging of domestic and "public" life. As the Exners well knew, the family's stability rested on the state's. It is therefore no coincidence that those who fought to maintain the family's identity in the 1930s—above all Ernst Frisch, Karl Frisch, and Franz Exner—were those who compromised and contorted themselves to serve the Nazi powers. They took charge of the Exners' legacy, but the historian need not leave it in their hands.

In the end, the Exner family's rise to prominence is the story of the creation of a new form of moral authority specific to Austria's liberal *Bildungs-*

36. Franz Exner, *Exnerei*, p. 22.

37. Franz Exner, *Kriminalbiologie*, pp. 237–48.

38. Franz Exner, *Exnerei*, 30.

bürgertum. The first Franz Exner won the loyalty of his students and the support of the imperial government on the basis of an authority founded not on birth, religion, or wealth. Indeed, his skeptical challenge undercut all traditional sources of authority, above all that of religious dogma. Exner traded instead on the finesse with which he navigated a world devoid of absolute certainty. He displayed the mental agility necessary to tread a straight course through an intellectual landscape with no firm footholds. Exner's probabilistic reasoning was a style of argument and a manner of comportment fine-tuned to a culture equally wary of discredited dogmatism and revolutionary relativism.

Yet the Exner family's history is even more remarkably a story of the transmission and diffusion of this new form of authority, in the face of powerful and enduring political opposition. Within the *Gymnasien* and universities, the liberals used the probability calculus to transmit a moral authority grounded in the quantification, and thus domestication, of uncertainty. The preceding chapters have traced the role that probability played as a pedagogical tool in Vienna, first in securing the liberals' power in the wake of the 1848 revolution, then in buttressing their authority against new incursions after 1879. The probability calculus was a lesson in a barely veiled program of "political *Bildung*." As a language free of absolutes, probability steeled students against dogmatism, preparing them to resist clerical control of education. At the same time, probability braced students against the revolutionary potential of radical skepticism by showing them that knowledge gleaned from experience need not be solipsistic. As a means of liberal self-fashioning, probability located an intellectual standpoint that transcended divisions of nation and class without falling back on absolutism. It was a display of the moral authority that came from the disciplined tolerance of uncertainty.

Nonetheless, academic institutions alone were not responsible for the wide diffusion of the values of Austrian liberalism. This process depended crucially on those "clubs" that were, as contemporaries noted, fully exempt from state surveillance. Families could nurture values that were threatened by elements of the wider society. At Brunnwinkl the Exners could transmit values through traditions that remained relatively independent of broader norms.

Austrian liberalism was thus a product not only of schools and universities, civic societies and cafés in the metropolis. It took root as well "auf dem Land," through the rise of bourgeois summer homes in the countryside. As exceptional as the Exners were in their intellectual endeavors at Brunnwinkl, they were entirely typical of families of the Austrian *Bildungsbür-*

gertum in seizing on a summer home as a strategy of self-preservation. Summer homes shaped and sheltered the identity of Austrian liberals. Essential to this identity was an open-ended yet self-disciplined manner of interacting with the unpredictable natural environment. This agility was the physical evidence of liberals' fitness to live in a postdogmatic world. It was the embodiment of that "taming of chance" which they accomplished in their scholarship by means of the probability calculus. At Brunnwinkl, the Exners fashioned a lifestyle through which to pass on to their children a moral authority founded on the acceptance of uncertainty.

The Exners' manner of living "in the open" in turn shaped the family's scientific achievements and those of their students. The defining feature of the Austrian summer home was its seclusion, its insulation from historical change. This myth of isolation made possible the dissolution of internal boundaries. The porous borders between inside and out erased distinctions between pests and specimens, making all local organisms objects of research. At the same time, this architecture confused the nineteenth-century physicist's demarcation between constant "phenomena" and fluctuating "disturbances." The Exners dealt resourcefully with this situation. Felix Exner imported new mathematical techniques from the field of statistics to identify "significant" fluctuations. Likewise, Karl von Frisch found that man-made boundaries were unnecessary if one worked with organisms that segregated themselves. For the natural sciences in Austria, Brunnwinkl was a site of convergence and exchange, providing a common solution to the problem of uncertainty and a shared ideal for the language of science.

The summer retreat was a central node in the social and intellectual networks that undergirded Vienna's *Bildungsbürgertum*. It was not by accident that the Exners vacationed next door to cultural luminaries like Theodor Billroth and Marie von Ebner-Eschenbach, nor that several members of the family first met their spouses during summers in the Salzkammergut. In its role as a bridge between the domestic world and the field of research and politics, Brunnwinkl was crucial to the diffusion of the moral value of uncertainty in imperial Austria. There the probabilistic attitude spread through the closely knit branches of the extended Exner family and beyond, into their network of friends and colleagues. From there this incarnation of liberal authority penetrated into a broad spectrum of pursuits, from physics, physiology, and psychology, to legal reform, women's education, and modernist art.

After the First World War, Austria's legacy of probabilistic liberalism fractured, its branches leading in two very different directions. One path isolated the liberals' pedagogical program and embedded it in a new political context. As we have seen, socialist educators like Otto Neurath and Otto Glöckel breathed new life into the values of Austrian liberal education, despite the divergence of their Austro-Marxist goals from those of liberalism. Aided by the philosophers of the Vienna Circle, they resuscitated probability as a form of reasoning that guarded against dogmatism while transcending solipsistic individualism.

A second group hewed a path closer to the original ideals of probabilistic liberalism but divorced them from their pedagogical context. This was the Austrian School of economics, led by F. A. Hayek and Ludwig von Mises. Due to their scholarship, Vienna remained one of the premier centers of classical liberal economics throughout the 1920s, maintaining a reputation forged by Carl Menger in the 1870s. Against the "historicist" economics for which Germany had become known, the Austrian school denounced all assumptions of historical determinism. Economic history was too complex and human motivations too multifaceted to admit of predictive theories, they argued. Against socialists at home and in Germany they attacked the Hegelian determinism at the core of Marxist theory, recalling the liberal anti-Hegelian tradition in Austria that stretched back to Franz Exner's critiques of the 1840s.[39]

To Schorske, and to most theorists of Central European history in the tradition of Jürgen Habermas, liberalism and the public sphere have appeared to be mutually constitutive historical phenomena. This study has instead emphasized the debt of Austrian liberalism, for better or worse, to the domestic world of the educated middle class. The Exners were typical of educated Viennese liberals in their conviction that family life offered opportunities to cultivate "many-sidedness," and so to approach asymptotically to a universal perspective. The challenges of family life, as they saw them, were those of a liberal society in miniature. The goal was a state of autonomy that would not devolve into solipsism and a form of relatedness that would not lapse into blind conformity. At Brunnwinkl, the Exners performed a sensitive political calculation, weighing mechanisms of communication against guarantees of individual freedom. Through the layout of the colony, the linden tree's functions and rituals, the decoration of the cottages, and the rhythm of their days, they invented suitable forms of autonomy and intimacy. Through the activities of observing, experimenting, and representing, they offered probabilistic science and aesthetic realism as models of liberal

39. See Hacohen, *Karl Popper*, especially chapter 10.

sociability—as means of achieving consensus through disciplined, sensitive eyes and ears, not through dogmatic conformity.

It was thus within the domestic sphere that the Exners began to map the boundaries of the self alongside the norms of universality, and science was among the tools of this cartography. Note how the Exners inverted the ideal of "public science" of their compatriot Karl Popper. They saw in their intimate relations not a threat to their autonomy but a means of achieving a perspectival objectivity. They saw no divide between public reason and private affect.

By demoting the family from a training ground for public life to a retreat from civic responsibility, Popper's alignment of science and democracy against the intimate sphere undermined the authority of *Bildungsbürger* elites. In *The Open Society,* he argued that "concrete social groups," circles of family and friends, were unable to sustain a critical rationality. Yet his, youth in Vienna had shown him possibilities to the contrary. Perhaps he had in mind the reputations of Vienna's academic dynasties, names like Gomperz, Benndorf, Przibram, Lieben, or Exner; perhaps he thought back to the lively intellectual environment of his own childhood home.[40] In any case, the twentieth century's most ardent defender of science's "public" face was forced to recognize its sometimes private roots. And so he added a caveat: rationality could not take root in the private sphere, "with the exception," he wrote, "of some lucky family groups."[41]

40. Hacohen, *Karl Popper,* chapter 1.

41. Karl Popper, *The Open Society and Its Enemies* (New York, 1950), p. 170.

APPENDIX

An Exner-Frisch Family Tree

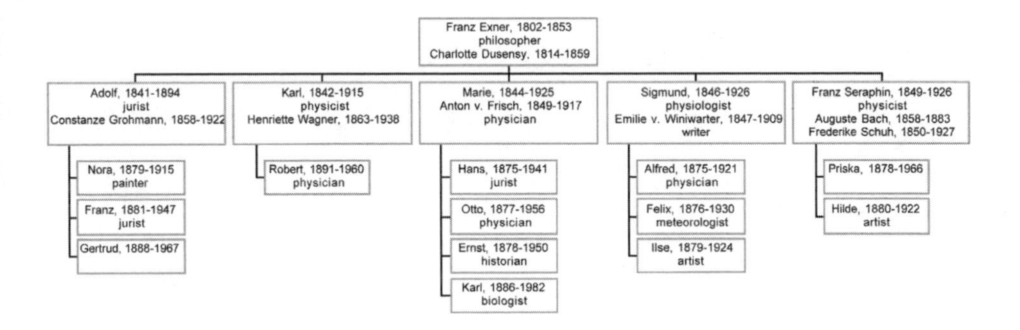

Bibliography

ARCHIVES

Akademisches Gymnasium, Vienna: Archiv.
Archive for the History of Quantum Physics (microfilm).
Institut für Geschichte der Medizin, Vienna: Bibliothek; Archiv.
Max-Planck-Gesellschaft, Dahlem, Germany: Nachlass F. A. Paneth.
Österreichische Akademie der Wissenschaften, Vienna: Aktenarchiv; Archiv des Instituts für Radiumforschung.
Österreichische Forschungstelle für Philosophie, Graz, Austria: Nachlass Alois Höfler.
Österreichische Nationalbibliothek, Vienna: Handschriftensammlung.
Österreichische Zentralbibliothek für Physik, Universität, Vienna: Bibliothek; Archiv.
Österreichisches Staatsarchiv, Vienna: Allgemeines Verwaltungsarchiv.
Preussischer Kulturbesitz, Staatsbibliothek, Berlin: Handschriftensammlung.
Private collection of Dr. Herwig Frisch, Vienna: Nachlass Franz Exner.
Private collection of Dr. P. C. Dijkgraaf, Den Helder, The Netherlands.
Richard-Wagner-Archiv, Bayreuth, Germany.
Stadt- und Landesarchiv, Vienna: Handschriftensammlung; Sammlung Wiesinger.
Universität für angewandte Kunst, Vienna: Sammlung Oskar Kokoschka; Archiv.
Universität, Vienna: Universitätsarchiv; Handschriftensammlung, Universitätsbibliothek.

PRIMARY SOURCES

Albert, Georg. *Ein Wort für das humanistische Gymnasium zur Erwiderung an Geheimrat Ostwald.* Leipzig and Vienna: Deuticke, 1908.

Baechtold, Jakob. *Gottfried Kellers Leben. Seine Briefe und Tagebücher.* 3 volumes. Berlin: Verlag W. Herz, 1897.

Bahr, Hermann. *Gegen Klimt.* Vienna: J. Eisenstein, 1903.

Beck v. Mannagetta, Paul Ritter. "Das österreichische Patentrecht." In *Das oesterreichische Staatswörterbuch.* Ed. E. Mischler and J. Ulbrich. Vienna: Alfred Hölder, 1896.

Benndorf, Hans. "Messungen des Potentialgefälles in Sibirien" ("Beiträge zur Kenntniss der atmosphärischen Elektricität II"). *Sitzungsberichte der Kaiserlichen Akademie der Wissenschaften zu Wien IIa* 108 (1899): 341–70.

———. "Beiträge zur Kenntnis der atmosphärischen Elektricität. VI and XXIII." *Sitzungsberichte der Kaiserlichen Akademie der Wissenschaften zu Wien IIa* 109 (1900): 923–40, 115 (1906): 425–56.

———. "Gedenkrede bei der Enthüllung von Exners Denkmal in der Wiener Universität." Vienna: self-published, 1937.

Billroth, Theodor. *Aphorismen zum Lehren und Lernen in der medizinischen Wissenschaften.* Vienna: C. Gerold's Sohn, 1886.

———. *Briefe von Theodor Billroth.* Eighth edition. Hannover and Leipzig: Hahsche Buchhandlung, 1910.

Bohm, David. *Causality and Chance in Modern Physics.* Philadelphia: University of Pennsylvania Press, 1957.

Boltzmann, Ludwig. "The Second Law of Thermodynamics." In *Theoretical Physics and Philosophical Problems: Selected Writings.* Ed. Brian McGuinness, pp. 13–32. Dordrecht: Reidel, 1974.

Brakel, J. van. "The Possible Influence of the Discovery of Radio-Active Decay on the Concept of Physical Probability." *Archives for the History of the Exact Sciences* 31 (1985): 369–85.

Brentano, Franz. *Ueber die Zukunft der Philosophie. Mit apologetisch-kritischer Berücksichtigung der Inaugurationsrede von Adolf Exner.* Vienna: Alfred Hölder, 1893.

Broglie, Louis de. *Une tentative d'interpretation causale et non-linéare de la mécanique ondulatoire.* Paris: Gauthier-Villars, 1956.

Brücke, Ernst. *Die Physiologie der Farben für die Zwecke der Kunstgewerbe auf Anregung der Direktion des kaiserlichen Oesterreichischen Museums für Kunst und Industrie.* Leipzig: Hirzel, 1866.

Carnap, Rudolf. *Logical Foundations of Probability.* Chicago: University of Chicago Press, 1950.

Chevalier, Ludwig. "Über den Unterricht in der philosophischen Propädeutik an österreichischen Gymnasien." *Jahresbericht des k.k. Staats-Untergymnasiums in Prag-Neustadt* 4 (1885): 3–39.

Die Mittelschule-Enquete im k.k. Ministerium für Kultus und Unterricht, Vienna 21–25 Januar 1908, Stenographisches Protokoll. Vienna: Alfred Hölder, 1908.

"Dr. Hans Przibram." *Österreichsiche Ex Libris Gesellschaft* 2 (1904): 8–10.

Ebner-Eschenbach, Marie von. "Exner, Emilie (Felicie Ewart)." In *Biographisches Jahrbuch und deutscher Nekrolog*, vol. 14, *Die Toten des Jahres 1909.* Ed. Anton Bettelheim, pp. 10–18. Berlin: Georg Reimer, 1912.

———. *Tagebücher.* Vol. 4 of *Kritische Texte und Deutungen.* Ed. K. K. Polheim and N. Gabriel. Tübingen: Max Niemeyer Verlag, 1995.

Ehrenfest, Paul and Tatyana. "Begriffliche Grundlagen der statistischen Auffassung in der Mechanik." *Encyclopädie der mathematischen Wissenschaften.* Vol. 4, pp. 3–90. Leipzig: Teubner, 1911.

Ehrenhaft, Felix. "Über die Messung von Elektrizätsmengen, die kleiner zu sein scheinen als die Ladung des einwertigen Wasserstoffions oder Elektrons oder von dessen Vielfachen abweichen." *Sitzungsberichte der Kaiserlichen Akademie der Wissenschaften zu Wien IIa* 119 (1910): 815–66.

Einstein, Albert. *Investigations on the Theory of the Brownian Movement.* Ed. R. fürth. 1926. New York: Dover, 1956.

Eisler, Max. *Österreichische Werkkultur.* Vienna: Kunstverlag Anton Schroll, 1916.

Exner, Adolf. *Die Lehre vom Rechtserwerb durch Tradition nach österreichischem und gemeinem Recht.* Vienna: Manz, 1867.

———. "Ueber Recht und Billigkeit." *Juristische Blätter* 3 (1874): 85–87, 105–7.

———. *Der Begriff der höheren Gewalt (vis major) im römischen und heutigen Verkehrsrecht.* 1883. Reprint, Darmstadt: Scientia Verlag Aalen, 1970.

———. Review of Rudolph Sohm, *Institutionen des römischen Rechts*. *Grünhuts Zeitschrift* 11 (1884): 622–25.

———. *Über politische Bildung*. Vienna: Adolf Holzhausen, 1892.

Exner, Emilie [Felicie Ewart, pseud.]. *Die Emancipation in der Ehe: Briefe an einen Arzt*. Hamburg: Voß, 1895.

———, ed. *Jugendschatz. Deutsche Dichtungen*. Vienna: R. v. Waldheim, 1897.

———. "Ein Flüchtling." *Velhagen und Klasings Monatshefte* 1, no. 6 (1898): 602–27.

———. *Goethe's Vater. Eine Studie*. 1899. Schutterwald: Wissenschaftler Verlag, 1999.

———. *Weibliche Pharmaceuten*. Vienna: E. Kainz & R. Liebhart, 1902.

———. *Der Brunnwinkl*. Vienna: self-published, 1906.

———. *Eine Abrechnung in der Frauenfrage*. Hamburg: L. Voss, 1906.

———. "Der Mittelschüler in Literatur und Wirklichkeit." *Österreichische Rundschau* 12 (1907): 123–30.

———. *Zwei Frauenbildnisse. Erinnerungen*. Vienna: self-published, 1908.

———. *Zur Kenntnis der Geschlechtsbestimmung beim Menschen*. Bonn: 1908.

Exner, Emilie [Felicie Ewart, pseud.], ed. *Jugendschatz. Deutsche Dichtungen*. Vienna: R. v. Waldheim, 1897.

Exner, Felix M. "Messungen der täglichen Temperaturschwankungen in verschiedenen Tiefen des Wolfgangsees." *Sitzungsberichte der Kaiserlichen Akademie der Wissenschaften zu Wien IIa* 111 (1900): 707–25.

———. "Notiz zu Brown's Molecularbewegung." *Sitzungsberichte der Kaiserlichen Akademie der Wissenschaften zu Wien IIa* 109 (1900): 843–47.

———. "Versuch einer Berechnung der Luftdruckänderungen von einem Tage zum nächsten." *Sitzungsberichte der Kaiserlichen Akademie der Wissenschaften zu Wien IIa* 111 (1902): 707–25.

———. "Grundzüge einer Theorie der synoptischen Luftdruckveränderungen." *Sitzungsberichte der Kaiserlichen Akademie der Wissenschaften zu Wien IIa* 115 (1906): 1171–1246.

———. "Über eigentümliche Temperaturschwankungen von eintätiger Periode im Wolfgangsee." *Sitzungsberichte der Kaiserlichen Akademie der Wissenschaften zu Wien IIa* 117 (1908): 9–26.

———. "Über eine erste Annäherung zur Vorauschberechnung synoptischer Wetterkarten." *Meteorologische Zeitschrift* 25 (1908): 57–67.

———. "Der Korrelationsfaktor und seine Verwendung in der Meteorologie." *Meteorologische Zeitschrift* 27 (1910): 263–66.

———. *Über die Korrelationsmethode*. Jena: Verlag Gustav Fischer, 1913.

———. "Über monatliche Witterungsanomalien auf der nördlichen Erdhälfte im Winter." *Sitzungsberichte der Kaiserlichen Akademie der Wissenschaften zu Wien IIa* 122 (1913): 1165–1241.

———. "Studien über die Ausbreitung kalter Luft auf der Erdoberfläche." *Sitzungsberichte der Kaiserlichen Akademie der Wissenschaften zu Wien Iia* 127 (1918): 795–847.

———. "Über natürliche Bewegungen in geraden und gewellten Linien." *Naturwissenschaftliche Wochenschrift* n.s. 19 (1920): 385–90.

———. "Zur Physik der Dünen." *Sitzungsberichte der Kaiserlichen Akademie der Wissenschaften zu Wien IIa* 129 (1920): 929–52.

———. "Dünen und Mäander, Wellenformen der festen Erdoberfläche, deren Wachstum und Bewegung." *Geografiska Annaler* 3 (1921): 327–35.

———. "Sind die Zyklonen Wellen in der Polarfront oder Durchbrüche derselben?" *Meteorologische Zeitschrift* 38 (1921): 21–23.

———. Review of Lewis F. Richardson, *Weather Prediction by Numerical Process*. *Meteorologische Zeitschrift* 40 (1923): 189–90.

———. *"Über ein Bewegungsprinzip in der Natur." Lecture delivered at the Akademie der Wissenschaften in Vienna on 30 May 1923*. Vienna: Österreichsiche Akademie der Wissenschaften, 1923.

———. "Monatliche Luftdruck- und Temperaturanomalien auf der Erde: Korrelationen des Luftdrucks auf Island mit dem anderer Orte." *Sitzungsberichte der Kaiserlichen Akademie der Wissenschaften zu Wien IIa* 133 (1924): 307–408.

———. *Dynamische Meteorologie*. Second edition. Vienna: Springer, 1925.

———. "Über die Zirkulation zwischen Rossbreiten und Pol." *Meteorologische Zeitschrift* 44 (1927): 46–53.

Exner, Franz S. *Die Psychologie der Hegelschen Schule*, 2 vols. Leipzig: Friedrich Fleischer, 1842–44.

———. *Leibniz Universal-Wissenschaft*. Prague: Borrosch & André, 1843.

———. "Entwurf der Organisation der Gymnasien und Realschulen in Oesterreich." 1849. Vienna: k.k. Schulbücher Verlag, 1875.

Exner, Franz Serafin. *Untersuchungen über die Härte an Krystallflächen*. Vienna: k.k. Hof- und Staatsdruckerei, 1873.

———. "Über transportable Apparate zur Beobachtung der atmosphärischen Elektrizität." *Repertorium der Physik* 23 (1886): 657–69.

———. "Über unsere Atmosphäre." In *Populäre Vorträge*, pp. 129–63. Vienna: Verein zur Verbreitung naturwissenschafticher Kenntnisse, 1891.

———. "Zur Characteristik der schönen und hässlichen Farben." *Sitzungsberichte der Kaiserlichen Akademie der Wissenschaften zu Wien IIa* 3 (1902): 901–21.

———. *Der Schlichten Astronomia*. Vienna: self-published, 1908.

———. *Über Gesetze in Naturwissenschaft und Humanistik*. Vienna: Hölder, 1909.

———. *Vorlesungen über die physikalischen Grundlagen der Naturwissenschaften*. Vienna: Franz Deuticke, 1919, second edition, 1922.

———. "Zur Erinnerung an Josef Loschmidt." *Die Naturwissenschaften* 9 (1921): 177–80.

Exner, Franz Serafin, and Sigmund Exner. "Die physikalischen Grundlagen der Blütenfarbungen." *Sitzungsberichte der Kaiserlichen Akademie der Wissenschaften zu Wien I* 119 (1910): 191–245.

Exner, Franz Serafin, and J. Tuma. "Studien zur chemischen Theorie des galvanischen Elementes." *Sitzungsberichte der Kaiserlichen Akademie der Wissenschaften zu Wien IIb* 97 (1889): 917–57.

———. "Ueber Quecksilber-Tropfelektroden." *Repertorium der Physik* 25 (1889): 597–614.

———. "Ueber Ostwald'sche Tropfenelektroden: Zweite Erwiderung." *Repertorium der Physik* 26 (1890): 91–101.

Exner, Franz. *Krieg und Kriminalität in Österreich*. Vienna: Holder-Pichler-Tempsky, 1927.

———. *Kriminalbiologie in ihren Grundzügen*. Hamburg: Hanseatische Verlagsanstalt, 1939.

———. *Einiges über die Exnerei*. Staltach: self-published, 1944.

Exner, Karl. "Über die Curven des Anklingens und des Abklingens der Lichtempfindungen." *Sitzungsberichte der Kaiserlichen Akademie der Wissenschaften zu Wien IIa* 62 (1870): 197–201.

———. "Über das Funkeln der Sterne und die Scintillation überhaupt." *Sitzungsberichte der Kaiserlichen Akademie der Wissenschaften zu Wien IIa* 84 (1881): 1038–81.

———. "Die neuen Instructionen für den Unterricht in der Physik." In *Stimmen über den österreichischen Gymnasiallehrplan vom 26. Mai 1884*. Ed. K. F. Kummer. Vienna: Carl Gerold's Sohn, 1886.

Exner, Robert. "Die Genesis der vier Kardinalfarben." *Psychiatrisch-Neurologische Wochenschrift* 40 (1938): 1–18.

Exner, Robert, and R. Routil. "Studien zur Farbentheorie." *Annalen des Naturhistorischen Museums in Wien* 57 (1949–50): 6–11.

Exner, Sigmund. "Untersuchungen über Brown's Molecularbewegung." *Sitzungsberichte der Kaiserlichen Akademie der Wissenschaften zu Wien II* 105 (1867): 116–23.

———. "Experimentelle Untersuchungen der einfachsten psychischen Processe." *Archiv für die gesammte Physiologie* 11 (1875): 403–32.

———. *Die Physiologie des Fliegens und Schwebens in den bildenden Künsten*. Vienna: Wm. Braumüller, 1882. Reprinted in Olaf Breidbach, ed., *Natur der Ästhetik, Ästhetik der Natur*. Vienna: Springer, 1997.

———. "Grosshirnrinde." In *Handbuch der Physiologie*. Ed. L. Hermann. Abt. IIa, *Allgemeine Physiologie*, p. 238. Leipzig: F. C. W. Vogel, 1879–83.

———. *Untersuchungen über die Localisation der Functionen in der Grosshirnrinde des Menschen*. Vienna: Braumüller, 1881.

———. *Physiologisches und Pathologisches in den bildenden Künsten*. Vienna: Selbstverlag des Vereines zur Verbreitung naturwissenschaftlicher Kenntnisse in Wien, 1889.

———. "Ueber allgemeine Denkfehler." *Deutsche Rundschau* 4 (1889): 54–67.

———. "Ernst v. Brücke und die moderne Physiologie." *Wiener klinische Wochenschrift* Nr. 3 (1890): 807–12.

———. *Die Physiologie der Facettirten Augen von Krebsen und Insecten, Eine Studie*. Leipzig and Vienna: Franz Deuticke, 1891.

———. "Die Moral als Waffe im Kampfe ums Dasein." *Almanach der Akademie der Wissenschaften* (1892): 243–73.

———. "Wilhelm Ritt. v. Hartel." *Wiener klinische Wochenschrift* Nr. 4 (1907).

———. *Studien auf dem Grenzgebiete des localisirten Sehens*. Bonn: Emil Strauss, 1898.

———. "Kennen, Können, und Erkenntnis in der ärtzlichen Kunst." *Wiener Zeitung* 241 (19 Oct. 1899).

———. "Physiologie der männlichen Geschlechtsfunktionen" (1903). Reprinted in *Handbuck der Urologie*, ed. Anton von Frisch and Otto Zuckerkandl. Vienna: Alfred Hoelder, 1903.

———. "Ein Versuch aus Goethes Farbenlehre und seine Erklärung." *Wiener Klinischer Wochenschrift* 25 (1912): 22–23.

———. *Entwurf zu einer physiologischen Erklärung der psychischen Erscheinungen*. 1894. Thun and Frankfurt: Verlag Harri Deutsch, 1999.

Fechner, Gustav Theodor. *Das Büchlein vom Leben nach dem Tode*. Third edition. Hamburg: Leopold Voss, 1887.

———. *Vorschule der Aesthetik*. Hildesheim: G. Olms, 1978.

Frank, Philipp. *The Law of Causality and Its Limits*. Ed. R. S. Cohen, trans. by M. Neurath and R. S. Cohen. Dordrecht: Kluwer, 1998.

Frankfurter, Solomon. *Graf Leo Thun-Hohenstein, Franz Exner und Hermann Bonitz. Beiträge zur österreichischen Unterrichtsreform*. Vienna: Hölder, 1893.

———. *Verlauf und Ergebnisse der Mittelschulenquete des Unterrichtsministeriums 21.-25. Jänner 1908 und andere Beiträge zur Geschichte der österreichischen Mittelschulreform*. Vienna: Carl Fromme, 1910.

Franzos, Karl Emil et al. "Die Suggestion und die Dichtung." *Deutsche Dichtung* 9 (1890–91): 71–130.

Freud, Sigmund. *Totem and Taboo: Some Points of Agreement between the Mental Lives of Savages and Neurotics*. Trans. by James Strachey. New York: Norton, 1950.

Frisch, Ernst. *Chronik von Brunnwinkl*. Vienna: self-published, 1906.

———. *Geschichte der Brunnwinklmühle 1615–1882*. Vienna: self-published, 1918.

———. *Sommer am Abersee*. Vienna: self-published, 1938.

Frisch, Hans. *50 Jahre Brunnwinkl*. Vienna: self-published, 1931.

———. *Die Gewaltherrschaft in Österreich, 1933 bis 1938. Eine Staatsrechtliche Untersuchung.* Leipzig and Vienna: Johannes Günther Verlag, 1938.
Frisch, Karl. "Der Farbensinn und Formensinn der Biene." *Zoologischer Jahrbücher* 35 (1914): 1–105.
———. *Aus dem Leben der Bienen*. Berlin: Springer, 1927.
———. *Du and Das Leben*. Berlin: Deutsche Verlag, 1936.
———. *Erinnerungen eines Biologen*. Berlin: Springer, 1973.
———. *fünf Häuser am See. Der Brunnwinkl: Werden und Wesen eines Sommersitzes*. Berlin: Springer, 1980.
fürth, Reinhold. *Schwankungserscheinungen in der Physik*. Braunschweig, 1920.
———. "Über einige Beziehungen zwischen klassischer Statistik und Quantenmechanik." *Zeitschrift für Physik* 81 (1933): 143–62.
Gimpl, Georg, ed. *Unter uns gesagt: Friedrich Jodls Briefe an Wilhelm Bolin*. Vienna: Löcker, 1990.
Goethe, Johann Wolfgang von. *Faust*. Trans. by Walter Arndt. New York: Norton, 1976.
Groos, Karl. *Die Spiele des Menschen*. Jena: Gustav Fischer, 1899.
Habermas, Jürgen. *The Structural Transformation of the Public Sphere*. Trans. Thomas Burger. Cambridge, Mass.: MIT Press, 1991.
Hainisch, Marianne. "Emilie Exner." *Der Bund* 4 (May 1909): 1–2.
Hanslick, Eduard. *Vom Musikalischen-Schönen: Ein Beitrag zur Revision der Aesthetik der Tonkunst.* Second edition. Leipzig: Weigel, 1858.
Hartel, Wilhelm von. *Festrede zur Enthüllung des Thun-Exner-Bonitz Denkmals*. Vienna: Verein deutscher Philologen und Schulmänner, 1893.
Haschek, Eduard. "Quantitative Beziehungen in der Farbenlehre." *Sitzungsberichte der Kaiserlichen Akademie der Wissenschaften zu Wien IIa* 136 (1927): 461–68.
Hatschek, Berthold. *Medicin, Naturwissenschaft und Gymnasialreform*. Prague: Calve, 1896.
Hayek, F. A. *The Fortunes of Liberalism: Essays on Austrian Economics and the Ideal of Freedom.* Ed. Peter G. Klein. London: Routledge, 1992.
Heilsberg, Alois. "Bemerkungen über modernen Betrieb des naturgeschichtlichen Unterrichtes am Gymnasium." In *Jahresbericht des k.k. Staatsgymnasiums im XIX. Bezirk von Wien für das Schuljahr 1902/3* (1903): 29–40.
Heilsberg, Franz. "Die Bedeutung des Zufallsbegriffes in der Geschichtswissenschaft." *XI. Programm der Kaiser Franz Josef-Staats-Realschule* (1908–9): 6–26.
Helmholtz, Hermann von. "Ueber Wirbelbewegungen." *Journal für die reine und angewandte Mathematik* 55 (1858): 25–55.
———. *Handbuch der physiologischen Optik*, 388–417. Leipzig: L. Voss, 1867.
———. *Science and Culture: Popular and Philosophical Essays*. Ed. David Cahan. Chicago: University of Chicago Press, 1995.
Herbart, J. F. *Systematische Pädagogik*, Bd. 2, *Interpretationen*. Ed. Dietrich Benner. Weinheim: Deutscher Studien Verlag, 1997.
Hering, Ewald. *Wissenschaftliche Abhandlungen*. Leipzig: Georg Thieme, 1931.
———. *Vier Reden*. Amsterdam: E. J. Bunset, 1969.
Hillebrand, Franz. "Purkinjesches Phänomen und Eigenhelligkeit." *Zeitschrift f. Sinnesphysiologie* 51 (1920): 46–95.
Höfler, Alois. *Zur Propadeutik-Frage*. Vienna: A. Hölder, 1884.
———. *Was die gegenwärtige Psychologie unserem Gymnasium sein und werden könnte*. Vienna: Deutsche Philologen und Schulmänner, 1893.
———. *Die neuen Instructionen für philosophische Propadeutik*. Linz: Feichtinger, 1901.
Höfler, Alois, and Alexius Meinong. *Philosophische Propadeutik*. 2 vols. Prague and Vienna: F. Tempsky, 1890.
Höfler, Alois, and S. Witasek. *Physiologische oder experimentelle Psychologie am Gymnasium?* Vienna: Alfred Hölder, 1898.

Hofmannsthal, Hugo von. "Preuße und Österreicher. Ein Schema." In *Gesammelte Werke. Reden und Aufsätze II (1914–1924)*. Ed. Bernd Schoeller, pp. 459–61. Frankfurt: Fischer Taschenbuch Verlag, 1979.

Hromoda, Adolf. *Briefe über den naturhistorischen Unterricht an der medicinischen Facultät und am Gymnasium*. Vienna: Carl Gerold's Sohn, 1897.

Hruza, Ernst. *Der romanistische Rechtsunterricht in Oesterreich*. Czernowitz: Pardini, 1886.

Humboldt, Alexander von. *Kosmos. Entwurf einer physicischen Weltbeschreibung*. Volume 1. Stuttgart: J. G. Cotta, 1874. Darmstadt: Wissenschaftliche Buchgesellschaft, 1993.

Huxley, Thomas H. "On the Method of Zadig: Retrospective Prophecy as a Function of Science." In *Science and Culture and Other Essays*, pp. 135–55. New York: D. Appelton and Co., 1893.

Itten, Johannes. *The Art of Color: The Subjective Experience and Objective Rationale of Color*. Trans. by Ernst van Haagen. New York: Reinhold, 1966.

Jerusalem, Wilhelm. "Zur Reform des Unterrichtes in der philosophischen Propädeutik." *Programm des Staats-Gymnasiums in Nikolsburg* 12 (1884–85): 3–32.

Jhering, Rudolf von. "Der Takt." Ed. Christian Helfer. *Nachrichten der Akademie der Wissenschaften in Göttingen* (1968): 73–98.

Jodl, Friedrich. *Das Problem des Moralunterrichts in der Schule: Zwei Vorträge*. Frankfurt am Main: Neuer Frankfurter Verlag, 1912.

———. *Vom Lebenswege: Gesammelte Vorträge und Aufsätze*. Ed. Wilhelm Börner. Stuttgart and Berlin: Cotta, 1916–17.

Jodl, Margarete. *Friedrich Jodl: Sein Leben und Wirken*. Stuttgart and Berlin: J. G. Cotta, 1920.

Kann, Robert A., ed. *Marie von Ebner-Eschenbach—Dr. Josef Breuer: Ein Briefwechsel, 1889–1916*. Vienna: Bergland, 1969.

Kant, Immanuel. *Foundations of the Metaphysics of Morals* and *What Is Enlightenment?* Trans. by Lewis White Beck. New York: Macmillan, 1990.

Kerber, Gabriele et al., eds. *Dokumente, Materialien, und Bilder zur 100. Wiederkehr des Geburtstages von Erwin Schrödinger*. Vienna: Fassbaender, 1987.

Köhler, Wolfgang. "Über unbemerkte Empfindungen und Urteilstäuschungen." *Zeitschrift für Psychologie* 66 (1913): 51–80.

Kohlrausch, K. W. F. "Der experimentelle Beweis für den statistischen Charakter des radioaktiven Zerfallsgesetzes." *Ergebnisse der exakten Naturwissenschaften* 5 (1926): 192–212.

Lehrbuch der Philosophie. 2 volumes. Vienna: k.k. Schulbücher-Verschleiß-Administration, 1835.

Leitmeier, Hans, ed. *Die österreichischen Alpen*. Leipzig and Vienna: Franz Deuticke, 1928.

Lewin, Kurt. "Kriegslandschaft." In *Kurt-Lewin-Werkausgabe*, vol. 4, *Feldtheorie*. Ed. Carl-Friedrich Graumann, pp. 315–26. Bern: Hans Huber, 1982.

Lichtenfels, Johann von. *Lehrbuch zur Einleitung in die Philosophie*. 2 volumes. Vienna: Wm Braumüller, 1852.

Lichtwark, Alfred. *Die Erziehung des Farbensinnes*. Berlin: Cassirer, 1901.

Loschmidt, Josef. "Zur Grösse der Luftmolecüle." *Sitzungsberichte der Kaiserlichen Akademie der Wissenschaften zu Wien II* 52 (1866): 395–407.

Löwe, Johann Heinrich. *Ueber den Unterricht in der philosophischen Propädeutik am Gymnasium*. Prague: A. G. Steinhauser, 1865.

Mach, Ernst. *Grundriß der Naturlehre für die oberen Classen der Mittelschulen. Ausgabe für Realschulen*. Vienna: F. Tempsky, 1891.

———. *Analyse der Empfindungen*. Second edition. Jena: G. Fischer, 1900.

Madelung, Erwin. "Quantentheorie in hydrodynamischer Form." *Zeitschrift für Physik* 40 (1926): 322–26.

Mahler, Margaret. *Selected Papers*. Vol. 2, *Separation and Individuation*. New York: Jason Aronson, 1979.

Maxwell, James Clerk. "Molecules." *Nature* (Sept. 25, 1873): 437–41.

Meinong, Alexius. *Über philosophische Wissenschaft und ihre Propädeutik*. Vienna: A. Hölder, 1885.

———. *Möglichkeit und Wahrscheinlichkeit*. Leipzig: Barth, 1915.

———. *Zum Erweise des allgemeinen Kausalgesetzes*. Vienna: A. Hölder, 1918.

Menger, Carl. "Die Eroberung der Universitäten." Reprinted in A. J. Peters, *"Klerikale Weltauffassung" und "Freie Forschung": Ein offenes Wort an Karl Menger*, pp. 7–9. Vienna: Georg Eichinger, 1907.

Meyer, Stefan, and Egon Schweidler. *Radioaktivität*. Leipzig and Berlin: Teubner, 1927.

Mill, J. S. *On Liberty*. Indianapolis: Hackett, 1978.

———. *The Subjection of Women*. Buffalo: Prometheus, 1986.

Mises, Richard von. *Wahrscheinlichkeit, Statistik, und Wahrheit*. Vienna: Springer, 1936.

Mittelschul-Enquete 1908 Referate und Korreferate. Vienna: Verlag des k.k. Ministeriums für Kultus und Unterricht, 1908.

Mozart, Josef. *Deutsches Lesebuch für die unteren Klassen der Gymnasien*. Third edition. Vienna: Carl Gerold & Sohn, 1859.

Musen Almanach der Hochschueler Wiens. Berlin and Leipzig: Heinrich Meyer, 1900.

Neurath, Otto. *International Picture Language: The Rules of Isotype*. London: Kegan Paul, 1936.

———. *Empiricism and Sociology*. Dordrecht: Reidel, 1973.

Ostwald, Wilhelm. "Über Tropfelektroden." *Zeitschrift für physikalische Chemie* 3 (1889): 354–58; 4 (1890): 570, 597–614.

———. *Naturwissenschaftliche Forderungen zur Mittelschulreform*. Vienna: Verein für Schulreform, 1908. Reprinted in *Die Forderung des Tages*, pp. 517–37. Leipzig: Akademische Verlagsgesellschaft, 1910.

———. *Goethe, Schopenhauer und die Farbenlehre*. Leipzig: Unesma, 1918.

———. *Farbenfibel*. Thirteenth edition. Leipzig: Unesma, 1928.

———. *Goethe der Prophete*. Leipzig: Hegner, 1932.

Pattai, Robert. *Reden und Gedanken*. Second edition. Vienna: self-published, 1909.

Popper, Karl. *The Open Society and Its Enemies*. New York, 1950.

Przibram, Karl. "Ladungsbestimmungen an Nebelteilchen II." *Sitzungsberichte der Kaiserlichen Akademie der Wissenschaften zu Wien IIa* 119 (1910): 1719–53.

Przibram, Karl, ed. and trans. *Letters on Wave Mechanics: Schrödinger, Planck, Einstein, Lorentz*. New York: Philosophical Library, 1967.

Reichenbach, Hans. *Experience and Prediction: An Analysis of the Foundations and the Structure of Knowledge*. Chicago: University of Chicago Press, 1938.

Riegl, Alois. *Altorientalische Teppiche*. Leipzig: T. O. Weigel Nachfolger, 1891.

———. *The Group Portraiture of Holland*. Trans. by E. M. Kain and D. Britt. Los Angeles: Getty, 1999.

Rückblick auf die ersten 25 Vereinsjahre der Philosophischen Gesellschaft an der Universität Wien. Vienna: Philosophische Gesellschaft der Universität Wien, 1913.

Saar, Ferdinand von. "Schloss Kostenitz." In Norbert Miller, *Camera Obscura*. Stuttgart: Mayer, 2000.

Schnitzler, Arthur. *Der Weg ins Freie*. Frankfurt: Insel Taschenbuch, 2002.

Schrödinger, Erwin. *Science, Theory, and Man*. 1935. New York: Dover, 1957.

———. *Gesammelte Abhandlungen. Collected Papers*. Vienna: Verlag der Österreichischen Akademie der Wissenschaften, 1984.

Schuster, P., and K. Kadletz. "Sechs Briefe Josef Loschmidts an Franz Serafin Exner (1840–1845)." In *Mitteilungen der Österreichische Gesellschaft für Wissenschaftsgeschichte* 14 (1994): 180–93.

Schweidler, Egon von. *Über Schwankungen der radioaktiven Umwandlung*. Liège, 1905.

Sitte, Camillo. "Über Farbenharmonie." *Centralblatt für das gewerbliche Unterrichtswesen in Österreich* 18 (1900): 196–227.

Smidt, Irmgard, ed. *Aus Gottfried Keller's Glücklicher Zeit: Der Dichter im Briefwechsel mit Marie und Adolf Exner*. Zurich: Th. Gut & Co., 1981.

Spengler, Oswald. *The Decline of the West*. Trans. by Charles Atkinson, abridged by Helmut Werner. Oxford: Oxford University Press, 1991.

Stahl, F. J. *Die Philosophie des Rechts*. Vol. 2, *Rechts- und Staatslehre auf der Grundlage christlichen Weltanschauung*. Second edition. Heidelberg: J. C. B. Mohr, 1845.

Steiner, Rudolf. *Nature's Open Secret: Introductions to Goethe's Scientific Writings*. Trans. by J. Barnes and M. Spiegler. Great Barrington, Mass.: Anthroposophic Press, 2000.

Stifter, Adalbert. *Der Nachsommer*. Frankfurt: Insel Taschenbuch, 1982.

Svevo, Italo. *Confessions of Zeno*. Trans. by Beryl de Zoete. New York: Vintage, 1989.

Systematischer Religions-Unterricht für Candidaten der Philosophie. 2 volumes. Vienna: k.k. Schulbücher-Verschleißes, 1821.

Unger, Josef. *Die Ehe in ihrer Welthistorischen Entwicklung: Ein Beitrag zur philosophie der Gechsichte*. Vienna, 1850.

———. *Das österreichische Erbrecht systematisch dargestellt*. Leipzig: Breitkopf und Härtel, 1864.

———. *Adolf Exner*. Vienna: k.k. Hof- & Universität Buchhändler, 1894.

Wahrmund, Ludwig. *Katholische Weltanschauung und freie Wissenschaft*. Munich: J. F. Lehmann, 1908.

Wiesner, J. *Die Nothwendigkeit der naturhistorischen Unterrichtes im medicinischen Studium, aus Anlass der bevorstehenden Reform der medicinischen Studien an den österreichischen Universitäten*. Vienna: Alfred Hölder, 1896.

Wilson, Francesca M. *A Lecture by Professor Cizek*. [N.p.]: Children's Art Exhibition Fund, 1921.

———. *The Child as Artist: Some Conversations with Professor Cizek*. [N.p.]: Children's Art Exhibition Fund, 1921.

Windelband, Wilhelm. *Die Lehren vom Zufall*. Berlin: A. W. Schade, 1870.

Winter, Eduard, ed. *Der Briefwechsel B. Bolzano's mit F. Exner*. Prague: Böhmische Gesellschaft der Wissenschaften, 1935.

Zimmermann, Robert. *Leibnitz und Herbart: Eine Vergleichung ihrer Monadologien*. Vienna: Braumüller, 1849.

———. *Philosophische PropaXdeutik*. Vienna: Wilhelm Braumüller, 1852, second edition, 1860, third edition, 1867.

Zweig, Stefan. *Die Welt von Gestern: Erinnerungen eines Europäers*. Frankfurt: Fischer, 2002.

SECONDARY SOURCES

Albisetti, James C. *Secondary School Reform in Imperial Germany*. Princeton: Princeton University Press, 1983.

Algazi, Gadi. "Scholars in Households: Refiguring the Learned Habitus, 1480–1550." *Science in Context* 16 (2003): 9–42.

Allison, Henry. *Kant's Transcendental Idealism*. New Haven: Yale, 1983.

Anderson, Harriet. *Utopian Feminism: Women's Movements in Fin-de-siècle Vienna*. New Haven: Yale University Press, 1992.

Anderson, Katharine. "The Weather Prophets: Science and Reputation in Victorian Meteorology." *History of Science* 37 (1999): 179–216.

Anderton, Keith. "The Limits of Science: A Social, Political, and Moral Agenda for Epistemology in Nineteenth-Century Germany." Ph.D. dissertation, Harvard University, 1993.

Arendt, Hannah. *The Human Condition*. Chicago: University of Chicago Press, 1998.

Arens, Katharine. *Structures of Knowing: Psychologies of the Nineteenth Century*. Dordrecht and Boston: Kluwer, 1988.

Ash, Mitchell. *Gestalt Psychology in German Culture, 1890–1967: Holism and the Quest for Objectivity*. Cambridge: Cambridge University Press, 1995.

Asman, Carrie. "Ornament and Motion: Science and Art in Gottfried Semper's Theory of Adornment." In *Natural History*. Ed. Jacques Herzog and Pierre de Meuron. Montreal: Canadian Centre for Architecture, 2002.

Auslander, Leora. *Taste and Power: Furnishing Modern France*. Berkeley: University of California Press, 1996.

Barkey, K., and M. von Hagen. *After Empire: Multiethnic Societies and Nation-Building*. Boulder, Colo.: Westview Press, 1997.

Bauer, Franz J. *Bürgerwege und Bürgerwelten: Familienbiographische Untersuchungen zum deutschen Bürgertum*. Göttingen: Vandenhoeck and Ruprecht, 1991.

Beller, Mara. "Experimental Accuracy, Operationalism, and the Limits of Knowledge, 1925–1935." *Science in Context* 2 (1988): 147–62.

Beller, Steven, ed. *Rethinking Vienna 1900*. New York: Berghahn Books, 2001.

Benjamin, Walter. *Illuminationen: Ausgewählte Schriften*. Frankfurt: Suhrkamp, 1977.

Benner, Dietrich. *Die Pädagogik Herbarts*. Weinheim: Deutscher Studien Verlag, 1986.

Berg, Christa et al., eds. *Handbuch der deutschen Bildungsgeschichte*. 6 volumes. Munich: C. H. Beck, 1987.

Bisanz-Prakken, Marian. *Heiliger Frühling: Gustav Klimt und die Anfänge der Wiener Secession, 1895–1905*. Vienna: Brandstätter, 1999.

Bittner, Lotte. "Geschichte des Studienfaches Physik an der Wiener Universität in der letzten Hundert Jahren." Ph.D. dissertation, University of Vienna, 1949.

Blackbourn, David. *The Long Nineteenth Century: A History of Germany, 1780–1918*. Oxford: Oxford University Press, 1997.

Blackbourn, David, and Geoff Eley. *The Peculiarities of German History: Bourgeois Society and Politics in Nineteenth-Century Germany*. Oxford: Oxford University Press, 1984.

Blackmore, John T. *Ernst Mach: His Work, Life, and Influence*. Berkeley: University of California Press, 1972.

———. *Ludwig Boltzmann: His Later Life and Philosophy, 1900–1906*. Vol. 2, *The Philosopher*. Boston Studies in the Philosophy of Science, 174. Dordrecht and Boston: Kluwer, 1995.

Blaukopf, Kurt. "Von der Ästhetik zur 'Zweigwissenschaft.' Robert Zimmermann als Vorläufer des Wiener Kreises." In *Kunst, Kunsttheorie und Kunstforschung im wissenschaftlichen Diskurs. In memoriam Kurt Blaukopf*. Ed. Martin Seiler und Friedrich Stadler, pp. 35–46. Vienna: Öbv & Hpt, 2000.

Bourdieu, Pierre. *Homo Academicus*. Trans. by Peter Collier. Stanford: Stanford University Press, 1984.

Boyer, John W. *Political Radicalism in Late Imperial Vienna: Origins of the Christian Social Movement, 1848–1897*. Chicago: University of Chicago Press, 1981.

———. *Culture and Political Crisis in Vienna: Christian Socialism in Power, 1897–1918*. Chicago: University of Chicago Press, 1995.

Breidbach, Olaf. "Bemerkungen zu Exners Physiologie des Fliegens und Schwebens." In *Natur der Ästhetik, Ästhetik der Natur*. Ed. Breidbach, pp. 221–23.Vienna: Springer, 1997.

———. *Die Materialisierung des Ichs. Zur Geschichte der Hirnforschung im 19. und 20. Jahrhundert*. Frankfurt: Suhrkamp, 1997.

Broman, Thomas. "Rethinking Professionalization: Theory, Practice, and Professional Ideology in Eighteenth-Century German Medicine." *Journal of Modern History* 67 (1995): 835–72.

Browne, Janet. *Charles Darwin: The Power of Place*. Princeton: Princeton University Press, 2002.

Bruckmüller, Ernst et al., eds. *Bürgertum in der Habsburgermonarchie*. Vienna: Böhlau, 1990.

Burks, Arthur W. "Peirce's Theory of Abduction." *Philosophy of Science* 13 (1946): 301–6.

Cahan, David. *An Institute for an Empire: The Physikalisch-Technische Reichsanstalt, 1871–1914*. Cambridge: Cambridge University Press, 1989.

———, ed. *Hermann von Helmholtz and the Foundations of Nineteenth-Century Science*. Berkeley: University of California Press, 1993.

Canales, Jimena. "Sensational Differences: Individuality in Observation, Experimentation and Representation (France 1853–1895)." Ph.D. dissertation, Harvard University, 2003.

Canales, Jimena, and Andrew Herscher. "Criminal Skins: Tattos, Criminal Anthropology and Modern Architecture in the Work of Adolf Loos." *Architectural History* 48 (2005): 235–56.

Canguilhem, Georges. *The Normal and the Pathological*. New York: Zone, 1991.

Christianson, J. R. *On Tycho's Island: Tycho Brahe and His Assistants*. Cambridge: Cambridge University Press, 2000.

Coen, Deborah R. "Determinism in Decay." M.Phil. dissertation, Cambridge University, 1998.

———. "Scientist's Errors, Nature's Fluctuations, and the 'Law' of Radioactive Decay, 1899–1926." *Historical Studies in the Physical and Biological Sciences* 32 (2002): 179–206.

———. "Living Precisely in Fin-de-Siècle Vienna." *Journal of the History of Biology* 39 (2006): 493–523.

Cohen, Gary B. *Education and Middle-Class Society in Imperial Austria, 1848–1918*. West Lafayette, Indiana: Purdue Universitiy Press, 1996.

Conze, Werner, and Jürgen Kocka, eds. *Bildungsbürgertum im 19. Jahrhundert*. Vol. 4. Stuttgart: Klett-Cotta, 1989.

Cooper, Alix. "Homes and Households." In *The Cambridge History of Science*. Volume 3, *Early Modern Science*. Ed. Katharine Park and Lorraine Daston, pp. 224–37. Cambridge: Cambridge University Press, 2006.

Craig, Gordon A. *The Triumph of Liberalism: Zürich in the Golden Age, 1830–1969*. New York: Collier, 1988.

Crary, Jonathan. *Techniques of the Observer*. Cambridge, Mass.: MIT Press, 1990.

———. *Suspensions of Perception: Attention, Spectacle, and Modern Culture*. Cambridge, Mass.: MIT, 1999.

Daston, Lorraine. *Classical Probability in the Enlightenment*. Princeton: Princeton University Press, 1988.

Daston, Lorraine, and Peter Galison. "The Image of Objectivity." *Representations* 40 (1992): 81–128.

Daston, Lorraine, and Otto Sibum, eds. "Scientific Personae." *Science in Context* 16 (2003).

Daston, Lorraine, and Fernando Vidal, eds. *The Moral Authority of Nature*. Chicago: University of Chicago Press, 2004.

Daum, Andreas. *Wissenschaftspopularisierung im 19. Jahrhundert: Bürgerliche Kultur, naturwissenschaftliche Bildung, und die deutsche Öffentlichkeit, 1848–1914*. Munich: R. Oldenbourg, 1998.

Davidoff, Leonore, and Catherine Hall. *Family Fortunes: Men and Women of the English Middle Class, 1780–1850*. London: Hutchinson, 1987.

Desrosières, Alain. *The Politics of Large Numbers*. Cambridge, Mass.: Harvard University Press, 1998.

Döcker, Ulrike. "Bürgerlichkeit und Kultur." In *Bürgertum in der Habsburgermonarchie*, vol 1. Ed. Ernst Bruckmüller et al. Vienna: Böhlau, 1990.

Doser, Barbara. "Das Frauenkunststudium in Österreich, 1870–1935." Ph.D. dissertation, Leopold-Franz-Universität, Innsbrück, 1988.

Dülmen, Andrea van. *Das irdische Paradies: Bürgerliche Gartenkultur der Goethezeit*. Cologne: Böhlau, 1999.

Eduard Suess, Forscher und Politiker. Horn: Österreichische Geologische Gesellschaft, 1981.

Ellenberger, Henri F. *The Discovery of the Unconscious*. New York: Basic Books, 1970.

Engelbrecht, Helmut. *Geschichte des österreichischen Bildungswesens*, vols. 3–4. Vienna: Österreichische Bundesverlag, 1986.

Fahr-Becker, Gabriele. *Wiener Werkstaette, 1903–1932*. Cologne: Taschen, 1995.

Felt, Ulrike. "'Öffentliche' Wissenschaft: Zur Beziehung von Naturwissenschaften und Gesellschaft in Wien von der Jahrhundertwende bis zum Ende der Ersten Republik." *Österreichische Zeitschrift für Geschichtswissenschaften* 7 (1996): 45–66.

Ferry, Luc. *Homo Aestheticus: The Invention of Taste in the Democratic Age*. Trans. by Robert de Loaiza. Chicago: Chicago University Press, 1993.

Fliedl, Gottfried, ed. *Kunst und Lehre am Beginn der Moderne: Die Wiener Kunstgewerbeschule, 1867–1918*. Salzburg and Vienna: Residenz Verlag, 1986.

Föger, Benedikt, and Klaus Taschwer. *Die Andere Seite des Spiegels: Konrad Lorenz und der Nationalsozialismus*. Vienna: Czernin, 2001.

Forman, Paul. "Weimar Culture, Causality, and Quantum Theory, 1918–1927: Adaptation by German Physicists and Mathematicians to a Hostile Intellectual Environment." *Historical Studies in the Physical Sciences* 3 (1971): 1–116.

Franz, Georg. *Liberalismus. Die deutschliberale Bewegung in der Habsburgischen Monarchie*. Munich: Georg Callwey, 1956.

Franz, Roland. "Stilvermeidung and Naturnachahmung. Ernst Haeckels 'Kunstformen der Natur' und ihr Einfluß auf die Ornamentik des Jugendstils in Österreich." *Stapfia* 56 (1998): 475–80.

Freier, Elke, and Walter Reineke, eds. *Karl Richard Lepsius (1810–1884)*. Berlin: Akademie-Verlag, 1988.

Frevert, Ute. *Men of Honour: A Social and Cultural History of the Duel*. Trans. by Anthony Williams. Cambridge: Polity Press, 1995.

Friedman, Robert Marc. *Appropriating the Weather: Vilhelm Bjerknes and the Construction of a Modern Meteorology*. Ithaca: Cornell University Press, 1989.

Friedrich, M. "Versorgungsfall Frau." *Jahrbuch des Vereins für die Geschichte der Stadt Wien* 47–48 (1991–92): 263–308.

Galison, Peter. *How Experiments End*. Chicago: University of Chicago Press, 1987.

———. "Aufbau/Bauhaus: Logical Positivism and Architectural Modernism." *Critical Inquiry* 17 (1990): 709–52.

———. *Image and Logic: A Material Culture of Microphysics*. Chicago: University of Chicago Press, 1997.

———. *Einstein's Clocks, Poincaré's Maps, Empires of Time*. New York: Norton, 2003.

Gall, Lothar. *Bürgertum in Deutschland*. Berlin: Siedler, 1989.

Gay, Peter. *Freud: A Life for Our Time*. New York: Norton, 1998.

Geehr, Richard S. *Karl Lueger: Mayor of Fin-de-Siècle Vienna*. Detroit: Wayne State University Press, 1990.

Geimer, Peter. "Das Gewicht der Engel. Eine Physiologie des Unmöglichen." In *Kultur im Experiment*. Ed. Henning Schmidgen, Peter Geimer, and Sven Dierig, pp. 170–90. Berlin: Kadmos Kulturverlag, 2004.

Gigerenzer, Gerd et al., eds. *The Empire of Chance: How Probability Changed Science and Everyday Life*. Cambridge: Cambridge University Press, 1989.

Gigerenzer, Gerd, and D. J. Murray. *Cognition as Intuitive Statistics*. Hillsdale, N.J.: Lawrence Erlbaum Associates, 1987.

Gillispie, Charles C. *Science and Polity in France: The Revolutionary and Napoleonic Years*. Princeton: Princeton University Press, 2004.

Ginzburg, Carlo. *Clues, Myths, and the Historical Method*. Trans. by J. and A. Tedeschi. Baltimore: Johns Hopkins University Press, 1989.

Goldstein, Catherine. "Mathematik im Frankreich des frühen 17. Jahrhunderts." In *Zwischen Vorderbühne und Hinterbühne: Beiträge zum Wandel der Geschlechterbeziehungen in der Wissenschaft vom 17. Jahrhundert bis zur Gegenwart*. Ed. Theresa Wobbe, pp. 41–72. Bielefeld: transcript Verlag, 2003.

Goldstein, Jan. "The Hysteria Diagnosis and the Politics of Anticlericalism in Late Nineteenth-Century France." *Journal of Modern History* 54 (1982): 209–39.

Goldstein, Jan, ed. *Foucault and the Writing of History*. Oxford: Blackwell, 1994.

Golinski, Jan. *Science as Public Culture: Chemistry and Enlightenment in Britain, 1760–1820*. Cambridge: Cambridge University Press, 1992.

Goller, Peter. *Die Lehrkanzeln für Philosophie an der philosophischen Fakultät der Universität Innsbruck, 1848–1945*. Innsbruck: Kommissionsverlag der Wagner'schen Kommissionsbuchhandlung, 1989.

Haas, Hannes. "Der Traum von Dazugehören—Juden auf Sommerfrische." In *Der Geschmack der Vergänglichkeit: Jüdische Sommerfrische in Salzburg*. Ed. Kriechbaumer, pp. 41–58. Vienna: Böhlau, 2002.

———. "Die Sommerfrische—Ort der Bürgerlichkeit." In *"Durch Arbeit, Besitz, Wissen und Gerechtigkeit."* Ed. Hannes Stekl et al., pp. 364–77. Bürgertum in der Habsburgermonarchie, 2. Vienna: Böhlau, 1992.

Habermas, Rebekka. *Frauen und Männer des Bürgertums: Eine Familiengeschichte (1750–1850)*. Göttingen: Vandenhoek & Ruprecht, 2000.

Hacking, Ian. *The Taming of Chance*. Cambridge: Cambridge University Press, 1990.

Hacohen, Malachi. "Karl Popper, the Vienna Circle and Red Vienna." *Journal of the History of Ideas* 59 (1998): 711–50.

———. *Karl Popper, the Formative Years, 1902–1945: Politics and Philosophy in Interwar Vienna*. Cambridge: Cambridge University Press, 2000.

Hagner, Michael. *Homo cerebralis: Der Wandel vom Seelenorgan zum Gehirn*. Berlin: Berlin Verlag, 1997.

Hajós, Géza. "Die 'Verhüttelung' der Landschaft—Beiträge zum Problem Villa und Einfamilienhaus seit dem 18. Jahrhundert." In *Landhaus und Villa in Niederösterreich, 1840–1914*. Ed. Österreichische Gesellscahft für Denkmal- und Orstbildpflege, pp. 9–56. Vienna: Böhlau, 1982.

Gruber, Helmut. *Red Vienna: Experiment in Working-Class Culture, 1919–1934*. Oxford: Oxford University Press, 1991.

Hacking, Ian. *The Emergence of Probability: A Philosophical Study of Early Ideas about Probability, Induction, and Statistical Inference*. London and New York: Cambridge University Press, 1975.

Hammerl, Christa et al., eds. *Die Zentralanstalt für Meteorologie und Geodynamik, 1851–2001*. Graz: Leykam, 2001.

Hanle, Paul. "Erwin Schrödinger's Statistical Mechanics." Ph.D. dissertation, Yale University, 1975.

———. "Indeterminacy before Heisenberg: The Case of Franz Exner and Erwin Schrödinger." *Historical Studies in the Physical Sciences* 10 (1980): 225–69.

Hannaway, Owen. "Laboratory Design and the Aim of Science: Andreas Libavius versus Tycho Brahe." *Isis* 77 (1986): 585–610.

Hansel, Karl, and Ingeborg Mauer. "Paul Krais, Wilhelm Ostwald und die Werkstelle für Farbkunde in Dresden." *Wissenschaftliche Zeitschrift der Technischen Universität Dresden* 49 (2000): 41–44.

Harrington, Anne. *Reenchanted Science: Holism in German Culture from Wilhelm II to Hitler*. Princeton: Princeton University Press, 1996.

Healy, Maureen. *Vienna and the Fall of the Habsburg Empire: Total War and Everyday Life in World War I*. Cambridge: Cambridge University Press, 2004.

Heidelberger, Michael. *Die Innere Seite der Natur: Gustav Theodor Fechners wissenschaftlich-philosophische Weltauffassung*. Frankfurt am Main: Klostermann, 1993.

Heilbron, John L. *The Dilemmas of an Upright Man: Max Planck and the Fortunes of German Science*. Cambridge, Mass.: Harvard University Press, 1996.

Hirschmüller, Albrecht. *Physiologie und Psychoanalyse in Leben und Werk Josef Breuers*. Bern: Huber, 1978.

Höbelt, Lothar. *Kornblume und Kaiseradler, Die deutschfreiheitlichen Parteien Altösterreichs, 1882–1918*. Vienna: Verlag für Geschichte und Politik, 1993.

Hoffmann, Christoph. "The Design of Disturbance: Physics Institutes and Physics Research in Germany, 1870–1910." *Perspectives in Science* 9 (2001): 173–92.

Höflechner, Walter. *Die Baumeister des kunftigen Glücks: Fragment einer Geschichte des Hochschulwesens in Oesterreich vom Ausgang des 19. Jahrhunderts bis in das Jahr 1938*. Graz: Akademische Druck- und Verlagsanstalt, 1988.

Hofer, Veronika. "Physiology Gains Space: On the Meaning of Sigmund Exner's Founding of the Phonogrammarchiv." Paper delivered at the History of Science Society annual meeting, Denver, Colo., 2001.

Hofmeister, Herbert. "Jhering in Wien." In *Rudolf von Jhering, Beiträge und Zeugnisse*. Ed. Okko Behrends, pp. 38–48. Göttingen: Wallstein Verlag, 1992.

Holton, Gerald. "Subelectrons, Presuppositions, and the Millikan-Ehrenhaft Dispute." In *The Scientific Imagination: Case Studies*. Ed. Gerald Holton, pp. 25–83. New York: Cambridge University Press, 1978.

Hon, Giora. "On the Concept of Experimental Error." Ph.D. dissertation, Cambridge University, 1985.

———. "Is the Identification of Experimental Error Contextually Dependent?" In *Scientific Practice*. Ed. Jed Buchwald, pp. 170–223. Chicago: University of Chicago Press, 1995.

Hörmann, Marianne. *Vally Wieselthier, 1895–1945*. Vienna: Böhlau, 1999.

Höttinger, Matthias. "Der Fall Wahrmund." Ph.D. dissertation, University of Vienna, 1949.

Jackson, Myles. "A Spectrum of Belief: Goethe's 'Republic' versus Newtonian 'Despotism.'" *Social Studies of Science* 24 (1994): 673–701.

Jagdzeit: Österreichs Jagdgeschichte. Eine Pirsch. Vienna: Historisches Museum der Stadt Wien, 1997.

Janik, Allan, and Stephen Toulmin. *Wittgenstein's Vienna*. New York: Touchstone, 1973.

Jansen, Sarah. "An American Insect in Imperial Germany: Visibility and Control in Making the Phylloxera in Germany, 1870–1914." *Science in Context* 13 (2000): 31–70.

Jean, Paul. *Levana*. Second edition. Stuttgart and Tübingen: J. G. Cotta, 1845.

Jelavich, Barbara. *Modern Austria: Empire and Republic, 1815–1986*. Cambridge: Cambridge University Press, 1987.

Johnston, William M. *The Austrian Mind: An Intellectual and Social History, 1848–1938*. Berkeley: University of California Press, 1972.

Jones, Ernest. *The Life and Work of Sigmund Freud*. New York: Basic, 1953.

Judson, Pieter. *Exclusive Revolutionaries: Liberal Experience, Social Politics, and National Identity in the Austrian Empire, 1848–1914*. Ann Arbor: University of Michigan Press, 1996.

———. *Wien Brennt! Die Revolution von 1848 und ihr liberales Erbe*. Vienna: Böhlau, 1998.

Jungnickel, Christa, and Russell McCormmach. *Intellectual Mastery of Nature: Theoretical Physics from Ohm to Einstein*. Two volumes. Chicago: University of Chicago Press, 1986.

Jurkowitz, Edward. "Helmholtz and the Liberal Unification of Science." *Historical Studies in the Physical and Biological Sciences* 32 (2002): 291–317.

Karlik, Berta, and Erich Schmid. *Franz Serafin Exner und sein Kreis*. Vienna: Österreichische Akademie der Wissenschaften, 1982.

Kern, Stephen. *The Culture of Time and Space: 1880–1918*. Cambridge, Mass.: Harvard University Press, 1983.

Kohler, Robert. *Landscapes and Labscapes: Exploring the Lab-Field Border in Biology*. Chicago: University of Chicago Press, 2002.

Koller, Gabrielle. "Die Kunstgewerbeschule des K.K. Österreichischen Museum für Kunst und Industrie, Wien, 1899–1905." Ph.D. dissertation, University of Vienna, 1983.

Konau, Ernst. *Rastlos zieht die Flucht der Jahre—: Josephine und Franziska von Wertheimstein, Ferdinand von Saar*. Vienna: Böhlau, 1997.

Kos, Wolfgang, and Elke Krasny, eds. *Schreibtisch mit Aussicht: Österrreichische Schriftsteller auf Sommerfrische*. Vienna: Verlag Carl Ueberreuter, 1995.

Koselleck, Reinhart. *The Practice of Conceptual History: Timing History, Spacing Concepts*. Stanford: Stanford University Press, 2002.

Kremer, Richard L. "Innovation through Synthesis: Helmholtz and Color Research." In *Hermann von Helmholtz and the Foundations of Nineteenth-Century Science*. Ed. David Cahan, pp. 205–58. Berkeley: University of California Press, 1993.

Kreidl, Alois. "Siegmund Exner zum 70. Geburtstag." *Wiener klinische Wochenschrift* 29 (1916): 426–27.

Kriechbaumer, Robert, ed. *Der Geschmack der Vergänglichkeit: Jüdische Sommerfrische in Salzburg*. Vienna: Böhlau, 2002.

Kriesl, Hans Max. *Gottfried Keller als Politiker*. Frauenfeld and Leipzig: Verlag Huber, 1918.

Krüger, Lorenz et al., eds. *The Probabilistic Revolution*. Two volumes. Cambridge, Mass.: MIT Press, 1987.

Kuklick, Henrika and Robert Kohler, eds. "Science in the Field." *Osiris* 11 (1996).

Kutzbach, Gisela. *The Thermal Theory of Cyclones: A History of Meteorological Thought in the Nineteenth Century*. Boston: American Meteorological Society, 1979.

Lacoue-Labarthe, P., and J.-L. Nancy. *The Literary Absolute*. Trans. by Bernard and Lester. Albany: State University of New York Press, 1988.

Landes, Joan B., ed. *Feminism, the Public and the Private*. Oxford: Oxford University Press, 1998.

Langewiesche, Dieter. *Liberalism in Germany*. Trans. by Christiane Banerji. Princeton: Princeton University Press, 2000.

Latour, Bruno. *Science in Action*. Cambridge, Mass: Harvard University Press, 1987.

Lattmann, Urs Peter, and Peter Metz. *Bilden und Erziehen*. Aarau: Sauerländer, 1995.

Lauscher, Friedrich, and Georg Skoda. "Zum Gedenken an Felix M. Exner." *Wetter und Leben* 33 (1981): 94–102.

Law, John, and Michael Lynch. "Lists, Field Guides, and the Descriptive Organization of Seeing: Birdwatching as an Exemplary Observational Activity." In *Representation in Scientific Practice*. Ed. Michael Lynch and Steve Woolgar, pp. 267–300. Cambridge, Mass.: MIT Press, 1990.

Lawrence, Christopher. "Incommunicable Knowledge: Science, Technology, and the Clinical Art in Britain, 1850–1914." *Journal of Contemporary History* 20 (1985): 503–20.

Leed, Eric J. *No Man's Land: Combat and Identity in World War I*. Cambridge: Cambridge University Press, 1979.

Lenger, Friedrich. *Werner Sombart, 1863–1941: Eine Biographie*. Munich: C. H. Beck, 1994.

Lenoir, Timothy. "The Eye as Mathematician." In *Hermann Helmholtz: Philosopher and Scientist*. Ed. David Cahan, pp. 109–53. Berkeley: University of California Press, 1993.

———. "The Politics of Vision: Optics, Painting, and Ideology in Germany, 1845–95." In *Instituting Science: The Cultural Production of Scientific Disciplines*. Stanford: Stanford University Press, 1997.

Lentze, Hans. "Graf Thun und die voraussetzungslose Wissenschaft." In *Festschrift Karl Eder*. Ed. Helmut Mezler-Andelberg. Innsbruck: Universitätsverlag, 1959.

———. *Die Universitätsreform des Ministers Graf Leo Thun-Hohenstein*. Vienna: Böhlau, 1962.

Lesky, Erna. *The Vienna Medical School of the Nineteenth Century*. Baltimore: Johns Hopkins University Press, 1976.

Lindenfeld, David F. *The Transformation of Positivism: Alexius Meinong and European Thought, 1880–1920*. Berkeley: University of California Press, 1980.

Loewenberg, Peter. "The Creation of a Scientific Community: The Burghölzi, 1902–1914." In *Fantasy and Reality in History*, pp. 46–89. New York: Oxford University Press, 1995.

MacKenzie, John M. *The Empire of Nature: Hunting, Conservation and British Imperialism*. Manchester: Manchester University Press, 1988.

Maisel, Thomas. *Alma Mater auf den Barrikaden*. Vienna: Universitätsverlag, 1998.

Malgrave, H. F. *Gottfried Semper, Architect of the Nineteenth Century*. New Haven: Yale University Press, 1996.

Manuel, Frank. *Utopian Thinking in the West*. Cambridge, Mass.: Harvard University Press, 1979.

———, ed. *Utopias and Utopian Thought*. Beacon: Boston, 1965.

McGrath, William G. *Dionysian Art and Populist Politics in Austria*. New Haven: Yale University Press, 1974.

McNeely, Ian F. *"Medicine on a Grand Scale": Rudolf Virchow, Liberalism, and the Public Health*. London: Wellcome Trust Centre for the History of Medicine at University College London, 2002.

Meehan, Johanna, ed. *Feminists Read Habermas: Gendering the Subject of Discourse*. New York: Routledge, 1995.

Michler, Werner. *Darwinismus und Literatur: Naturwissenschaftliche und literarische Intelligenz in Österreich, 1859–1914*. Vienna: Böhlau, 1999.

Miller, Norbert. *Camera Obscura: Schloss Kostenitz*. Stuttgart: Mayer, 2000.

Mönninger, Michael. *Vom Ornament zum Nationalkunstwerk: Zur Kunst und Architekturtheorie Camillo Sittes*. Braunschweig and Wiesbaden: Seweg 1998.

Mooney, Chris. *The Republican War on Science*. New York: Basic Books, 2005.

Mosse, George L. *The Crisis of German Ideology*. New York: Schocken, 1981.

Muhlack, Herbert. "Bildung zwischen Neuhumanismus und Historismus." In *Bildungsbürgertum im 19. Jahrhundert. Teil II: Bildungsgüter und Bildungswissen*. Ed. Reinhart Koselleck, pp. 80–105. Industrielle Welt, Bd. 41. Stuttgart: Klett-Cotta, 1990.

Munz, Tania. "The Bee Battles: Karl von Frisch, Adrian Wenner, and the Honey Bee Dance Language Controversy." *Journal of the History of Biology* 38 (2005): 535–70.

Nebeker, Frederik. *Calculating the Weather*. San Diego: Academic Press, 1995.

Nemeth, Elisabeth, and Friedrich Stadler, eds. *Encyclopedia and Utopia: The Life and Work of Otto Neurath (1882–1945)*. Dordrecht and Boston: Kluwer, 1996.

Nye, Mary Jo. *Molecular Reality: A Perspective on the Scientific Work of Jean Perrin*. New York: American Elsevier, 1972.

———. "Aristocratic Culture and the Pursuit of Science: The De Broglies in Modern France." *Isis* 88 (1997): 397–421.

Odelberg, Wilhelm, ed. *Les Prix Nobel en 1973*. Stockholm: Nobel Prize Foundation, 1974.

Ogris, Werner. "Die historische Schule der österreichischen Zivilistik." In *Festschrift Hans Lentze*. Ed. Nikolaus Grauss and Werner Ogris, pp. 449–96. Innsbruck and Munich: Universitätsverlag Wagner, 1969.

———. "Die Entwicklung der österreichischen Privatrechtswissenschaft im 19. Jahrhundert." In *Handbuch der Quellen und Literatur der neueren europäischen Privatrechtsgeschichte*. Ed. Helmut Coing. Munich: Beck, 1973.

Olesko, Kathryn M. *Physics as a Calling: Discipline and Practice in the Königsberg Seminar for Physics*. Ithaca, N.Y.: Cornell University Press, 1991.

Olin, Margaret. *Forms of Representation in Alois Riegl's Theory of Art*. University Park: Pennsylvania State University Press, 1992.

Österreichische Gesellschaft für Denkmal- und Ortsbildpflege. *Landhaus und Villa in Niederösterreich, 1840–1914*. Vienna: Böhlau, 1982.

Otis, Laura. *Organic Memory: History and the Body in the Late Nineteenth and Early Twentieth Centuries*. Lincoln: University Nebraska Press, 1994.

Palme, Johanna. *Sommerfrische des Geistes: Wissenschaftler im Ausseerland*. Bad Aussee: Alpenpost, 1999.

Patka, Erika, ed. *Kunst—Anspruch und Gegenstand: Von der Kunstgewerbeschule zur Hochscule für angewandte Kunst in Wien, 1918–1991*. Vienna: Residenz Verlag, 1991.

Penck, Albrecht. *Friedrich Simony: Leben und Wirken eines Alpenforschers. Ein Beitrag zur Geschichte der Geologie in Österreich*. Vienna: Hölzel, 1898.

Perrot, Michelle, ed. *A History of Private Life*. Trans. by Arthur Goldhammer. Vol. 4. Cambridge, Mass.: Harvard University Press, 1990.

Pichler, Dietlind. *Bürgertum und Protestantismus: Die Geschichte der Familie Ludwig in Wien und Oberösterreich (1860–1900)*. Bürgertum in der Habsburgermonarchie, 10. Vienna: Böhlau, 2003.

Plakolm-Forsthuber, Sabine. *Künstlerinnen in Österreich, 1897–1938: Malerei, Plastik, Architektur*. Vienna: Picus, 1994.

Plato, Jan von. *Creating Modern Probability: Its Mathematics, Physics, and Philosophy in Historical Perspective*. Cambridge: Cambridge University Press, 1994.

Pollack, Henry N. *Uncertain Science . . . Uncertain World*. Cambridge University Press, 2003.

Porter, Theodore M. *The Rise of Statistical Thinking, 1820–1900*. Princeton: Princeton University Press, 1986.

Pycior, H. M., et al., eds. *Creative Couples in the Sciences*. New Brunswick: Rutgers University Press, 1996.

Pynsert, Robert B., ed. *Decadence and Innovation: Austro-Hungarian Life and Art at the Turn of the Century*. London: Weidenfeld and Nicolson, 1989.

Rabinbach, Anson. *The Crisis of Austrian Socialism: From Red Vienna to Civil War, 1927–1934*. Chicago: University of Chicago Press, 1983.

———. *The Human Motor: Energy, Fatigue, and the Origins of Modernity*. Berkeley: University of California Press, 1992.

Ramhardter, Günther. *Geschichtswissenschaft und Patriotismus: Österreichische Historiker im Weltkrieg 1914–1918*. Munich: Oldenbourg, 1973.

Reiter, Wolfgang L. "Zerstört und vergessen: Die Biologische Versuchsanstalt und ihre Wissenschaftler/innen." *Österreichische Zeitschrift für Geschichte* 10 (1999): 585–614.

Richards, Joan. *Mathematical Visions: The Pursuit of Geometry in Victorian England*. Boston: Academic Press, 1988.

Richards, Robert. *The Romantic Conception of Life: Science and Philosophy in the Age of Goethe* Chicago: University of Chicago Press, 2002.

Ringer, Fritz. *The Decline of the German Mandarins*. Cambridge, Mass.: Harvard University Press, 1969.

Rinofner, Sonja, ed. *Zwischen Orientierung und Krise. Zum Umgang mit Wissen in der Moderne*. Vienna: Böhlau, 1998.

Ritter, Harry. "Austrian-German Liberalism and the Modern Liberal Tradition." *German Studies Review* 7 (1984): 227–48.

Rosner, Robert W. *Chemie in Österreich, 1740–1914: Lehre—Forschung—Industrie*. Vienna: Böhlau, 2004.

Rossbacher, Karlheinz. *Literatur und Liberalismus. Zur Kultur der Ringstrassenzeit in Wien.* Vienna: J&V, 1992.

———. *Literatur und Bürgertum. fünf Wiener jüdische Familien von der liberalen Ära bis zum Fin de Siècle.* Vienna: Böhlau, 2003.

Schaffer, Simon. "Astronomers Mark Time." *Science in Context* 2 (1988): 115–45.

———. "Physics Laboratories and the Victorian Country House." In *Making Space for Science: Territorial Themes in the Shaping of Knowledge.* Ed. Crosbie Smith and Jon Agar, pp. 149–80. New York: St. Martin's Press, 1998.

———. "On Astronomical Drawing." In *Picturing Science, Producing Art.* Ed. Peter Galison and Caroline Jones, pp. 441–74. New York: Routledge, 1998.

Schorske, Carl E. *Fin-de-Siècle Vienna: Politics and Culture.* New York: Vintage, 1981.

Schweiger, Werner J. *Wiener Werkstätte: Design in Vienna, 1903–1932.* New York: Abbeville Press, 1984.

Schwinges, Rainer C., ed., *Humboldt International. Der Export des deutschen Universitätsmodells im 19. und 20. Jahrhundert.* Basel: Schwabe, 2001.

Shapin, Steven. "The Mind Is Its Own Place: Science and Solitude in Seventeenth-Century England." *Science in Context* 4 (1991): 191–218.

Shapin, Steven, and Simon Schaffer. *Leviathan and the Air Pump: Hobbes, Boyle and the Experimental Life.* Princeton: Princeton University Press, 1985.

Sheehan, James. *German Liberalism in the Nineteenth Century.* Chicago: Humanities Press, 1978.

Sheynin, O. B. "On the History of the Statistical Method in Meteorology." *Archive for the History of the Exact Sciences* 31 (1984–85): 53–95.

Silverman, Debora. *Art Nouveau in Fin-de-Siècle France.* Berkeley: University of California Press, 1989.

Smidt-Dörrenberg, Irmgard. *Gottfried Keller und Wien.* Vienna: Museumsverein Josefstadt, 1977.

Smith, Barry. *Austrian Philosophy: The Legacy of Franz Brentano.* Chicago: Open Court, 1994.

Smith, Crosbie. *The Science of Energy: A Cultural History of Energy Physics in Victorian Britain.* Chicago: University Chicago Press, 1998.

Sommerhalder, Mark. *Pulsschlag der Erde!: Die Meteorologie in Goethes Naturwissenschaft und Dichtung.* Bern and New York: P. Lang, 1993.

Stadler, Friedrich. *Studien zum Wiener Kreis.* Frankfurt: Suhrkamp, 1997.

Stekl, Hannes, ed. *Bürgerliche Familien: Lebenswege im 19. und 20. Jahrhundert.* Bürgertum in der Habsburgermonarchie, 8. Vienna: Böhlau, 2000.

Stekl, Hannes, et al., eds. *Durch Arbeit, Besitz, Wissen und Gerechtigkeit.* Bürgertum in der Habsburgermonarchie, 2. Vienna: Böhlau, 1992.

Stewart, Larry. *The Rise of Public Science: Rhetoric, Technology and Natural Philosophy in Newtonian Britain.* Cambridge: Cambridge University Press, 1992.

Stigler, Stephen M. *The History of Statistics: The Measurement of Uncertainty before 1900.* Cambridge, Mass.: Harvard University Press, 1986.

Stöltzner, Michael. "Vienna Indeterminism: Mach, Boltzmann, Exner." *Synthese* 119 (1999): 85–111.

———. "Franz Serafin Exner's Indeterminist Theory of Culture." *Physics in Perspective* 4 (2002): 267–319.

———. "Vienna Indeterminism. Causality, Realism and the Two Strands of Boltzmann's Legacy." Ph.D. dissertation, University of Bielefeld, 2003.

Stürmer, Michael, et al. *Wagen und Wägen: Sal. Oppenheim jr. & Cie.: Geschichte einer Bank und einer Familie.* Munich: Piper, 1989.

Sulloway, Frank. *Freud: Biologist of the Mind.* 1979. Cambridge, Mass.: Harvard University Press, 1992.

Sutton, Geoffrey. *Science for a Polite Society: Gender, Culture, and the Demonstration of Enlightenment*. Boulder, Colo.: Westview, 1995.

Terrall, Mary. "Heroic Narratives of Quest and Discovery." *Configurations* 6 (1998): 223–42.

Teske, Armin. *Marian Smoluchowski, Leben und Werk*. Warsaw: Polish Academy of Sciences, 1977.

Tuchman, Arleen Marcia. *Science, Medicine and the State in Germany: The Case of Baden, 1815–1871*. Oxford: Oxford University Press, 1993.

Turner, R. Stephen. *In the Eye's Mind: Vision and the Helmholtz-Hering Controversy*. Princeton: Princeton University Press, 1994.

Wagner, Manfred. *Alfred Roller in seiner Zeit*. Salzburg: Residenz Verlag, 1996.

Wanner, Heinz, et al., "North Atlantic Oscillation—Concepts and Studies." *Surveys in Geophysics* 22 (2001): 321–82.

Whalen, Robert W. *Bitter Wounds: German Victims of the Great War*. Ithaca: Cornell University Press, 1984.

White, Paul. "Science at Home." In *Thomas Huxley: Making the "Man of Science."* Cambridge: Cambridge University Press, 2002.

Whitman, James Q. *The Legacy of Roman Law in the German Romantic Era*. Princeton: Princeton University Press, 1990.

Wieacker, Franz. *A History of Private Law in Europe (with Particular Reference to Germany)*. Trans. by Tony Weir. Oxford: Clarendon Press, 1995.

Winter, Alison. *Mesmerized: Powers of Mind in Victorian Britain*. Chicago: University of Chicago Press, 1998.

Winter, Eduard. *Religion und Offenbahrung in der Religionsphilosophie B. Bolzanos*. Breslau: Müller und Seiffert, 1932.

Wobbe, Theresa, ed. *Zwischen Vorderbühne und Hinterbühne: Beiträge zum Wandel der Geschlechterbeziehungen in der Wissenschaft vom 17. Jahrhundert bis zur Gegenwart*. Bielefeld: transcript Verlag, 2003.

Wonders, Karen. "Hunting Narratives of the Age of Empire: A Reading of Their Iconography." *Environment and History* 11 (2005): 269–91.

Yeo, Richard. *Science in the Public Sphere: Natural Knowledge in British Culture, 1800–1860*. Aldershot: Ashgate, 2001.

Zahra, Tara. "Your Child Belongs to the Nation: Nationalization, Germanization, and Democracy in the Bohemian Lands, 1900–1945." Ph.D. dissertation, University of Michigan, 2005.

Zeps, Michael J. *Education and the Crisis of the First Republic*. Boulder: East European Monographs, 1987.

Ziche, Paul. "Erklärung und Erklärbarkeit. Sigmund Exners Wissenschaftsphilosophie des Physiologie-Psychologie-Verhältnisses." Paper delivered at the International Workshop Sigmund Exner, Vienna, 5–6 March 2004.

Index